Catalysis –
Science and Technology

Volume 4

EDITORS:

PROF. DR. J. R. ANDERSON (Melbourne/Australien)
PROF. DR. M. BOUDART (Stanfort/USA)

CONTRIBUTORS:

PROF. DR. H.-P. BOEHM, PROF. DR. G. ERTL,
PROF. DR. H. KNÖZINGER, DR. P. N. RYLANDER

CATALYSIS - Science and Technology

Volume 4

With 106 Figures and 21 Tables

Akademie-Verlag · Berlin
1983

Die
Originalausgabe erscheint
im Springer-Verlag
Berlin · Heidelberg · New York

Vertrieb
ausschließlich für die DDR
und die sozialistischen Länder

Erschienen im Akademie-Verlag, DDR-1086 Berlin, Leipziger Straße 3—4

Lizenznummer: 202 · 100/532/83
Gesamtherstellung: VEB Druckerei „Thomas Müntzer“, 5820 Bad Langensalza
Umschlaggestaltung: Eckhard Steiner
Bestellnummer: 763 255 8 (6776) · LSV 1215
Printed in GDR
DDR 132,— M

General Preface to Series

In one form or another catalytic science reaches across almost the entire field of reaction chemistry, while catalytic technology is a cornerstone of much of modern chemical industry. The field of catalysis is now so wide and detailed, and its ramifications are so numerous, that the production of a thorough treatment of the entire subject is well beyond the capability of any single author. Nevertheless, the need is obvious for a comprehensive reference work on catalysis which is thoroughly up-to-date, and which covers the subject in depth at both a scientific and at a technological level. In these circumstances, a multi-author approach, despite its well-known drawbacks, seems to be the only one available.

In general terms, the scope of *Catalysis: Science and Technology* is limited to topics which are, to some extent at least, relevant to industrial processes. The whole of heterogeneous catalysis falls within its scope, but only biocatalytic process which have significance outside of biology are included. Ancillary subjects such as surface science, materials properties, and other fields of catalysis are given adequate treatment, but not to the extent of obscuring the central theme.

Catalysis: Science and Technology thus has a rather different emphasis from normal review publications in the field of catalysis: here we concentrate more on important established material, although at the same time providing a systematic presentation of relevant data. The opportunity is also taken, where possible, to relate specific details of a particular topic in catalysis to established principles in chemistry, physics, and engineering, and to place some of the more important features into a historical perspective.

Because the field of catalysis is one where current activity is enormous and because various topics in catalysis reach a degree of maturity at different points in time, it is not expedient to impose a preconceived ordered structure upon *Catalysis: Science and Technology* with each volume devoted to a particular subject area. Instead, each topic is dealt with when it is most appropriate to do so. It will be sufficient if the entire subject has been properly covered by the time the last volume in the series appears. Nevertheless, the Editors will try to organize the subject matter so as to minimize unnecessary duplication between chapters, and to impose a reasonable uniformity of style and approach. Ultimately, these aspects of the presentation of this work must remain the responsibility of the Editors, rather than of individual authors.

The Editors would like to take this opportunity to give their sincere thanks to all the authors whose labors make this reference work possible. However, we all stand in debt to the numerous scientists and engineers whose efforts have built the discipline of catalysts into what it is today: we can do no more than dedicate these volumes to them.

Preface

A very wide range of catalytic conversions find industrial use in organic process chemistry. The scale of the operations varies enormously from very high volume processes to specialty chemical preparations. Many of these processes are functional group conversions or class reactions, and the more important of these will receive detailed treatment in specific chapters throughout this series. Nevertheless, the scope is very broad, and it is all too easy for the non-specialist to become lost in a large volume of detail. To try to avoid this, the first chapter in this volume, by Dr. Paul N. Rylander provides a working summary of the more important catalytic conversions of this type. In doing this, he also gives some valuable comments about catalyst selection, together with an indication of the reaction conditions used in practice, the more important of the problems usually encountered, and comments about the most important of the mechanistic features.

It has long been recognized that an understanding of the chemical nature of solid surfaces is fundamental to an understanding of catalytic processes which may take place upon them. This question may be approached in two distinct ways. One is *via* surface crystallography which focuses attention upon *long range order*. The second concentrates upon the concept of the surface functional group where attention is mainly upon the chemistry characteristic of a particular *localized* atomic arrangement at the surface. In practice, of course, there exists a continuum between these idealized extremes. The second chapter of this volume, by Professor H.-P. Boehm and Professor H. Knözinger, gives a comprehensive account of the chemistry of surface functional groups, including their characterization, estimation, synthesis, and interconversion. In many cases, of course, surface group functionality is central to the behavior of catalytically active sites. However, their importance extends beyond this. Heterogeneized homogeneous catalysts are now of considerable importance, and the provision of chemically suitable surface functional groups is the key which makes it possible to bind catalytically active molecular species to support surfaces.

The wish to elucidate catalytic reaction mechanisms has a history which is as long as that of catalytic science itself. One of the principal motives for this has been the predictive insight which mechanistic understanding provides. In general, this has gone hand-in-hand with the development of chemical theory and with the introduction of more sophisticated techniques by means of which catalyst surfaces may be characterized and the energetics and dynamics of surface processes studied. Even so, there is only a very small number of catalytic processes for which there is a detailed understanding of the molecular dynamics and energetics. The third chapter in the present volume, by Professor G. Ertl, is devoted to this subject. The fact is that a detailed understanding is limited to a relatively small number of rather simple systems characterized by the methods of surface science. No doubt others will be added to the list in the course of time and as a result of further work. However, probably the most important conceptual benefit which studies of this sort have to offer to catalytic science as a whole is a framework of mechanistic concepts, principles, and insights, which should be part of the intellectual equipment of anyone working in the field of heterogeneous catalysis. But there is another important practical benefit deriving from surface science studies presented in this last chapter of Volume 4. In several instances, investigators working with large single crystals of metals have measured turnover rates for a given catalytic reaction which agree almost completely with those reported by others working with supported metal catalysts of the type used in industry. In a way, surface science studies provide standards by which the quality of work done by catalytic scientists can be assessed.

Contents

List of Contributors

Professor Hanns-Peter Boehm
Institut für Anorganische Chemie
Universität München
Meiserstr. 1
D-8000 München 2, FRG

Professor Gerhard Ertl
Institut für Physikalische Chemie
Universität München
Sophienstr. 11
D-8000 München 2, FRG

Professor Helmut Knözinger
Institut für Physikalische Chemie
Universität München
Sophienstr. 11
D-8000 München 2, FRG

Dr. Paul N. Rylander
Engelhard Industries Div.
429 Delancy Street
Newark, N.J. 07105, USA

Chapter 1

Catalytic Processes in Organic Conversions

Paul N. Rylander

Engelhard Industries Div.
429 Delancy Street
Newark, N.J. 07105, USA

1. Introduction

Catalytic conversions of organic compounds are the heart of modern chemical industry. Few bulk or specialty organic chemicals are produced nowadays that have not been touched by catalysts either directly or indirectly in the synthetic sequence. This review of catalyzed organic reactions is limited to heterogeneous catalysts and mainly to chemicals of industrial importance having functionality beyond those of simple olefins and aromatics. Its aims are to illustrate the diversity of chemical transformations that can be achieved, to discuss the interplay of catalyst and chemical properties of the organic reactants, to show how various intrinsic problems can be minimized, and to suggest the type of catalyst suitable for various reactions.

Organization of this review is based on the type of reaction being catalyzed, rather than the class of compound produced with the intent of emphasizing general relationships between catalyst, functionality, and reaction. Space limitations preclude details of catalyst functioning, but many leading references to this area are included for interested readers.

2. Hydrogenation

One of the most useful experimental and industrial means of achieving controlled transformations of organic compounds is through selective catalytic hydrogenations. In total number of applications there are probably more examples of industrial hydrogenation than any other type of reaction. The reason for this exceptional usefulness of hydrogenation is not hard to find. Most functional groups can be reduced readily in high regio- and stereoselectivity, often under mild conditions. An attractive feature of catalytic hydrogenation is that the characteristic properties of various catalysts toward each functional group is likely to remain invariant, with due allowance, of course, for overall structure. A consequence of this is that development of a new use is not apt to require development of a new catalyst;

a relatively few "standard catalysts" suffice for a great variety of substrates. Satisfactory catalysts and conditions for a new process can often be found with relatively little experimental work, given a suitable precedent.

In the sections that follow industrial examples of hydrogenation of various common functional groups (olefins, aromatic nitro, heterocyclic aromatics, nitriles, and carbonyl compounds) are described together with problems often encountered in the hydrogenation of these functions. Several recent texts on hydrogenation are available [1–3].

A. Hydrogenation of Carbon Monoxide

Hydrogenation of carbon monoxide, which includes methanation and Fischer-Tropsch synthesis, has been one of most important and must studied of reactions, and currently with the prospect of returning to coal based economies, this interest is intensifying [4]. Focus in this section is on a single aspect of carbon monoxide hydrogenation, the synthesis of methanol.

World-wide production of methanol in 1979 was in excess of 13 million metric tons, made from carbon monoxide and hydrogen by a reaction first commercialized in the early 1920's

$$CO + 2\,H_2 \rightleftharpoons CH_3OH$$

This methanol synthesis stands as a monument to catalytic chemists, for, as Stiles [5] pointed out, methanol is thermodynamically the least stable of a number of products that have been formed in good yield catalytically from carbon monoxide and hydrogen under conditions similar to those used in methanol synthesis.

Formation of methanol is highly exothermic and reactors are designed to remove heat and operate within a fairly narrow temperature band, commensurate with sufficient catalyst activity at the lower range, and favorable equilibrium at the upper range. Methanol synthesis reactors are usually either tubular with incoming gas being heat exchanged against reacting gases or water, or more commonly are multistage or multitray reactors with cold gas (cold-shot) being mixed with reacted gases as they emerge from each section [5–6]. Metallic iron and nickel are good catalysts for reduction of carbon monoxide to methane, and for this reason the steel reactors are often copper-lined. Iron can also be transported as iron pentacarbonyl from one portion of the process to another. Some processes which operate at low carbon monoxide partial pressures permit the use of carbon-steel, as the iron carbonyl is not formed at these pressures.

Early processes operated at high pressures (24–30 MPa; 593–653 K) over Zn—Cr oxide catalysts [1]. But the development of more active catalysts of copper oxide, zinc oxide, and chronium or aluminum oxides permitted the use of lower pressure (4–10 MPa; 533 K), with its attendent operating economies. These lower pressure catalysts are much more sensitive to poisoning, especially sulfur and chlorine, and reactors are designed to allow facile catalyst change-out. A survey of these industrial catalysts with catalyst composition, operating conditions and space-time yields, has been made [8].

These zinc-copper oxide catalysts are at least three orders of magnitude more active than each of the separate catalyst components [8]. The initial step of methanol synthesis is thought to be nondissociative chemisorption and activation of CO on Cu^+ centers and of hydrogen on the surrounding ZnO centers, hydrogen being split heterolytically. Attack by proton on the oxygen atom of CO and by hydride on the carbon atom gives $Cu—CH_2OH$. Hydrogenolysis of this species is thought to be rate limiting. One cause of catalyst deactivation has been explained as reduction of Cu^+ to Cu metal, whereas oxygen, water, and carbon dioxide have a rate enhancing effect, due to their tendency to keep copper in the active Cu^+ form [8].

Some amounts of carbon dioxide, up to 6 vol. %, have been found desirable in the syn gas feed. The carbon dioxide can serve a number of functions. Through the endothermic reverse water-gas shift reaction, it helps adjust the carbon monoxide-hydrogen ratio as well as consuming some of the heat liberated in the methanol synthesis [9–10]. With lower pressure, lower temperature synthesis catalysts, carbon dioxide helps to maintain copper in the active Cu^+/ZnO state; in the absence of carbon dioxide, catalysts gradually lose activity with a color change from black to pink, indicative of reduction to inactive copper metal.

B. Hydrogenation of Olefins

Hydrogenation of olefins is frequently practiced. The reduction is one of great versatility and has been applied to a variety of compounds over wide ranges of conditions. In general, most olefins can be reduced at ambient conditions, but higher temperatures and pressures are usually employed to make more effective use of catalyst and equipment. Problems encountered in olefin hydrogenation may arise as a consequence of prior double-bond migration, *cis-trans* isomerization, or ineffective competition when other reducible functions are present. Palladium and nickel catalysts are widely used; platinum is effective when bond migration is to be avoided; and rhodium and ruthenium may be useful for avoiding hydrogenolysis of vinyl and allylic substituents.

1. Hydrogenation of Natural Oils

The largest single application of hydrogenation is in the partial hydrogenation of natural oils to margarines, shortenings, salad oils, toppings and various other edible products. Hydrogenation capacity for edible oils in 1980 was about 9 billion pounds annually in the U.S. There is also a large market for products derived by more complete hydrogenation, but the difficult selectivity problems described below do not exist in these deeper hydrogenations.

A variety of different oils, such as cottonseed, soybean, sunflower and rapeseed oil, are used in preparation of edible products. All of these oils differ in detail, but have structural similarities. They are all glycerides of a long chain saturated and unsaturated fatty acids. By partial hydrogenation, oxidation stability of these materials is improved and plastic properties suitably altered.

Oxidation instability in oils is caused by the presence of various homoconjugated dienes and trienes, a type of material especially prone to oxidation. The oxidation may be catalyzed by trace metals. An aim of partial hydrogenation is to remove multiple unsaturation preferentially. Prior isomerization of the unsaturated bonds into conjugation is a necessary prelude to effective selective removal of multiple unsaturation, and there is a close parallel between isomerizing activity of various catalysts and selectivity [11].

As a consequence of migration, isomerization of the naturally occurring *cis* isomer into *trans* may or may not occur, depending on the conformation of the chain at the time of migration. Geometrical isomerization can occur also without migration of the double-bond. The tendencies of various catalysts to promote double-bond isomerization and geometrical isomerization closely parallel, and are related to reversibility of adsorption of the "half-hydrogenated" states.

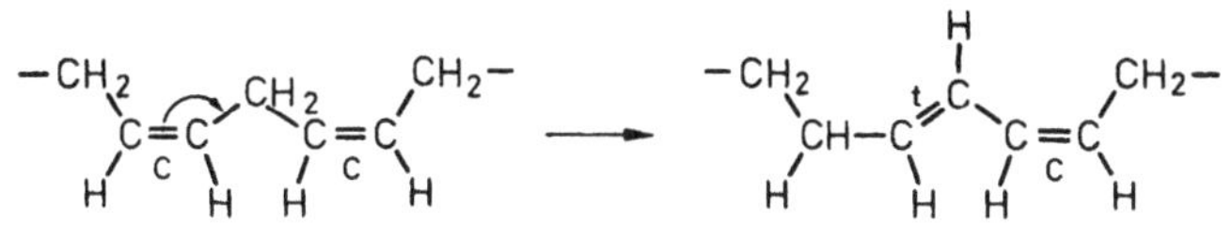

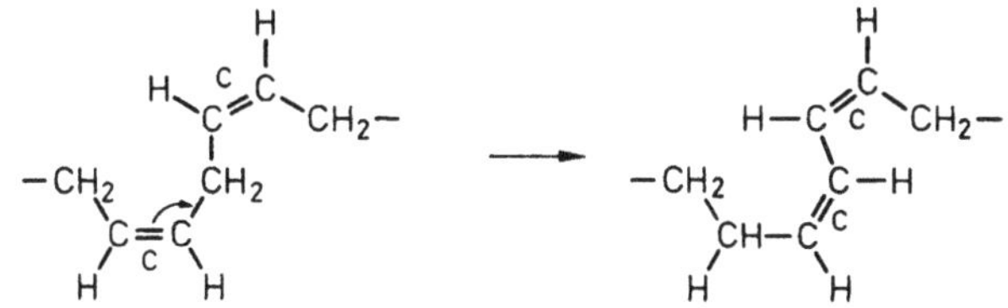

There has been extensive study of the hydrogenation of fats over a great variety of base and noble metal catalysts, but industrial practice overwhelmingly uses some from of nickel.

The demands of oil hydrogenation are such that batch processing is the preferred mode of operation. Continuous hydrogenation is made difficult by intermittent production and frequently changing feed stocks, by difficulty in adjusting to declining catalyst activity, and by difficulty in controlling hydrogen availability at the catalyst surface. Hydrogen availability is an important factor in determining product composition. Hydrogen "rich" catalysts tend to diminish double-bond migration, geometrical isomerization, and selectivity, whereas the reverse is true of hydrogen "poor" catalysts. Hydrogen availability is influenced by reaction parameters as well as by the intrinsic activity of the catalyst; it is increased by decreased temperature and decreased catalyst loading, and is increased by increased pressure and agitation [12–13]. Hydrogenation of oils to edible products is usually carried out at 0.2–0.3 MPa and 413 to 453 K.

Higher pressure (20 MPa), high temperature (523 K) hydrogenation of natural oils over copper chromite catalysts, is an industrial source of long chain saturated alcohols. Remarkably, catalysts (Cd, Cu, Cr) have been developed which allow selective hydrogenation of the ester function, resulting in long chain unsaturated alcohols. Catalysts suitable for the latter mode have been reviewed [14].

Saponification of natural fats gives glycerine and a mixture of long chain fatty acids containing various degrees of unsaturation, which are often hydrogenated to saturated acids of low iodine value (preferably < 1). Since no selectivity problem is involved, the major technical difficulty is prevention of catalyst poisoning. Nickel is usually the catalyst of choice, but it is attacked by the free carboxylic acid and tends to deactivate through formation of nickel soaps. Pressures and temperatures in the range of 1.50 to 3.25 MPa, 420–475 K, are often used and additional catalyst and/or increased temperature are employed if the catalyst deactivates before completion.

Palladium is an ideal catalyst to use in hydrogenation of highly purified fatty acids, since it is not deactivated by the carboxylic acid, but mainly by extrinsic impurities. However, the cost of the product is low enough that the requisite purification is generally not warranted.

Special selectivity problems are involved in the hydrogenation of castor oil, which has an allylic 12-hydroxy function. The value of the product, glyceryl tris(12-hydroxystearate), is diminished as hydroxyls are lost. The product is used in waxes, polishes, and paper coating. The reduction is usually carried out over nickel at temperatures around 410 K, with care being taken to minimize hydrogenolysis. Modified palladium catalysts give higher selectivities, but the increased value of the product must be balanced against extra catalyst costs [15].

Methyl esters of ricinoleic acid can be reduced to the unsaturated diol, ricinoleyl alcohol, in 70% yield over Cu-Cd catalysts, at 490 K, 26 MPa [16].

C. Hydrogenation of Aromatic Nitro Compounds

An important class of industrial hydrogenation is reduction of the nitro function in an aromatic nitro compound. The reactions are highly exothermic and provisions must be made for heat removal. Among side-reactions that may be ecountered are incomplete reduction to the hydroxylamine, partial ring reduction with perhaps some hydrolysis or hydrogenolysis of the amino group, coupling to give azo derivatives, and loss of other functions. These problems can usually be conquered. At times, these various side-reactions are the goal of the reduction.

The ideal of 100% yield has recently acquired new significance; quite tolerable yield losses, from an economic standpoint, may impose severe disposal problems in environmental conscious societies. In the same vein, manufactures, that for years reduced nitro compounds chemically, have switched to catalytic reductions to avoid disposal of reaction residues.

1. Anilines

Aniline is manufactured by hydrogenation of nitrobenzene in either fixed or fluized-bed in vapor phase or liquid phase [17]. The vapor phase catalysts, nickel sulfide, or supported copper, manganese or iron, are of generally low hydrogenation activity, but function well at the elevated temperatures, 543–748 K, of reduction. Heat liberated in this strongly exothermic reaction is removed by excess hydrogen and by internal cooling systems. Liquid phase processes use either reduced nickel or noble metal catalysts. Selectivities exceed 99% in all reductions.

2. Diaminotoluene

Hydrogenation of dinitrotoluene, in contrast to nitrobenzene, is carried out exclusively in the liquid phase, due to the unstable nature of the substrate and product. The resulting diamine is converted to toluenediisocyanate, used for manufacture of flexible foams. Hydrogenations are done either with Raney nickel at about 400 K and 7 MPa, or with 5% palladium-on-carbon at milder conditions.

Reductions of dinitrotoluene over palladium should be carried out under kinetic control, both to maximize catalyst efficiency and to prevent dissolution of metal. When the mass transport of hydrogen to the catalyst surface is low relative to the rate of hydrogen consumption, palladium hydride will exist in an α phase or a mixture of $\alpha + \beta$ phases, instead of the β phase. Under the former conditions, there will be a dissolution of palladium if dissolved oxygen is present in the system, but no dissolution occurs when only the β phase is present [18].

An interesting, but not yet commercialized, route to toluenediisocyanate is the reduction and carbonylation of the nitro compound in a single step using carbon monoxide as reducing agent in the presence of palladium catalysts. The relative cost of carbon monoxide and hydrogen is an important factor affecting the relative merits of these two processes:

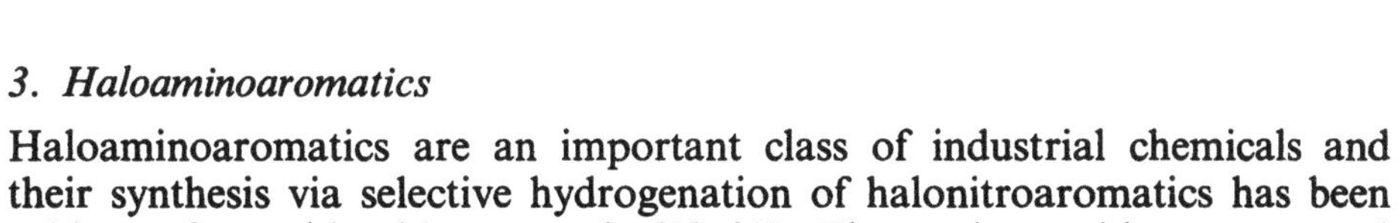

3. Haloaminoaromatics

Haloaminoaromatics are an important class of industrial chemicals and their synthesis via selective hydrogenation of halonitroaromatics has been subject of considerable research [19–20]. The main problem connected with the reaction is that the nitro group activates the halogen toward hydrogenolysis, with the tendency toward hydrogenolysis increasing in the order $F < Cl < Br < I$. Most catalytic systems will effect some loss of halogen, but with attention to the catalyst and conditions, dehalogenation can be kept to very low levels.

Platinum-on-carbon is the preferred catalyst and may be used without inhibitors, but with inhibitors, such as morpholine [19], the reaction conditions are less demanding. Inhibitors can be built directly into the catalyst and catalysts, such as 5% Pt-S_x-on-carbon, give nearly quantitative yields of haloaniline. The catalyst can be reused repeatedly.

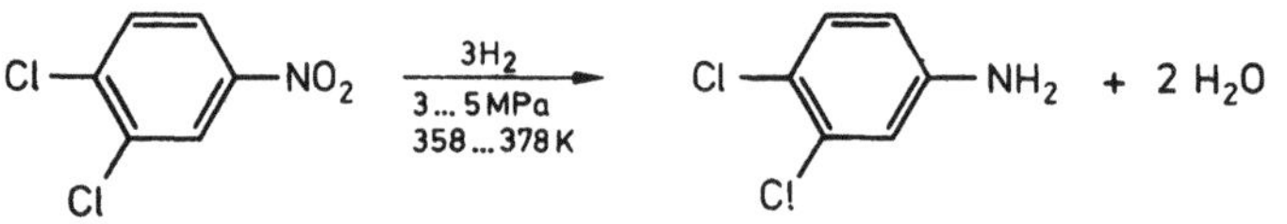

Various bases are often used stoichiometrically to promote catalytic dehydrohalogenations, but, paradoxically, the same bases in lesser amounts are effective in inhibiting loss of halogen during reduction of halonitroaromatics. For example, small amounts of magnesium oxide [21] or hydroxide, calcium hydroxide [22], and sodium acetate [23] all work well in this regard.

Selective inhibitors can be used sometimes with remarkable effect. Palladium is usually the catalyst of choice for achieving an aromatic or aliphatic dehydrohalogenation, but if palladium is used with added inhibitors of the type X—PH(=O)—OH where X = H, alkyl, or phenyl, little or no hydrogenolysis occurs even with iodonitroaromatics [24].

In hydrogenation of halonitroaromatics, the rate ratio of nitro reduction to halogen hydrogenolysis is increased by high hydrogen availability at the catalyst surface, a condition brought about by lower temperatures, higher pressures, vigorous agitation, and lower catalyst loadings. Loop type reactors which provide good temperature control and high hydrogen availability at the catalyst surface are useful in reductions of this type.

4. Phenylhydroxylamines

Phenylhydroxylamines are intermediates in the catalytic reduction of aromatic nitro functions to the aniline, but ordinarily they do not accumulate enough to make the reaction synthetically useful. The yield of hydroxylamine may become very good, however, if a promoter such as dimethylsulfoxide (DMSO) is added to the system [25–26].

The use of extrinsic promoters to influence selectivity markedly is a much neglected area of catalytic hydrogenation.

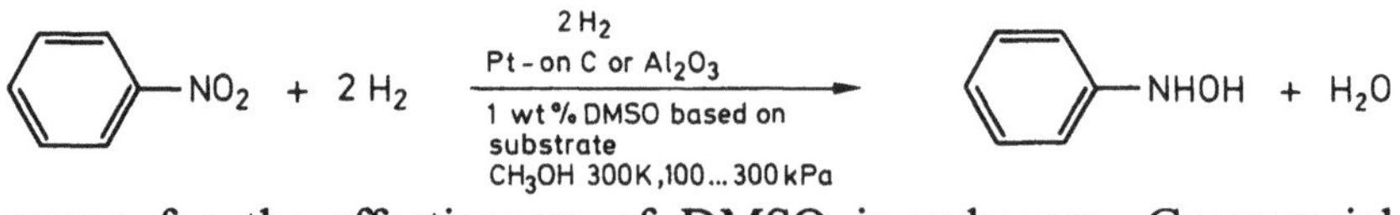

The reason for the effectiveness of DMSO is unknown. Commercial DMSO contains small amounts of dimethylsulfide and the sulfide could be the actual promoter. Pure DMSO behaves the same as sulfide-contaminated material, but this does not rule out a self-limiting reduction of DMSO to some sulfide. Probably DMSO functions as a ligand to change the relative strength of adsorption of nitrobenzene and phenylhydroxylamine.

An industrial synthesis of *p*-aminophenol involves reduction of nitrobenzene in sulfuric acid solution over platinum-on-carbon catalysts. Intermediate phenylhydroxylamine, rearanges in the acid solution as it is formed

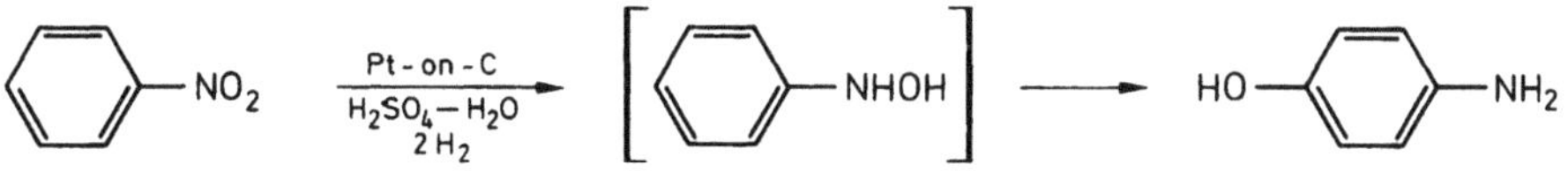

[27–28]. Since the hydroxylamine is easily reduced to aniline, reaction conditions are chosen so as to favor rearrangement instead of hydrogenation. The yield is sensitive to acid concentration, temperature, hydrogen pressure, mode of addition, solvents, wetting agents, such as water-soluble quaternary ammonium salts, or promoters such as DMSO. Early workers [29] thought good yields could be obtained only at subatmospheric pressure, but this restriction is not now necessary.

This process for *p*-aminophenol competes with reduction of *p*-nitrophenol, which preferably uses palladium-on-carbon.

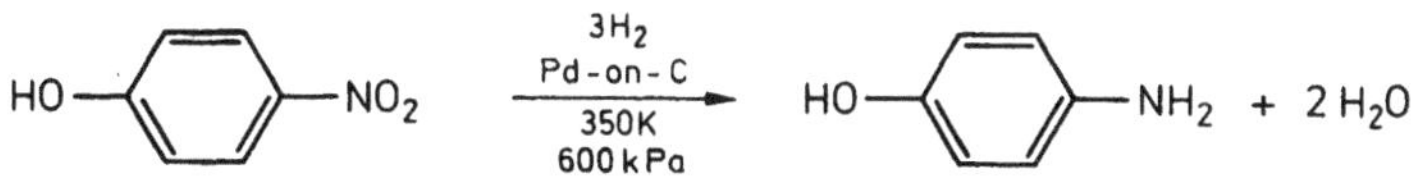

D. Aromatics

A number of important chemicals are made by hydrogenation of various aromatic systems. All of these reductions are highly exothermic and some provision has to be made for temperature control.

1. Cyclohexane

Cyclohexane is reduced industrially (493–670 K, 2.5 MPa) to cyclohexane over either supported platinum or nickel, in quantitative yield [30–31]. The reaction is reversible and the temperature must be controlled, so as not to rise excessively. Liquid phase processes are also operated. Excessive temperatures affect the equilibrium adversely, as well as causing isomerization to methylcyclopentane. Very pure material is obtained if acidic isomerizing activity is absent from the catalyst. The U.S. production of cyclohexane, most of which is obtained by hydrogenation, exceeds 3 billion pounds. It is used, primarily, in the manufacture of Nylon 6 and 66.

2. Cyclohexanecarboxylic Acids and Esters

A process for the manufacture of caprolactam, Nylon-6 monomer, consists of batch hydrogenation of benzoic acid over palladium to afford cyclohexanecarboxylic acid, which is then converted to caprolactam by treatment with nitrosylsulfuric acid [32–35]. This interesting rearrangement involves a ketene intermediate. The high heat of hydrogenation is moderated by recycling a portion of the hydrogenated product. Nickel fails in this reduction, due to attack by the acid. An advantage of this process is that it produces less by-product ammonium sulfate than conventional routes.

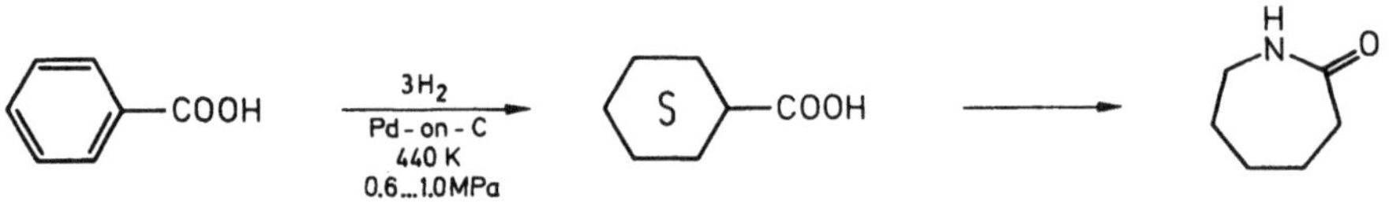

Aromatic esters are reduced more readily than aromatic acids. An example is the synthesis of 1,4-dimethylolcyclohexane, used in polyesters, polyurethanes, and polycarbonates, from dimethylterephthalate, by hydrogenation over palladium at 433–473 K, 30–40 MPa to cyclohexane 1,4-dicarboxylic acid dimethyl ester, followed by reduction over copper chromite at similar conditions [36].

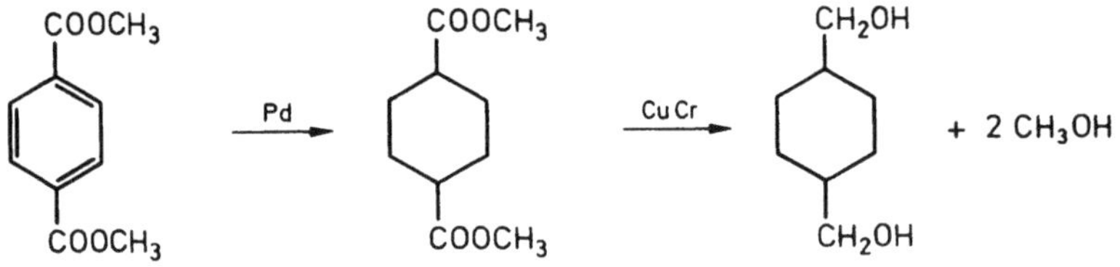

In the first reaction, a high pressure is used to obtain a satisfactory rate. In the second reaction, high pressure is used for rate as well as to shift the equilibrium in favor of the alcohols. The latter reaction is reversible and low pressure usually favors the ester [37].

The sequence of reductions shown above is expediant. The reverse sequence, reduction of dimethylterephthalate to the aromatic diol followed by ring reduction to the saturated diol, could not be done; palladium is an exceptionally fine catalyst for hydrogenolysis of benzyl functions, and the result would be near quantitative yields of dimethylcyclohexane.

3. Perhydrogenated Rosins

Rosin contains about 13% dehydroabietic acid, I, and similar compounds

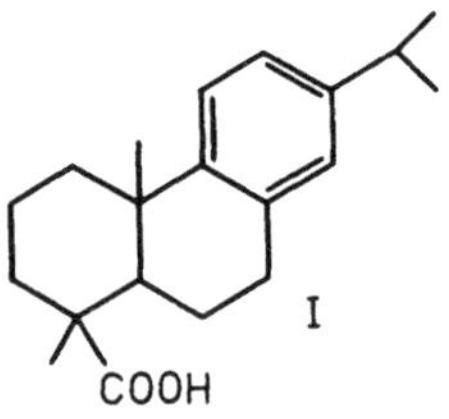

I

with highly hindered, difficult-to-reduce aromatic rings, as well as a variety of olefinic material that form aromatic rings by disproportionation during the hydrogenation. Full saturation of these rosins gives an oxidation-stable solid with desirable products for a variety of uses.

Palladium is the preferred catalyst for this hydrogenation (473 K, 33 MPa), not because of its intrinsic activity, but because it decarboxylates and de-carbonylates the substrate less, thereby diminishing self poisoning of the catalyst by carbon monoxide inhibition [38]. The phenomenom of self poisoning is quite common and its avoidance often dictates the choice of catalyst.

4. Cyclohexylamines

Hydrogenation of anilines yields a mixture of cyclohexylamines and dicyclohexylamines in high combined yield. The diamine probably arises through a step-wise hydrogenation of the ring to give an aminocyclohexene, which may undergo further hydrogenation to the cyclohexylamine, or isomerization to the imine. The amine and imine may combine to give an addition product that undergoes hydrogenolysis, or elimination followed by hydrogenation to give dicyclohexylamines [39].

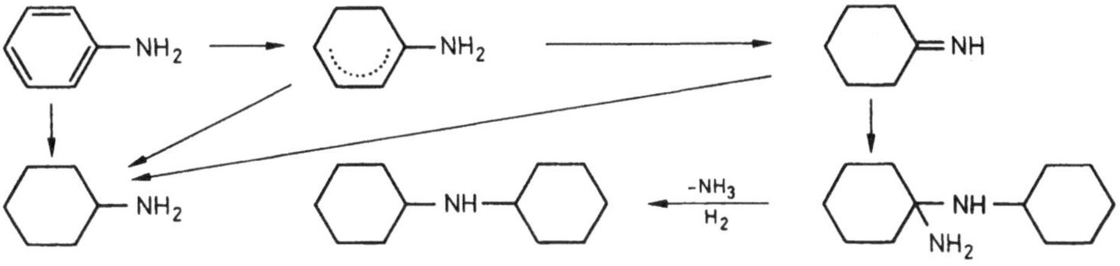

The ratio of cyclohexylamine to dicyclohexylamine depends on catalytic metal, solvent, catalyst support, and temperature. The reaction sequence is too complex to permit an a priori ordering of catalysts but the order for increasing dicyclohexylamine formation is found to be the same as the ordering for increasing tendency toward hydrogenolysis: Ru $<$ Rh $\ll$ Pd $\gtrsim$ Pt [40].

One of the most common sets of competing reactions in catalytic hydrogenation is some sort of hydrogenation versus some sort of hydrogenolysis. In general, polar solvents tend to favor hydrogenolysis relative to hydrogenation and this generality is applicable here; dicyclohexylamine formation decreases with decreasing dielectric constant of the solvent [41–42]. Also, in general, hydrogenolysis is favored by decreased pressure, by increased temperature, and in accord the percentage of dicyclohexylamine increases with increasing temperature, decreases with increasing pressure [43].

Industrial hydrogenations of anilines (ca. 373 K, 3–30 MPa) are often carried out in the presence of ammonia to decrease dicyclohexylamine formation. In the presence of ammonia, cyclohexylamine competes unfavorably in addition to the intermediate imine. However, ammonia may have other functions as well, for diamine formation can also be decreased by other bases such as calcium oxide [44] or lithium hydroxide.

Metal-on-carbon tends to give more dicyclohexylamine than metal-on-alumina [42]. The tendency of carbon to favor more coupling than alumina, seems to be generally true wherever coupling can occur.

Cyclohexylamine is the starting material for an artificial sweetner, and dicyclohexylamine is used to make antioxidants. Bis(4-aminocyclohexyl)methane, formed by hydrogenation of bis(4-aminophenylmethane), is a monomer for an important group of synthetic fibers [45].

4. Cyclohexanone

Cyclohexanone is obtained industrially by either homogeneously catalyzed (Mn, Co) oxidation of cyclohexane [46–47] to a mixture of cyclohexanonecyclohexanol (KA oil), or by partial hydrogenation of phenol by either batch or fixed-bed processing. In the first process, the products are separated and the cyclohexanol dehydrogenated around 720 K over copper or zinc oxide catalysts to cyclohexanone.

Cyclohexanone has been obtained by hydrogenation of phenol to cyclohexanol over nickel at 423 K, 1.5 MPa, followed by dehydrogenation. This older two stage process has been displaced by a single stage process, in which, by using Pd on carbon or alumina with an alkaline promoter, cyclohexanone is obtained directly in high yield. The reaction is done in both slurry and fixed-bed systems [48].

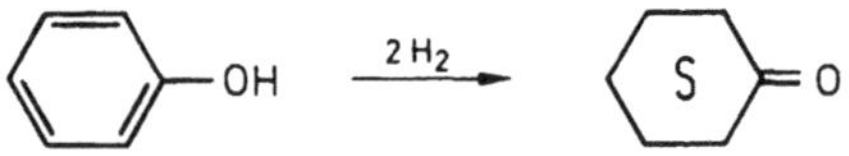

Palladium would appear a priori to be an excellent choice for conversion of phenol to cyclohexanone, since it has a high double-bond isomerization activity and readily can isomerize intermediate cyclohexenols to cyclohexanone, and, moreover, it has a relatively low activity for hydrogenation of the ketone to cyclohexanol. However, this view of palladium functioning has been questioned [49].

Over a billion pounds of cyclohexanone is used annually for production of caprolactam.

6. Tetralin

Tetralin is produced industrially by hydrogenation of naphthalene in fixed-bed reactors over nickel sulfide or nickel-molybdenum catalysts at 680 K, 2–6 MPa. Higher pressures give decalin. The highest yield of tetralin can be obtained over palladium catalysts. Palladium seems to be the preferred catalyst for the hydrogenation of only one of two rings, generally as shown in partial hydrogenation of biphenyl [50], *p*-phenylphenol [51], diphenylmethane, diphenylethane, and dibenzofuran [52].

E. Nitriles

Hydrogenation of nitriles provides a facile synthesis of the corresponding amines. This route to amines has grown in importance with the advent of ammoxidation reactions which introduces the nitrile function in molecules containing an activated methyl group. Reductive hydrolysis of nitriles in aqueous acid, is a convenient route to aromatic aldehydes [53].

Both aliphatic and aromatic nitriles are readily reduced, but the product is likely to be a mixture of primary, secondary, and with aliphatic nitriles, tertiary amines. Secondary amines arise through addition of primary amines to an intermediate imine [54].

$$RCH_2C{\equiv}N \xrightarrow{H_2} RCH_2CH{=}NH \xrightarrow{H_2} RCH_2CH_2NH_2$$

$$RCH_2CH{=}NH + RCH_2CH_2NH_2 \longrightarrow RCH_2CH(NH_2)NHCH_2CH_2R \xrightarrow{H_2} (RCH_2CH_2)_2NH + NH_3$$

Tertiary amines arise similarly. When primary amine is the desired product, the usual procedure is to add ammonia, which adds to the intermediate imine competitively and prevents coupling of the primary amine.

The catalysts have, for reasons not fully understood, a marked influence on product composition. For instance, with lower aliphatic nitriles, palladium or platinum give high yields of tertiary amines, whereas rhodium gives excellent yields of secondary amines. This generality holds, with some exceptions, despite changes in reaction conditions and catalyst support. When an amine is added initially, the coupled products are unsymmetrical amines. Rhodium is used industrially in this manner, for the preparation of unsymmetrical secondary amines, some of which contain functions readily susceptible to hydrogenolysis by other catalysts.

$$RNH_2 + R'CN \xrightarrow[H_2]{Rh} RNHCH_2R' + NH_3$$

It was found that the tendency toward coupled products is increased with increasing temperature and decreased with increasing pressure. Catalysts do not respond to pressure equally; in the range 0.3–10.0 MPa the affect of pressure was moderate with most catalysts, but, with rhodium oxide and lower nitriles, the product went from mostly secondary amine at low pressure to quantitative primary amine at high pressure (10 MPa) [55].

1. Hexamethylenediamine

Most (94%) of the hexamethylenediamine made in the United States is obtained by hydrogenation of adiponitrile. A small amount comes from ammonolysis of hexamethylenediol. Production in the United States in 1980 exceeded 1.2 billion pounds.

The difficulties inherent in the hydrogenation of adiponitrile are those found in hydrogenation of any nitrile, that is, mainly various kinds of coupling products, such as hexamethyleneimine, derived by addition of the amine to intermediate imines. In dinitriles, these coupling reactions can lead to polymeric materials that tend to poison the catalyst. With this particular nitrile, some small amount of coupling through carbon-carbon bond formation may occur to give 1,2-diaminocyclohexane, a compound that causes color problems.

The tendency to couple depends importantly on the catalyst. Cobalt, a catalyst often used in this type of reduction, has a low tendency to couple, which is further diminished by the standard technique of using large excesses of ammonia. Reductions are carried out under vigorous conditions, 3.5 MPa, 423 K.

Many preparations and procedures for this reduction have been patented [56–62]. They feature a variety of cobalt and nickel catalysts often with metal promoters, such as lead or manganese, or the use of various alkalis or bases to minimize coupling and to extend life. A flow sheet for continuous catalytic hydrogenation of adiponitrile is given in Kirk-Othmer [63].

$$NCCH_2CH_2CH_2CH_2CN \xrightarrow{4H_2} NH_2CH_2CH_2CH_2CH_2CH_2CH_2NH_2$$

A large scale route to adiponitrile is by reduction of dicyanobutene (obtained by chlorination of butadiene followed by displacement with cyanide) over palladium carbon, at 550 K, and nominal pressures [65]. Palladium is especially useful here in avoiding premature nitrile reduction. Polymeric material forms if any appreciable amine is produced while olefin is still present. Both displacement of chloride by cyanide and rearrangement of the 3,4-dichlorobutene-1 are catalyzed by Cu^+. An allylcopper type intermediate has been proposed to account for the effectiveness of copper.

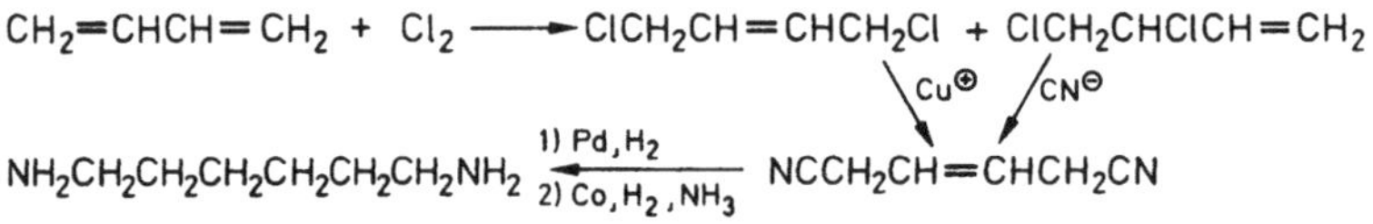

F. Carbonyl Compounds

Carbonyl functions, *i.e.*, aldehydes and ketones, are readily hydrogenated to the corresponding alcohol, and if the carbonyl is aromatic or otherwise activated, the reduction can be easily continued to the desoxy compound. Nickel, palladium, or ruthenium are often suitable catalysts.

1. Hydrogen Peroxide Formation

An important route to hydrogen peroxide involves auto-oxidation of an anthrahydroquinone to a hydroperoxide, which on hydrolysis forms hydrogen peroxide and an anthraquinone. The hydrogen peroxide is water-extracted and purified, the anthraquinone is selectively hydrogenated to reform the anthrahydroquinone, and the cycle is repeated. In practice, the anthraquinone carriers substituents such as 2-ethyl, *t*-butyl, or 2-amyl, for solubility reasons, and the reaction is carried out in a complex mixture of solvents, often referred to as the "working solution" [66].

The selective hydrogenation is best carried out with supported palladium catalysts which are either finely divided powders for slurry operation or pellets or spheres for fixed-bed operation. Hydrogenolysis of the oxygen function or ring saturation during hydrogenation causes deterioration of the working solution and must be minimized. These requirements make palladium especially suitable in this reaction, for it is very active for hydro-

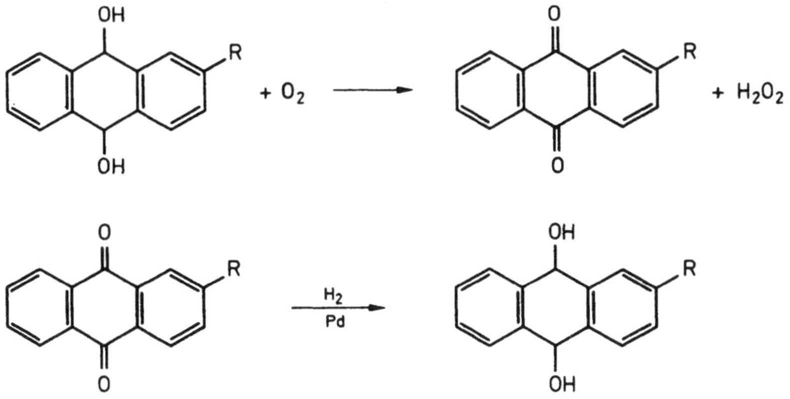

genation of aromatic carbonyl and relatively inactive for saturation of aromatic rings, as well as being non-pyrophoric.

2. Sorbitol

World consumption of sorbitol in 1978 has been estimated to be in the order of 300 thousand metric tons. All of it was produced by hydrogenation of D-glucose or in a mixture with mannitol by hydrogenation of fructose. Sorbitol is used in a variety of food stuffs, pharmaceuticals, and cosmetics *per se*, as well as being converted to polyurethanes, synthetic resins, adhesives and vitamin C.

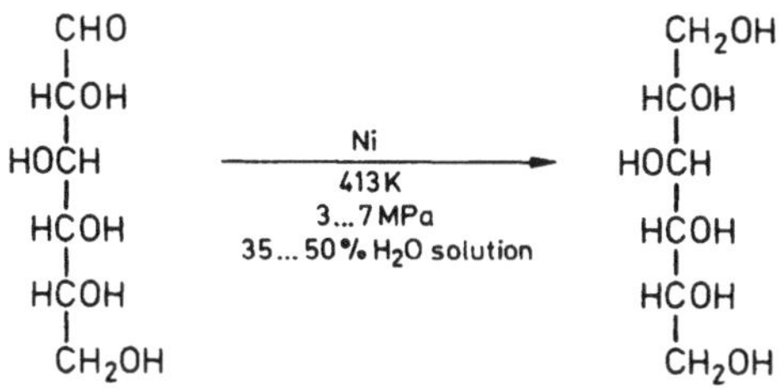

The demands on the hydrogenation are considerable. The temperature must be high enough to give adequate rates, but not so high as to produce a variety of by-products derived mainly by cyclization. Product quality also depends on pH which is maintained at 5–6. Alkaline media causes two undesired reactions; it epimerizes the alpha carbons through the Lobry de Bruyn-van Ekenstein rearrangement to produce mannitol as well as sorbitol, and through alkaline catalysis of the Cannizzaro reaction, gluconic acid is produced. On the other hand, nickel catalysts are damaged if the solution becomes too acid.

Most sorbitol plants use Raney or supported nickel catalysts [67–69] or nickel promoted by molybdenum [70], but more recently ruthenium catalysts have been used. Ruthenium is not so easily attacked by acid as nickel, and is especially useful when the glucose still contains some polysaccharides [71–72]. Ruthenium has been used in phosphoric acid solution to obtain polyhydric alcohols in excellent yield directly from materials such as cellulose, sawdust, and cotton [73–74].

3. 2-Ethylhexanol

Aldehydes and ketones bearing alpha hydrogens can undergo aldol condensations to afford α, β-unsaturated carbonyl compounds, some of which, either *per se* or has hydrogenated derivatives, have large scale industrial uses. An important member of this group is 2-ethylhexanol obtained from *n*-butyraldehyde through conversion to 2-ethylhexenal and hydrogenation. World production of 2-ethylhexanol is about 2 million metric tons per

$$2\ CH_3CH_2CH_2CHO \xrightarrow[OH^{\ominus}]{} CH_3CH_2CH_2CH{=}\underset{\displaystyle C_2H_5}{\underset{|}{C}}CHO \xrightarrow[Ni]{400\ldots460K,\ 8MPa} CH_3CH_2CH_2CH\underset{\displaystyle C_2H_5}{\underset{|}{C}H}CH_2OH$$

year. The hydrogenation is usually carried out over supported nickel in two stages, reducing first the olefin, then the aldehyde. This mode of operation gives a cleaner product than one stage hydrogenation and makes temperature control easier [75].

G. Acetylenes

The acetylene function is one of great synthetic utility because of the ability of the function to add to a carbon-carbon bond with the group still intact. Frequently it is converted to a *cis*-olefin preferably over palladium with or without promotes such as lead (Lindlar catalyst) or cadmium. This reaction sequence is used in synthesis of a variety of complex molecules such as vitamins A and K, prostagladins, and sex pheromones [76]. The largest volume industrial hydrogenation of acetylenes are of yneols and ynediols to the corresponding olefinic and saturated carbinol and diol.

1. Butanediol-1,4

Butanediol-1,4 has long been made by hydrogenation of butynediol over a variety of base metal catalysts, such as nickel copper, manganese on silica [77–80]. Nickel, the major component, is effective for hydrogenation of both acetylenic and olefinic bonds, copper is effective for hydrogenation of aldehydes formed by isomerization, and the small amount of manganese stabilizes the structure. Hydrogenations are carried out under vigorous conditions.

A remarkable synergism has been found with palladium-ruthenium on carbon catalysts when used in this reaction. Ruthenium, an excellent catalyst for hydrogenation of aliphatic aldehydes, is believed to function by removing inhibiting aldehydes as they are formed during the hydrogenation. The two metals together gave rates up to 10 times that of palladium alone, even though ruthenium was totally inactive for acetylene hydrogenation under the mild conditions of the experiment [81].

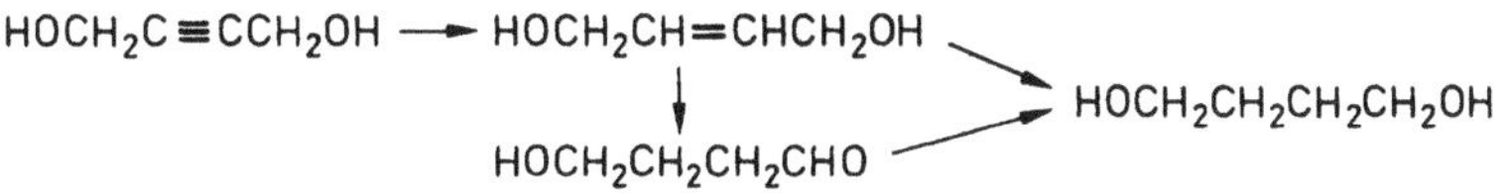

The growing importance of butanediol as a monomer has led to a variety of new catalytic approaches to this compound including hydrogenation of maleic acid or its esters, of butyrolactone, of polymeric peroxides, or of diacetoxybutene from acetoxylation of butadiene. (See also 3.C.)

3. Dehydrogenation

Dehydrogenation is a very important industrial reaction, although the number of different applications is quite limited relative to hydrogenation [82]. Most industrial dehydrogenations fall in one of the following categories: a. the accepted product is a complex mixture, as in reforming of naphtha to gasoline, or as in the dehydrogenation of linear paraffins to mixed olefins, for use in detergent alkylates; b. an unsaturated system is extended, as in conversion of ethylbenzene to styrene or benzene to diphenyl; c. there is little choice of product, as in dehydrogenation of butane to butenes or butadiene; d. there is a preferential point of attack, as in conversion of methanol, ethanol or cyclohexanol to the corresponding carbonyl compound; e. aromatization occurs.

Dehydrogenations are endothermic reactions and at room temperature the equilibrium lies far toward the hydrogenated species. For this reason, fixed or fluidized bed dehydrogenations are carried out at elevated temperatures to shift the equilibrium toward the dehydrogenated product, and/or with a hydrogen acceptor present to remove hydrogen as it is formed. Acceptors such as oxygen may function directly in the dehydrogenation as well. Slurry reactors can operate at lower temperature if hydrogen escapes from the products.

A. Styrene

About 9 billion pounds of styrene, an important monomer, is produced in the United States annually. Styrene is manufactured mainly by the dehydrogenation of ethylbenzene in the presence of super-heated steam. Reactors operate over narrow temperature ranges, the temperature being set high enough to achieve good rates of dehydrogenation, but low enough to avoid cracking of styrene and ethylbenzene. The dehydrogenation is endothermic, and with normal steam to hydrocarbon ratios, the temperature will fall about 1 K for each percent conversion. In adiabatic reactors, reheating at intermediate stages is required. Pressure drop through the reactor is an important consideration, and low pressure drops are essential if low pressure is to be maintained. Radial flow reactors are useful in this regard.

Dilution by steam has several desirable features. Steam at 993 K, introduced directly into the reactor, provides part of the heat for the endothermic

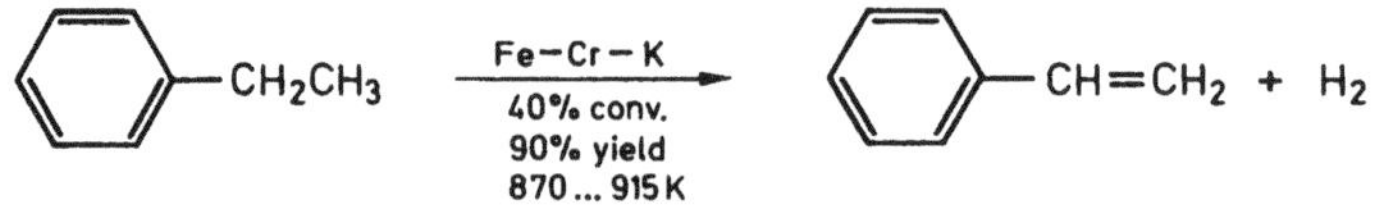

dehydrogenation. Steam, as an oxidant, equilibrates iron oxide catalysts to an oxidation state which is highly selective to styrene [86]. Through the water gas shift reaction, steam continuously removes carbonaceous deposits, permitting 1 to 2 years of continuous operation with a single catalyst charge, and steam also reduces the partial pressure of hydrocarbons which favors conversion to styrene. Common commercial catalysts are often iron oxide promoted with potassium and containing small amounts of chromium oxide as a structural stabilizer [87]. A great many ethylbenzene dehydrogenation catalysts have been developed. They contain, in general, some major active ingredient, such as iron, cobalt, or magnesium, a stabilizer chosen from a wide variety of metals, an alkali metal, and a support material.

Potassium is a potent promoter for iron catalysts, and increases the rate of dehydrogenation by an order of magnitude. During use, potassium migrates to the cooler center of the pellet, deactivating the catalyst on the periphery because of lack of promoter, and deactivating in the center because of excessive potassium. In plant practice, the reactor inlet temperature is raised slowly to compensate for decreasing activity, but finally the catalyst must be replaced [86]. Conversions in practice are only about 40%. As styrene approaches its equilibrium value, its rate of formation decreases, but the rate of formation of by-products remains constant, hence yields of styrene fall as conversion increases.

Commercial iron-potassium catalysts are so active that intraparticle diffusion limits the rate of dehydrogenation in normal operation. The particle size used commercially (4–6 mm) is a compromise between the disadvantages of diffusion limitations and excessive pressure drops.

Ethylbenzene is made by alkylation of benzene with ethylene over strong acids, such as $AlCl_3$-HCl or BF_3 complexes, or in vapor phase at high temperature over acid catalysts such as silica alumina, phosphoric acid on silica, or molecular sieves as crystalline aluminosilicate zeolites [88]. Zeolite ZSM-5 catalysts are a recent development in ethylbenzene manufacture [89].

These are aluminosilicates with elliptical pore openings of 5 by 7 angstroms. This pore size exhibits high shape selectivity and allows molecules of ethylbenzene to diffuse freely while restricting polyalkylated molecules. Selectivity to ethylbenzene is very high, for any diethylbenzenes formed in the process are recycled. Swing reactors permit periodic coke burnoffs, which does not cause catalyst deterioration [90].

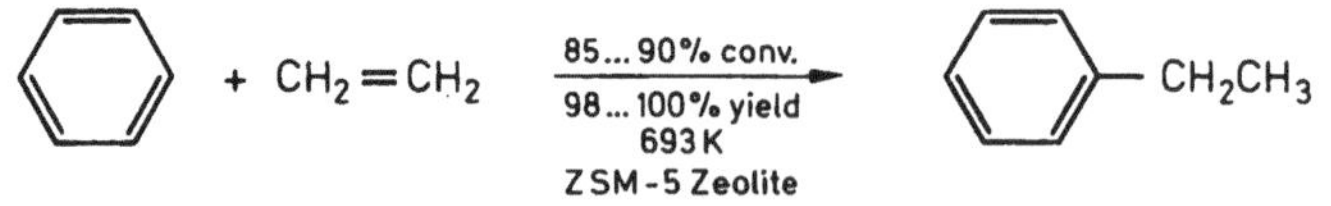

With changing economics of feed stocks, alkylation of benzene may be superceded someday by direct alkylation of toluene with methanol over a catalyst such as alkali-cation exchanged X and Y zeolites [91–92].

Other potential routes to styrene have been reviewed by Kaeding [93].

B. Formaldehyde

World-wide production of formaldehyde is around 9,000 metric tons annually. Most formaldehyde is derived from methanol by removal of hydrogen [94]. The reaction is endothermic, and to supply the necessary heat, air is added to combine with liberated hydrogen. Either insufficient air or a large excess can be used to avoid the explosive range. These two modes of operation are referred to as oxidative-dehydrogenation and oxidation, respectively. Each operates in different temperature ranges and with different type catalysts.

With large excesses of methanol, silver needles or gauzes are used near 873 K. The catalyst is arranged in a thin bed and the contact time is very short, the order of 0.01 second or less. The silver needles operate well above the Tammann temperature and physically change, undergoing some coalescence. Because of this, suitable operating beds are formed only when the silver needles meet specified shape and size requirements. Modern processes have been improved to the point where little methanol remains unchanged, obviating the usual removal of excess methanol.

When large excesses of air used, the catalysts are base metal oxides, such as iron-molybdenum oxides, sometimes promoted with materials such as chromium and copper. The catalysts are operated at 673–723 K, and produce formaldehyde in 91–93 % yield. The upper temperature limit is set to minimize catalyst damage by volitilization of molybdenum oxide. The reaction in this mode is more exothermic than that with an air deficiency and requires use of a multitubular reactor with external cooling. The yield with base metal catalysts is slightly higher than those obtained with silver, but this advantage is offset by more expensive equipment and operation. The majority of plants use silver catalysts.

Oxygen was at one time thought to function only by removing hydrogen formed in dehydrogenation of methanol over silver.

$$CH_3OH \xrightarrow{Ag} CH_2O + H_2$$

$$H_2 + 1/2\,O_2 \longrightarrow H_2O$$

Recent studies have shown a much more complex function [95]. Methanol is very poorly adsorbed on oxygen-free metallic silver, but in the presence of adsorbed oxygen, methanol reacts by

$$CH_3OH_{(g)} + O_{(ads)} \longrightarrow CH_3O_{(ads)} + OH_{(ads)}$$

to give methoxide, which can afford formaldehyde

$$CH_3O_{(ads)} \longrightarrow CH_2O_{(g)} + H_{(ads)}$$

as well as undergo other reactions.

Effectiveness of base metal oxides in converting methanol to formaldehyde has been linked to both their acid-base properties and to their oxidizing ability [96]. Methanol can be attacked at either its basic oxygen or acidic hydrogen. When attacked and activated at the oxygen atom by acidic oxides, such as V_2O_5, MoO_3, WO_3, or U_3O_8, formaldehyde is formed, but relatively basic oxides, such as TiO_2, SnO_2, Fe_2O_3, ZnO, NiO, or Co_3O_4, attacking the acidic hydrogen produce carbon dioxide. The principle is valid also in composite oxides and formaldehyde will be produced mainly as long as the composite is acidic enough.

However, acidity alone in insufficient and effective formaldehyde catalysts must have an oxidizing ability as well. Strongly acid catalysts, such as WO_3-P_2O_5 (rich in P), lacking either oxidizing function or basic properties produces methyl ether rather than formaldehyde.

C. Butyrolactone

A route to butyrolactone involves the dehydrogenation of butanediol over supported copper catalysts at 500–520 K. Presumably formation of butyrolactone involved dehydrogenation of one hydroxyl to an aldehyde, rapid cyclization to a hemi-acetal, and dehydrogenation of the hydroxyl.

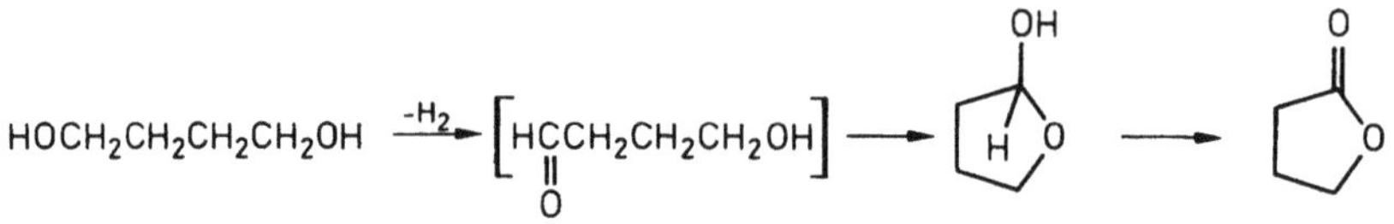

Interestingly, to diminish reliance on acetylene sources, a reverse sequence of reactions has been developed in Japan to permit synthesis of butanediol by hydrogenation of butyrolactone, which in turn is derived by hydrogenation of maleic anhydride [91–99]. The hydrogenation can be adjusted to yield tetrahydrofuran as the major product by changing the operating conditions, or by using Co-Re.

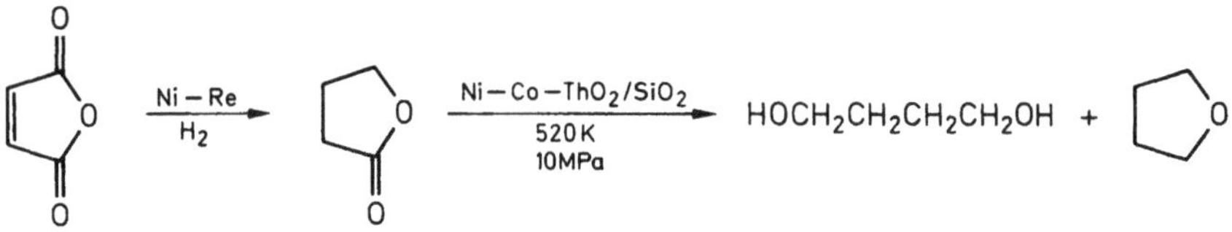

D. Acetone

For several decades following the first world war, the dominant production of acetone was through dehydrogenation of isopropanol. It is still an important process but is being replaced by the cumene oxidation process for phenol which produces acetone as a by-product.

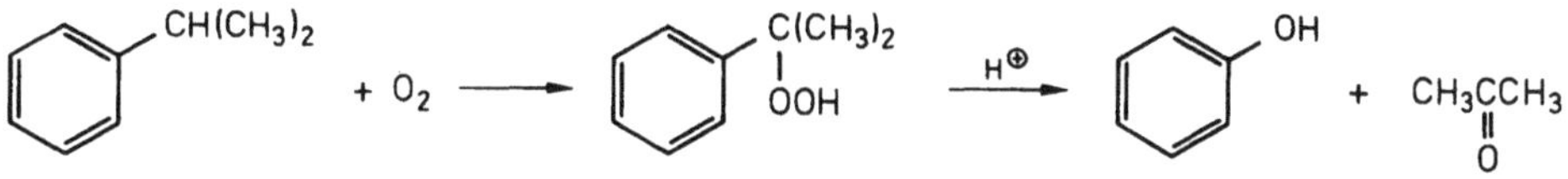

The lowest practical temperature permitted in isopropanol dehydrogenation is determined by equilibrium considerations. The dehydrogenation is endothermic and temperatures around 600 K are necessary for nearly complete conversion. Commercial reactors operate at considerably higher temperature, 500–870 K, to achieve better rates and space-time yields. In an oxidative-dehydrogenation mode, catalysts such as copper or silver are used, whereas zinc oxide or brass catalysts are frequently used in dehydrogenation. Yields are high, 98%. Acidic catalysts are avoided to minimize dehydration and aldol type condensation products.

4. Oxidation

Many large scale chemical processes are oxidative in nature and, despite a diversity of reaction types, have certain characteristics in common. All products of partial oxidation are derived by kinetic control of the reaction; thermodynamics favor ultimately formation of carbon dioxide. All oxidations are exothermic and provisions must be made to limit the temperature to prevent catalyst damage as well as over-oxidation. Heat and mass transfer effects may have significant influence on the reaction; these effects can be decreased by use of tubular or fluidized-bed reactors. Oxidations can also proceed non-catalytically and care must be taken to avoid explosive compositions.

Some common mechanistic factors in oxidation of hydrocarbons have been reviewed as well as a listing given of characteristic catalysts for various types of oxidation [100].

A. Acetoxylation

One of the most interesting catalytic reactions of recent years is acetoxylation, in which a hydrogen atom is replaced by an acetate group in an oxidizing atmosphere with the formation of water. The reaction is applicable to vinyl, allyl, benzyl, and aryl hydrogens.

1. Vinyl Acetate

Interest in this reaction began with the discovery of Moiseev [101] that vinyl acetate was formed by interaction of ethylene, acetic acid, a base, and stiochiometric palladium chloride, which was reduced to Pd^0 in the process. Air oxidation of the reduced palladium, catalyzed by copper ions, can be readily effected, thereby making the reaction catalytic with respect to palladium. Several large scale plants were built, based on liquid phase reactions, but they never operated satisfactorily, due to corrosion and other problems. All have been abandoned in favor of vapor-phase processes.

This latter mode of operation eliminates catalyst separation and corrosion problems, and minimizes the formation of acetaldehyde which was always an important accompaniment of liquid phase processing. In fact, the amount of acetaldehyde obtained could be controlled so that the process would provide its own acetic acid, through oxidation of acetaldehyde [102].

Vapor phase acetoxylation of ethylene is carried out in fixed-bed tubular reactors over palladium supported on carriers resistant to attack by acetic acid and containing alkali acetates and other promoters whose function is to increase selectivity, rate, and life, and minimize over-oxidation to carbon dioxide.

The ratio of reactants is adjusted to operate outside of the explosion limits, about 10% oxygen. This requirement on oxygen concentration limits the per pass conversions to about 25%, based on acetic acid. Pure oxygen, rather than air, is used to aid in recovery of ethylene and to eliminate the need for large nitrogen purge.

$$CH_2{=}CH_2 + HO\overset{}{\underset{\|\atop O}{C}}CH_3 \xrightarrow[410 \ldots 450\,K]{400 \ldots 800\,kPa} CH_2{=}CHO\overset{O\atop\|}{C}CH_3 + H_2O$$

91 ... 94% yield

The exothermic heat of reaction is removed by cooling water on the shell side of the tubular reactor and is recoverable at low heat levels. Corrosion problems are not severe and stainless steel reactors can be used. Catalyst life has exceeded $2^1/_2$ years [103]. The temperature is set as high as possible, consistent with good selectivity and life, and the pressure is raised as high as possible without approaching the dewpoint. Higher pressures are beneficial in separation of unchanged ethylene, in increasing productivity, and in lowering energy requirement for gas recycle.

Nakamura and Yasui [104] picture vinyl acetate being formed through hydrogen abstraction of both ethylene and acetic acid, followed by a combination of these dissociatively adsorbed species. Abstraction of hydrogen from ethylene occurs even in the absence of oxygen, but abstraction from acetic acid occurs only when oxygen is present. Co-catalysts, such as alkali metals, promote the abstraction of hydrogen from acetic acid and weaken palladium-oxygen bonds in dissociatively adsorbed acetic acid.

Catalyst stability depends on operating conditions [105]. Catalyst activity is maintained by those conditions under which palladium (II) acetate does not exist, that is, by lower partial pressures of acetic acid and oxygen, higher temperatures, and higher potassium acetate concentrations. It has been suggested that formation of palladium acetate is the cause of palladium aggregation with its consequent loss in activity.

2. *Other Acetoxylations*

Manufacture of vinyl acetate is by far the most important application of acetoxylation so far. The reaction has been considered for a number of other important materials, such as phenol [106], naphthols [101], and benzyl alcohol [108]. In each case, the acetoxylation catalyst was palladium, and the product was hydrolyzed to permit recycle of the acetic acid. Two routes to butanediol, an important monomer, involve acetoxylation of propylene or butadiene. Propylene is acetoxylated to allyl acetate, over a catalyst similar to that used for vinyl acetate, followed by hydroformylation over homogeneous rhodium catalysts, reduction of the aldehyde to an alcohol, and hydrolysis to give 1,4-butanediol [109].

A shorter route involves liquid phase diacetoxylation of butadiene to 1,4-diacetoxy-2-butene over a palladium-tellurium on carbon catalyst. The catalyst is remarkable because of the specificities involved. Without tellurium yields and activity are very low. Carbon as a support is markedly better than alumina or silica gel [110]. Hydrogenation and hydrolysis completes the synthesis

$$CH_2{=}CHCH{=}CH_2 + 2\ CH_3COOH + O_2 \longrightarrow CH_3\underset{\underset{O}{\|}}{C}OCH_2CH{=}CHCH_2O\underset{\underset{O}{\|}}{C}CH_3$$

trans/cis = 75/14

B. Ethylene Oxide

Ethylene oxide is used relatively little in itself, but it is an important intermediate, especially in the manufacture of ethylene glycol. The world-wide 1980 production of ethylene oxide was in excess of six million metric tons. Production today is almost entirely by direct oxidation of ethylene with air or oxygen over a supported silver catalyst according to the equation

$$H_2C{=}CH_2 + {}^1\!/_2\,O_2 \xrightarrow[523\,K]{0.5\ \dots\ 1.0\,MPa} H_2C{-}CH_2 \text{ (epoxide ring, bridged by } O\text{)}$$

Silver achieves expoxidation of ethylene uniquely; all other catalysts are relatively or totally ineffective and the reaction is limited, among lower hydrocarbons, to ethylene. Propylene or butylene do not form epoxides, but styrene, which has no β-hydrogen, is reported to give good yields of styrene oxide [113].

Side-reactions are the oxidation of both ethylene oxide and ethylene, especially the latter, to carbon dioxide and water with the liberation of large amounts of heat. Much of the effort in catalyst and process development has been directed toward minimizing these side-reactions which cause a yield loss *per se*, as well as catalyst deterioration through excessive temperature rises [112].

Catalysts are 10–15% silver supported on a low surface alumina, and may contain small amounts of a metal promoter such as calcium or barium. Over-oxidation is minimized by chloride inhibition, the chloride being fed continuously as a few parts per million of organic chlorides, such as 1,2-dichloroethylene. The role of modifiers has been explained in terms of electronic interactions between silver crystallites and the support, which alter the work function of electrons at the metal surface [114]. Chlorine modifiers are thought to change not only the relative concentration of atomic and molecular oxygen, but also to increase the probability that molecular oxygen reacts to give ethylene oxide [115].

Early plants used air as the oxidant, but later, despite added investment, pure oxygen was often used, as it simplifies recycle, allows higher reactor productivity and lowers over-all cost. The relative merits of air and oxygen have been discussed at length, as well as the respective operating conditions. The choice often depends on plant size and local circumstances [116–117]. When oxygen is used, an inert gas, such as methane, is recycled, which decreases ethylene losses, and decreases the flamability limits.

The highly exothermic oxidation is carried out in a tubular bundle with external cooling. Bundles in excess of 10,000 tubes are commonplace. Feed streams contain about 5–10% each of ethylene and oxygen with the remainder being inerts. Ethylene conversions are limited to about 10% and selectivities to ethylene oxide are 70–80%.

Some theorists project 85.6% selectivity as a theoretical maximum, based on mechanistic considerations and indeed this high selectivity is seldom reached. However, selectivities in the mid 90's have been demonstrated but only at low conversions, 3–4% [118–119]. The value of 85.6%, which is numerically equal to 6/7, is derived from the proposal that ethylene is not adsorbed on silver, but on nondissociatively adsorbed oxygen, which proceeds to give ethylene oxide and an adsorbed oxygen atom.

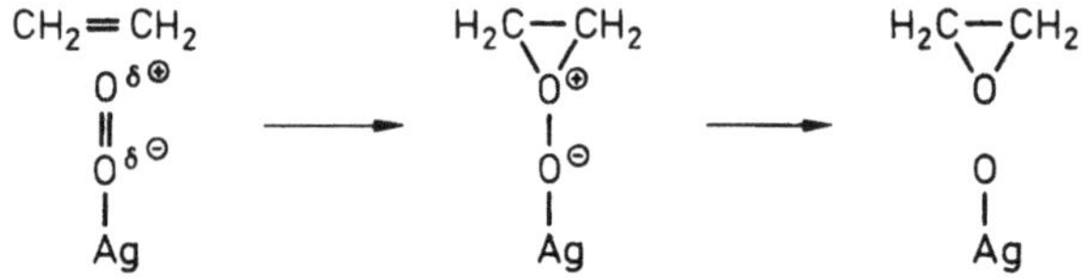

The adsorbed oxygen atoms thus produced react with ethylene to give carbon dioxide and water.

$$6\ O_{(ads)} + CH_2{=}CH_2 \longrightarrow 2\ CO_2 + 2\ H_2O$$

Overall, of every seven ethylene molecules, six go to ethylene oxide, one to carbon dioxide

$$7\ C_2H_4 + 6\ O_{2(ads)} \longrightarrow 6\ C_2H_4O + 2\ CO_2 + 2\ H_2O$$

The above oxygen adsorption process is assumed to occur when no more ensembles of four virgin silver atoms are available. These ensembles are consumed in a dissociative adsorption of oxygen. This mode of adsorption could be completely inhibited by introduction of chlorine atoms equal to one fourth the number of surface silver atoms. Each further chlorine atom introduced above this amount reduced by two the number of oxygen atoms subsequently adsorbed.

The dissociative mode of adsorption of oxygen is thought to lead to production of carbon dioxide; chlorine thus inhibits preferentially an unwanted mode of oxygen adsorption.

C. Maleic Anhydride

World-wide maleic anhydride capacity in 1980 exceeded $1^1/_2$ million metric tons, the U.S. being by far the largest producer. Most maleic anhydride still is produced by catalytic oxidation of benzene [122].

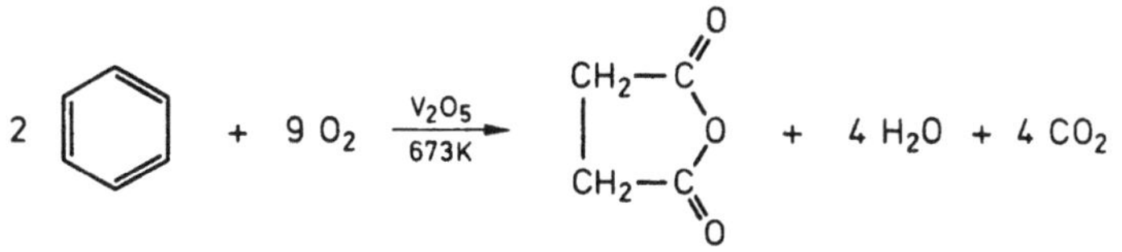

But this process wastes one third of the original carbons, and newer processes which use butane or butenes as starting materials are finding favor [123]. In both processes, much over-oxidation occurs so that the actual heat liberated is greater than that calculated for the reactions as written.

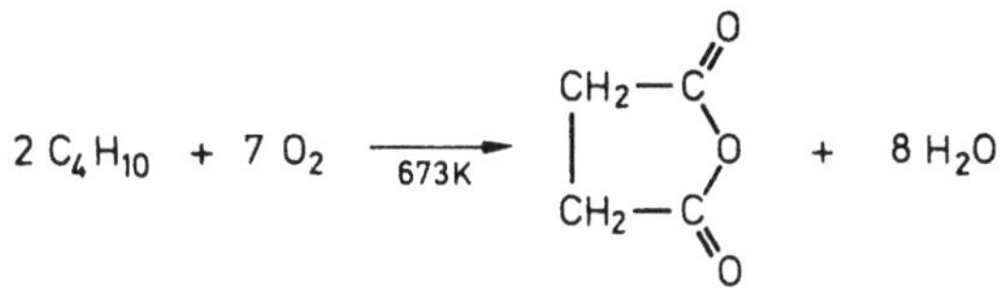

50 ... 60% yield

The relative merits of benzene and butane as feed stocks have been discussed in detail. The choice depends largely on the price differential which varies with supply and demand of each [124]. Maximum profitability can be obtained through a convertible plant concept that can utilize either feed stock [125].

Catalysts patents in this area are numerous. The literature up to 1972 has been reviewed extensively by Hucknall [126], and more recently, Varma and Saraf [127] have reviewed catalysts for the oxidation of C_4 hydrocarbons. Most catalysts are mainly vanadium with a variety of other components, often molybdenum or phosphorous. Loss of phosphorous from the catalyst is one cause of deactivation; this loss can be decreased markedly by use of alkaline promoters.

Kinetics of the oxidation of butene over a vanadium phosphate catalyst [127] were consistent with a two-stage redox model proposed by Mars and Van Krevelent [128], which suggests that oxidation takes place between butene in the gas phase and an oxygen ion of the oxide lattice.

All plants for oxidation of benzene to maleic anhydride use multitublar fixed-bed units that are cooled by molten salt (650 K). The reactors may have thousands of tubes, usually made of mild steel. The reaction is carried out in a non-explosive range with large excesses of air at low pressure, 0.1 to 0.2 MPa, and weight hourly space velocities of about 0.1 based on benzene. The effluent gas contains about 1% maleic anhydride. Conversions are around 95% with 75% selectivities.

A fluidized-bed process for oxidation of linear C_4 hydrocarbons is operated in Japan with a catalyst based on phosphorous-vanadium oxides. The advantages claimed are those of fluidized bed operation generally, easier temperature control and cheaper reactors both to build and maintain. The process is operated within the explosive range.

D. Phthalic Anhydride

Phthalic anhydride is an important industrial chemical whose world-wide production capacity in 1980 exceeded six billion pounds [129–130]. Phthalic anhydride was produced before 1945, by oxidation of naphthalene, but since then *o*-xylene has gained favor until now it is the preferred feed-stock. One pound of *o*-xylene can theoretically afford 1.40 pounds of phthalic anhydride, whereas only 1.16 pounds can be obtained from naphthalene. Also, the lower heat of reaction of *o*-xylene reduces heat transfer and over-oxidation problems.

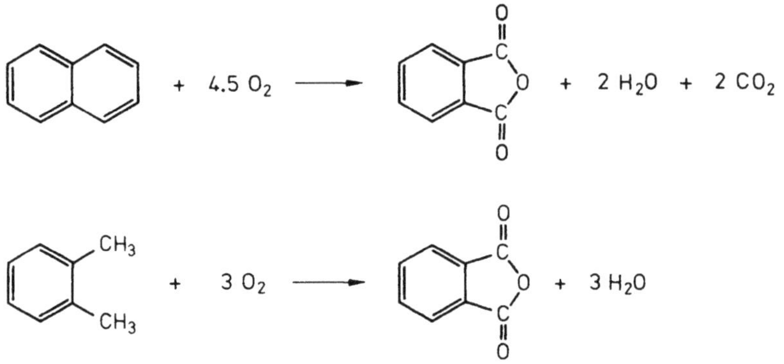

Most modern catalysts contain vanadium pentoxide and potassium sulfate on silica or titania supports [131]. Potassium-promoted catalysts are used to oxidize either *o*-xylene or naphthalene in fixed-beds and naphthalene in fluidized beds. Some catalysts require the addition of small amounts of sulfur dioxide for optimum functioning. Two theories have been offered to explain the role of this much-studied promoter in vanadium oxide-alkali metal sulfate systems [131]. It has been suggested [132] that SO_3 can act as an oxygen carrier by carrying oxygen to reduced vanadium sites, or by reducing vanadium pentoxide to form oxygen which oxidizes the hydrocarbon [133]. Sulfur trioxide reacts with the alkali metals to form pyrosulfate, which in turn forms a melt with vanadium pentoxide.

The major difference between catalysts used for naphthalene feeds and those for *o*-xylene is in the support material. Different supports have been found to have relatively little effect in naphthalene oxidation, but showed marked differences in *o*-xylene oxidation [134]. The current trend is use of nonporous supports coated with a thin layer of the active component, a technique that makes most of the active component accessible, and tends to reduce over-heating and over-oxidation. Traditionally, supports were spheres or pellets, but advantages, including lower pressure drop, higher yields, and longer life, have been claimed for inert ring-shaped forms [135].

Recent developments in catalyst and reactor systems have been reviewed by Foster and Wainwright [136].

E. Acrolein and Acrylic Acid

Acrylic acid, mainly in the form of its esters, is produced in the United States in excess of 500 thousand metric tons annually. This production comes almost entirely from the oxidation of propylene. The oxidation, which may be achieved in either one or two stages, proceeds stepwise through acrolein [131].

$$CH_2{=}CHCH_3 + O_2 \xrightarrow[0.1\ldots0.2\,MPa]{640\,K} CH_2{=}CHCHO + H_2O$$

$$CH_2{=}CHCHO + 1/2\,O_2 \xrightarrow{540\,K} CH_2{=}CHCOOH$$

Single stage processes have the disadvantage of requiring a single catalyst and condition for two separate reactions with different kinetics. However, commercial installations of this type claim selectivities to acrylic acid of 86% [138].

In two stage processes, it is possible to optimize the catalyst and conditions for each step. In this way, selectivities to acrylic acid of 90% can be achieved. A great many patents have been issued for the oxidation of propylene to acrolein; most of these patents claim molybdenum oxide, many claim bismuth oxides as well, and most claim minor amounts of other components including tungsten, iron, phosphorus, tellerium, tin, antimony, cadmium, and copper. Cuprous oxide catalysts were the earliest commercial catalysts, but these appear to have been superceded largely by bismuth molybdate. Bismuth molybdate catalysts have activities and selectivvities much greater than either component alone.

Details of the mechanism of catalytic oxidation of propylene over bismuth molybdate catalysts has received much attention [139–143]. It is generally agreed that the rate determining step is oxidative dehydrogenation involving abstraction of an allylic hydrogen atom. Distribution of ^{13}C in acrolein indicates an allylic intermediate that is rapidly isomerized by exchange of ends prior to addition of an oxygen atom [144]. Some workers have evidence indicating a second hydrogen is abstracted before the incorporation of oxygen [145–146], whereas others favor incorporation of oxygen before loss of a second hydrogen. Several mechanistic schemes may operate simultaneously, as well as various surface initiated homogeneous reaction.

The origin of oxygen in acrolein has been determined by the use of various ^{18}O labeled bismuth molybdates. The oxygen atom of acrolein clearly comes from the $(Bi_2O_2)_n^{2+}$ layer whether or not gaseous oxygen is present. It is likely that oxide ions are replenished by gaseous oxygen reacting with $(MoO_2)_n^2$ layers [149].

Oxidation of propylene is carried out in tubular reactors with a mixture of steam, propylene and air, at 680–780 K and nominal pressures. Steam reduces the oxygen concentration below the explosive limits, helps clean the catalyst, and absorbs part of the liberated heat. The reaction is run with an excess of air over propylene to maintain the catalyst in a proper oxidation state. To help maintain a more even temperature distribution, the catalyst bed may be unevenly packed with catalysts and inerts, with the catalyst more dilute at the front end of the bed.

If acrolein is not the desired end product, the gaseous effluent from the first reactor is led directly into a second reactor where it is oxidized at 530 to 570 K to acrylic acid. The lower operating temperature in the second reactor is a reflection of the relative ease of oxidizing propylene and acrolein. Catalysts for acrolein oxidation are similar to first stage catalysts but contain different promoters.

F. Ammoxidation

Ammoxidation refers to a reaction in which a methyl group is converted to a nitrile on treatment with ammonia and oxygen. An outstandingly successful use of this reaction is in conversion of propylene to acrylonitrile.

1. Acrylonitrile

Acrylonitrile is an important organic compound used widely as a monomer or comonomer, and as a building block in a variety of organic syntheses. World annual capacity exceeds 4.5 million metric tons.

Earlier, acrylonitrile was made by reaction of hydrogen cyanide with acetylene, or ethylene oxide, or acetaldehyde, but these processes have now been replaced by ammoxidation of propylene [150, 151].

$$CH_3CH{=}CH_2 + NH_3 + 3/2\,O_2 \xrightarrow[200\,kPa]{670\ldots770\,K} CH_2{=}CHCN + 3\,H_2O$$

The reaction is carried out usually in a fluidized-bed and to a smaller extent in fixed-bed. The catalysts have undergone numerous refinements in the last two decades; the yields of by-product hydrogen cyanide and acetonitrile have been decreased, and conversion increased to a point where reaction is nearly complete on a once through basis [152]. Bismuth and molybdenum are frequent components of acrylonitrile catalysts which may also contain a variety of other elements including nickel, cobalt, iron, tin, antimony, phosphorous, and uranium, on supports such as silica.

The frequent occurrence of bismuth and molybdenum in these and propylene oxidation catalysts have made bismuth-molybdates the subject of extensive studies [153]. Oxidation of propylene will occur readily at 595 K over Bi-Mo catalysts, but with ammonia present, no propylene conversion occurs until about 675 K, perhaps because of blocking of propylene chemisorption sites by ammonia [154]. The rate determining step is generally considered to be an α-hydrogen abstraction to form an allylic intermediate, a process which equilibrates the ends of the molecule. The observed kinetics, first order in propylene, and zero order in oxygen and ammonia, are consistent with this suggestion. Subsequent steps are less well understood. A number of mechanistic proposals have been made and compared [155].

2. Other Ammoxidations

Aromatic methyl substituents readily undergo ammoxidation to form nitriles. A process for nicotinamide and nicotinic acid involves ammoxidation of *β*-picoline followed by ammonia catalyzed hydrolysis [156]. The sequence is much superior to direct vapor-phase oxidations of *β*-picoline to nicotinic acid, which gives low yields. Kinetics of vapor-phase ammoxidation of 4-picoline over vanadium, chromium, aluminum oxides have determined [151]. This ammoxidation has been achieved also over V-Ti-oxide catalysts.

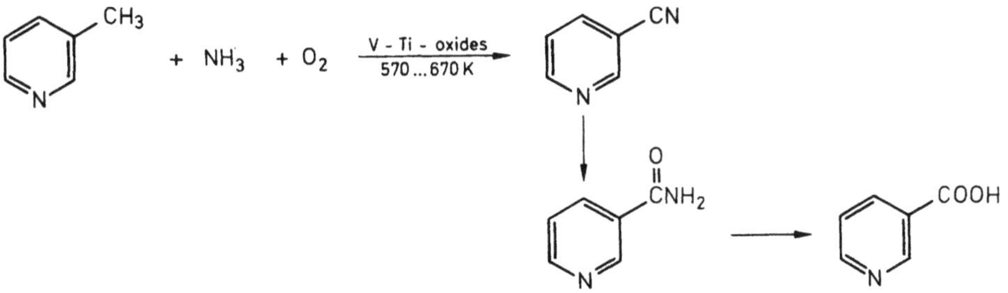

An ammoxidation process has been developed for the manufacture of terephthalic acid from *p*-xylene [159]. A major advantage of the process is elimination of aldehydic impures which are formed in oxidation routes. Phthalonitrile and isophthalonitriles are made similarly [160]. Hydrogenation of the dinitriles provides a convenient route to the diamines.

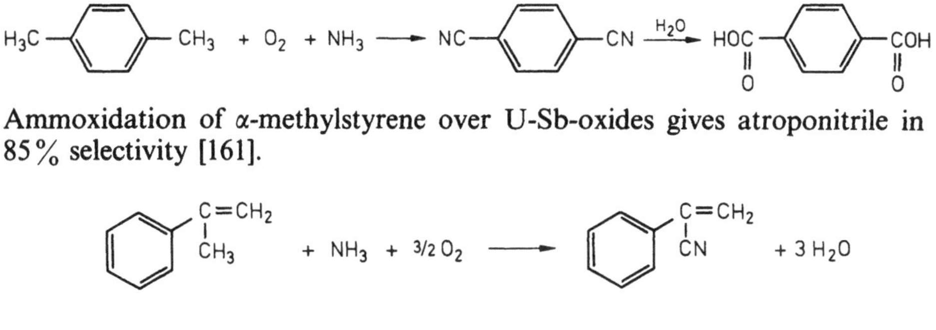

Ammoxidation of α-methylstyrene over U-Sb-oxides gives atroponitrile in 85% selectivity [161].

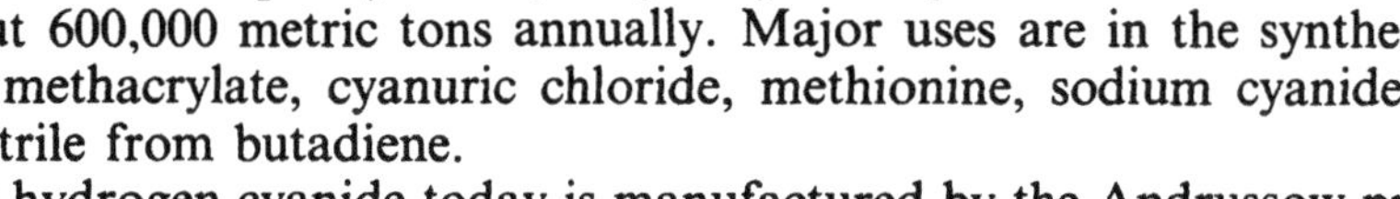

3. Hydrogen Cyanide

The world-wide capacity for hydrogen cyanide production is estimated to be about 600,000 metric tons annually. Major uses are in the synthesis of methyl methacrylate, cyanuric chloride, methionine, sodium cyanide, and adiponitrile from butadiene.

Most hydrogen cyanide today is manufactured by the Andrussow process or one of its many variations. The reaction in hindsight can be regarded as an ammoxidation of methane. Perhaps if it had always been viewed in this light, other ammoxidation reactions would have been discovered much sooner.

$$CH_4 + NH_3 + 1.5\ O_2 \xrightarrow{Pt} HCN + 3\,H_2O$$

The reaction is carried out over Pt (90%) Rh (10%) gauzes at temperatures around 1400 K, and nominal pressures with contact times the order of a few milliseconds [162–163]. The feed composition is within the range $CH_4/NH_3 = 0.7 \sim 1.7$ mole ratio and $air/(CH_4 + NH_3) = 2.80 \sim 3.25$ mole ratio [164]. In practice, some of the methane is burned to provide heat for the endothermic ammoxidation reaction.

Methane and ammonia in the absence of air will also produce hydrogen cyanide.

$$CH_4 + NH_3 \xrightarrow{Pt} HCN + 3H_2$$

The reaction is carried out in sintered tubes coated internally with alumina and noble metal. Heat is provided at 1475–1575 K by externally fired furnaces. This process provides an off-gas with more concentrated hydrogen cyanide than the Andrussow process, and is especially useful in locations where a demand for hydrogen exists. The Andrussow process advantages are those of lower investment and maintenance costs, and high yields based on natural gas.

Effluents are quenched quickly to temperatures below 675 K, to minimize decomposition and hydrolysis of hydrogen cyanide. The heat released is used to generate steam.

Ammoxidation of methane is carried out on the same Pt-Rh gauzes that are used in the manufacture of nitric acid by oxidation of ammonia. In the latter reaction, loss of metal adds an important operating cost to the manufacture [165], but in ammoxidation of methane, metal loss is negligible, despite about equal concentrations of oxygen in the feed in the two processes, and despite the temperature of ammoxidation being 200 K higher. However, the concentration of oxygen leaving the hydrogen cyanide process is much lower, due to its consumption by oxidation of methane [166].

In both ammonia oxidation and commercial synthesis of hydrogen cyanide, the catalyst undergoes an induction period during which the activity increases sharply and an increase in the surface area and roughness of the gauzes occurs, accompanied by formation of randomly oriented facets. Typically, complete activation of a hydrogen cyanide catalyst requires 60 to 80 hours on stream after which activity slowly declines giving an economical catalyst life of several thousand hours. Pan [164] examined the morphological changes which occur on use. At 200 × magnification, the new gauze had a very smooth surface, activated gauze (77 hours on stream) had developed whiskers, whereas at 2000 hours the whiskers had broken and the gauze wires were pitted. Surface area increased during these tests, but the area did not parallel activity. It is believed that formation of whiskers at the activation stage is responsible for both increased area and increased activity, but as time goes on, further increases in area are due mainly to carbon deposits with a consequent decrease in the number of active sites and activity [164].

The commercial rate of hydrogen cyanide formation is the order of 2×10^{19} mol s^{-1} cm^{-2} based on the original surface area of the gauze. This rate may be actually higher due to an over-estimation of how much of the gauze surface is used. The rates are approaching the maximum possible mass transfer rate [166].

G. Oxychlorination

Oxychlorination refers to processes in which oxidation of hydrogen chloride and chlorination occurs simultaneously. As applied to the synthesis of chlorobenzene, the reaction is $2\,\Phi H + 2\,HCl + {}^1/_2\,O_2 \rightarrow 2\,\Phi Cl + H_2O$ Oxychlorinations can be conducted in liquid phase in the presence of Friedel-Craft catalysts, such as ferric chloride, with the product as solvent, or in aqueous hydrochloric acid containing copper chloride, or in fixed- or fluidized-beds.

1. Vinyl Chloride

The most important application of oxychlorination is in synthesis of vinyl chloride. The capacity for vinyl chloride production in the United States in 1979 was in excess of 3,500,000 metric tons. Vinyl chloride, like acetic acid, acetaldehyde, and vinyl acetate, was prepared earlier from acetylene,

$$CH{\equiv}CH + HCl \xrightarrow{HgCl_2-C} CH_2{=}CHCl$$

but this process has now all but disappeared.

Synthesis of vinyl chloride by oxychlorination consists of three separate steps which are carefully intergrated in modern plants. The last step is an endothermic pyrolysis of 1,2-dichloroethane

$$ClCH_2CH_2Cl \xrightarrow[3MPa]{770 \ldots 870\,K} CH_2{=}CHCl + HCl$$

Pyrolysis is carried out in tubular reactors with contact times the order of 5 seconds, conversions of 50–60%, and selectivities to vinyl chloride of 98–99%.

Dichloroethane is derived by direct exothermic addition of chlorine to ethylene, usually carried out in liquid phase with Friedl-Craft type catalysts, such as $FeCl_3$

$$CH_2{=}CH_2 + Cl_2 \xrightarrow[FeCl_3]{EDC\ Solvent} ClCH_2CH_2Cl \quad ca\ 100\%$$

Hydrogen chloride derived in the pyrolysis unit is separated and led to the oxychlorination unit where it reacts with ethylene

$$CH_2{=}CH_2 + 2\,HCl + 1/2\,O_2 \xrightarrow[CuCl_2]{500 \ldots 580\,K} ClCH_2CH_2Cl + H_2O$$

Oxychlorination is carried out in gas phase in fixed- or fluidized-beds over catalysts such as copper chloride on inert material such as pumice or silica gel. These catalysts may contain alkali metal chlorides to reduce the volatility of the copper chloride, and rare earth metal chlorides to counteract the negative effects of alkali on rates. Oxychlorination catalysts and their mode of functioning have been reviewed [167–169]. The reaction is carried out at pressures ranging from 0.1 MPa–1.3 MPa, higher pressures being used with fixed-beds. Both oxygen and air are used; oxygen, despite its cost, offers an economic advantage especially at higher pressures. Reaction temperature is kept below 580 K, to minimize by-product formation.

Much effort has been devoted to minimizing the energy requirements of the process, a need that becomes more pressing with increasing energy costs. An ultimate goal is to combine all three steps into a single reaction, but this ideal situation has yet to be realized. Considerable progress has been made toward combining the oxychlorination step with the pyrolysis step, an innovation which would result in savings of both energy and investment. A major problem is development of an oxychlorination catalyst that is stable at temperatures needed to effect the pyrolysis.

Several processes have been developed as alternative ways of reusing hydrogen chloride resulting from the pyrolysis step. These include electrolysis to chlorine and hydrogen, oxidation to chlorine with nitrogen oxides and sulfuric acid (nitrosyl sulfuric acid) as catalyst, or oxidation over supported copper chloride, alkali, and rare earth chlorides.

2. Chlorination by Substitution

Oxychlorination can also be applied to systems where the chlorination is a substitution reaction rather than an addition. Industrial syntheses of a mixture of chloromethanes from methane and of chlorobenzene from benzene employ this reaction with various catalysts containing copper chloride. Conversions of benzene are usually limited to around 15% to minimize formation of dichlorobenzenes and to avoid excessive temperatures in the catalyst bed.

5. Metathesis

One of the most surprising reactions of the last few decades is olefin metathesis or olefin disproportionation in which olefinic carbon-carbon bonds are catalytically cleaved and recombined in a highly specific manner. For example,

$$2\,CH_3CH{=}CH_2 \rightleftharpoons CH_3CH{=}CHCH_3 + CH_2{=}CH_2$$

A number of both heterogeneous [170–171] and homogeneous [172–173] catalysts, all of which contain a transition metal, such as tungsten oxide, molybdenum oxide, or rhenium oxide supported on high surface silica or alumina, are known to catalyze this reaction. An increasing body of evidence suggests that metathesis proceeds through a carbene mechanism [174].

The reaction has found as yet limited industrial use but its potential remains high. Plants have been built for production of butene and ethylene from propylene [175]. The butenes are predominantly *trans*- and *cis*-2-butene. The process can be operated with tungsten oxide on silica or alumina at 573 to 773 K. The feed must be free of poisons, such as sulfur or oxygenated compounds, and life is extended if acetylenes and diolefins are first removed from the feed by selective hydrogenation over palladium. The reaction is cyclic and coke is removed by burn-off with air mixed with an inert gas.

The process can be operated in reverse if butenes and ethylenes are a glut and propylene in short supply. In this case, isomerization activity is incorporated in the catalyst to convert butene-1 to butene-2 [176].

In the Shell Higher Olefin Process, C_6–C_{18} linear olefins are produced from ethylene by a sequence of oligomerization to a mixture of alpha-olefins, catalyzed double-bond migration to a near equilibrium mixture of isomers, and metathesis of the resulting mixture. If detergent range olefins, C_{11}–C_{14}, are wanted, they are separated and the remainder recycled. All three reactions take place at moderate conditions, 353–413 K, 350–1700 kPa [177].

Neohexene, used in the synthesis of musk, is produced by disproportionation of diisobutylene with ethylene [176].

Disproportionation of cyclic olefins has been extensively studied and leads to high molecular weight compounds [178]. Disproportionation of a cyclic olefin should yield a mixture of cyclic oligomeric fractions, but the high molecular weight fraction is acyclic. Trace acyclic olefins probably account for the acyclic polymer since one molecule of an acyclic olefin can convert a cyclic polymer containing many thousands of cycloalkene units to an open chain structure. Polypentenamers have been obtained on an industrial scale by metathesis of cyclopentene.

A French company, CdF Chimie, is reported to polymerize norbornene to polynorbornene over a ruthenium based catalyst, and Huls of Germany make a polyoctenamer of high trans content [179].

6. Ammonolysis

A variety of important compounds are prepared by displacement of halogen or hydroxyl by ammonia. Amines derived from hydroxyls having an alpha hydrogen can be formed by reductive alkylation as well as displacement, i.e., the hydroxyl undergoes dehydrogenation to a carbonyl, followed by imine formation, and hydrogenation to the amine.

A. Aniline and Toluidines

There are three major routes to commercial aniline; catalytic or chemical reduction of nitrobenzene, ammonolysis of chlorobenzene, or of phenol. Ammonolysis of chlorobenzene is done at 475 K, under pressure with aqueous ammonia catalyzed by homogeneous cuprous chloride. Ammonolysis of phenol is carried out at 2 MPa and 600 K in adiabatic fixed-bed reactors.

$$C_6H_5OH + NH_3 \rightleftharpoons C_6H_5NH_2 + H_2O$$

There is only a small amount of heat liberated in this reversible reaction. The upper temperature limit is set by avoidance of ammonia cracking to hydrogen and nitrogen. Aniline as well as ammonia can function as a nucleophile, and some diphenylamine (1–2 %) also forms in the process; the amount formed is near the equilibrium value at the conditions of the reaction.

Two key developments aided in commercialization of this process. An active long-life catalyst was found as well as an efficient distillation for the separation of aniline and phenol in the product. The catalysts are described as Lewis-acid type alumina or silica-, titania, or zirconia-aluminas with surface areas of 100–150 $m^2 gm^{-1}$ combined with additional promoters. A feature of the catalyst is said to be limitation on alkali content [180–181].

Toluidenes can be made similarly from cresols.

B. Aliphatic Amines

Methanol reacts with ammonia in the presence of amination catalysts to give a mixture of methyl-, dimethyl-, and trimethylamines. The ratio of products can be adjusted at will by recycling any product formed in excess to obtain equilibrium mixtures. Over-yields based on both methanol and ammonia are in excess of 95 %. The catalysts are similar to those just described for production of aniline.

Higher alcohols react similarly, but the yields may be lower, due to dehydration. General conditions are 600–850 K, 5–30 MPa. A variant of the reaction is amination in the presence of hydrogen over dehydrogenation-hydrogenation catalysts such as copper or nickel; the reaction proceeds through an intermediate carbonyl compound. Hydrogen is not consumed in the reaction, but is added to keep the catalyst clean.

Fatty amines are not made by amination, but through hydrogenation of a fatty nitrile, readily available from dehydration of the corresponding ammonium salt of naturally occurring fatty acids. Excellent yields of primary amines can be obtained by hydrogenation of the nitrile over cobalt or nickel in the presence of ammonia at 410 K, 3.5 MPa.

References

1. Freifelder, M.: Catalytic Hydrogenation in Organic Synthesis, Procedures and Commentary, New York: Wiley Interscience, 1978
2. Kieboom, A. P. G., van Rantwijk, F.: Hydrogenation and Hydrogenolysis in Synthetic Organic Chemistry, Delft: Delft Univ. Press, 1977
3. Rylander, P. N.: Catalytic Hydrogenation in Organic Syntheses, New York: Academic Press, 1979
4. Vannice, M. A.: Catalytic Rev.-Sci. Eng. **14** (2), 153 (1976)
5. Stiles, A. B.: Amer. Inst. Chem. Eng. **23**, 362 (1977)
6. Cappelli, A., Collina, A., Dente, M.: Ind. Eng. Chem. Proc. Res. Develop. **11**, 184 (1972)
7. Williams, R. J. J., Cunnigham, R. E.: Ind. Eng. Chem. Prod. Res. Develop. **13**, 49 (1974)
8. Herman, R. G., Klier, K., Simmons, G. W., Finn, B. P., Bulko, J. B., Kobylinski, T. P.: J. Catal. **56**, 407 (1979)
9. Hydrocarbon Process., **56** (Nov.), 182 (1977)
10. Satterfield, C. N.: Heterogeneous Catalysis in Practice, New York: McGraw-Hill Book Co., 1980, p. 297
11. Rylander, P. N.: J. Am. Oil Chemists' Soc. **47**, 482 (1970)

12. Rylander, ref. 3, p. 7
13. Zajcew, M.: J. Am. Oil Chemists' Soc. **37**, 473 (1960)
14. Cherkaev, V. G., Bliznyak, N. V., Bag, A. A.: Tr. Vses. Nauchno-Issled. Inst. Sint. Nat. Dushistykh Veshchestv **1968**, 234; Chem. Abstr. **71**, 38200 (1969)
15. Zajcew, M.: J. Am. Oil Chemists' Soc. **35**, 175 (1958)
16. Pantula, A. J., and Achaya, K. T.: J. Am. Oil Chemists' Soc. **41**, 511 (1964)
17. Hydrocarbon Process. **58** (Nov.), 136 (1979)
18. Bird, A. J., Thompson, D. T.: Noble Metal Catalysis in Industrial Hydrogenations, in: Catalysis in organic syntheses, Jones, W. H. (ed.) New York, Academic Press, 1980, pp. 61–106
19. Kosak, J. R.: Ann. N.Y. Acad. Sci., **172**, 175 (1970)
20. Malone, R. J.: Chem. Eng. Progr. **76** (June), 53 (1980)
21. Spiegler, L.: U.S. Patent 3,073,865, Jan. 15, 1963
22. Dietzler, A. J., Keil, T. R.: U.S. Pat. 3,051,753, Aug. 28, 1962
23. German Patent 1,159,956
24. Kosak, J. R.: U.S. Pat. 4,020,107, April 26, 1977
25. Rylander, P. N., Karpenko, I. M., Pond, G. R.: Ann. N.Y. Acad. Sci. **172**, 266 (1970)
26. Rylander, P. N., Karpenko, I. M., and Pond, G. R.: U.S. Patent 3,715,397 Feb. 6, 1973
27. Benner, R. G.: U.S. Pat. 3,383,416 May 14, 1968
28. Benwell, N. R. W., Brit. Pat. 1,181,969, Feb. 18, 1970
29. Spiegler, L.: U.S. Pat. 3,073, 865, Jan. 15, 1963
30. Hydrocarbon Process. **58** (Nov.), 149 (1979)
31. Hydrocarbon Process **58** (Nov.), 150 (1979)
32. Hydrocarbon Process. **53** (Nov.), 145 (1979)
33. Taverna, M., and Chita, M.: Hydrocarbon Process. **49** (Nov.), 137 (1970)
34. Giuffre, L., Tempesti, E., Fornaroli, M., Sioli, G., Mattone, R., and Airoidi, G.: Hydrocarbon Process. **52** (Sept.), 199 (1973)
35. Mattone, R., Sioli, G., Giuffrè, L.: Hydrocarbon Process. **54** (Jan.), 85 (1975)
36. Weissermel, K., Arpe, H.-J.: Industrial Organic Chemistry, New York, Verlag Chemie, N.Y., 1978, p. 350
37. Adkins, H., Burks, R. E., Jr.: J. Am. Chem. Soc. **70**, 4174 (1948)
38. Montgomery, J. B., Hoffmann, A. N., Glasebrook, A. L., and Thigpen, J. I.: Ind. Eng. Chem. **50**, 313 (1958)
39. Greenfield, H.: J. Org. Chem. **29**, 3082 (1964)
40. Rylander, P. N., Hasbrouck, L., Karpenko, I.: Ann. N.Y. Acad. Sci. **214**, 100 (1973)
41. Nishimura, S., Shu, T., Hara, T., and Takagi, Y.: Bull. Chem. Soc. Jap. **39**, 329 (1966)
42. Nishimura, S., Kono, Y., Otsuki, Y., Fukaya, Y.: Bull. Chem. Soc. Jap. **44**, 240 (1971)
43. Greenfield, H.: Ann. N.Y. Acad. Sci, **214**, 233 (1973)
44. Kalina, M., Pasek, J.: Kinet. Katal. **10** (3) 574 (1969); Chem. Abstr. **71**, 54119 (1969)
45. Barkdoll, A. E., England, D. C., Gray, H. W., Kirk, W., Jr., Whitman, G. M.: J. Am. Chem. Soc. **75**, 1156 (1953)
46. Kochi, J. K.: Organometallic Mechanisms and Catalysis, New York: Academic Press, 1978, p. 2
47. Hydrocarbon Process. **58** (Nov.), 144 (1979)
48. Ref. 3, p. 194
49. Takagi, Y., Ishii, S., Nishimura, S.: Bull. Chem. Soc. Jap. **43**, 917 (1970)
50. Rylander, P. N., Steele, D. R.: U.S. Patent 3,387,048 June 4, 1968
51. Rylander, P. N., Vaflor, X.: Am. Chem. Soc. Northeast Reg. Meeting 6th, Burlington, VT, 1974
52. Rylander, P. N., Steele, D. R.: Engelhard Ind. Tech. Bull. **7**, 153 (1967)
53. Tinapp, P.: Chem. Ber. **102**, 2770 (1969)
54. Greenfield, H.: Ind. Eng. Chem. Prod. Res. Develop. **6**, 142 (1967)
55. Rylander, P. N., Hasbrouck, L., Karpenko, I.: Ann. N.Y. Acad. Sci. **214**, 100 (1973)
56. German Offen 2,053,799; Chem. Abstr. **75**, 64572a (1971)
57. British Patent 1,214,035; Chem. Abstr. **74**, 124788k (1971)
58. German Offen 2,101,911; Chem. Abstr. **75**, 109829b (1971)
59. German Offen 2,164,341; Chem. Abstr. **72**, 140721a (1972)

60. German Offen 2,038,104; Chem. Abstr. **76**, 112660r (1972)
61. Hydrocarbon Process. **56** (Nov.), 169 (1977)
62. Bartalini, G., and Gluggioli, M.: U.S. Pat. 3,821,305
63. Kirk-Othmer, Encyclopedia of Chemical Technology, New York: Wiley-Interscience, 3rd Ed., vol. 2, p. 363
64. Parshall, G. W.: Homogeneous Catalysis, New York: Wiley-Interscience, 1980, p. 219
65. Hatch, L. F., and Matar, S.: Hydrocarbon Process. **57** (Aug.), 164 (1978)
66. Kirk-Othmer, Encyclopedia of Chemical Technology, New York: Wiley-Interscience, 3 rd Ed., vol. 13, 1981, p. 16
67. Fedor, W. S., Millar, J., Accola, A. J., Jr.: Ind. Eng. Chem. **52** (Apr.), 282 (1960)
68. Wright, L. W.: Chemtech **4**, 42 (1974)
69. Kruse, W. M., and Wright, L. W.: U.S. Patent 3,963,789 June 15, 1976
70. Albert, R., Stratz, A., Vollheim, G.: Chem. Ing. Tech. **52**, 582 (1980)
71. Boyers, G. G.: U.S. Patent 2,866,847 Jan. 13, 1959
72. Wisniak, J., Simon, R.: Ind. Eng. Chem. Prod. Res. Develop. **18** (1), 50 (1979)
73. Sharkow, V. I.: Angew. Chem. Intern. Ed. Engl. **2**, 405 (1963)
74. Balandin, A. A.: Vasyunina, N. A., Chepigo, S. V., and Barysheva, G. S. Dokl. Akad. Nauk SSSR **128**, 941 (1959)
75. Cornils, B., Mullen, A.: Hydrocarbon Process. **59** (Nov.), **93** (1980)
76. Henrick, C. A.: Tetrahedron **33**, 1845 (1977)
77. Kirk-Othmer Encyclopedia of Chemical Technology, New York: Wiley Interscience, 2nd Ed., vol. 10, p. 671
78. Low, F. G.: French Demande 2,029,788; Chem. Abstr. **75**, 19684w (1971)
79. Wetherhill, F. E.: U.S. Pat. 3,449,445 June 10, 1969
80. Adam, K., Haarer, E.: German Pat. 1,285,992; Chem. Abstr. **70**, 67588n (1969)
81. Rylander, P. N., Cohn, G. J.: Liquid Phase Hydrogenations with Platinum-Ruthenium and Palladium-Ruthenium. In: Actes du deuxième congress international de catalyse, Editions Technique, Paris, France, 1961, pg. 977
82. Hydrocarbon Process. **58** (Nov.), 182 (1979)
83. Hydrocarbon Process. **58** (Nov.), 243 (1979)
84. Hydrocarbon Process. **58** (Nov.), 244 (1979)
85. Hydrocarbon Process. **58** (Nov.), 245 (1979)
86. Lee, E. H.: Catal. Rev.-Sci. Eng. **8**, 284 (1973)
87. Kearby, K. K.: U.S. Patent 2,426,829 Sept. 2, 1947
88. Hatch, L. F., Matar, S.: Hydrocarbon Process. **57** (Nov.), 291 (1978)
89. Chen, N. Y., Garwood, W. E.: J. Catal. **52**, 453 (1978)
90. Walsh, D. E., Rollmann, L. D.: J. Catal. **49**, 369 (1977)
91. Unland, M. L., Barker, G. E.: U.S. Pat. 4,115,424 Sept. 19, 1978
92. Itoh, H., Miyamoto, A., Murakami, Y.: J. Catal. **64**, 284 (1980)
93. Kaeding, W. W.: Catal. Rev.-Sci. Eng. **8**, 307 (1973)
94. Hydrocarbon Process. **58** (Nov.), 174, 175 (1979)
95. Wachs, I. E., Madix, R. J.: Surf. Sci. **76**, 531 (1978)
96. Ai, M.: J. Catal. **54**, 426 (1978)
97. Yamaguchi, M., Kageyama, Y.: German Pat. 2,144,316; Chem. Abstr. **76**, 126380 (1972)
98. Hiari, H., Miyata, K.: Japan Pat. 7242,832; Chem. Abstr. **77**, 111 105t (1973)
99. Murata, A., Kobayashi, H.: German Pat. 2,043,349; Chem. Abstr. **76**, 126378s (1972)
100. Dadyburjor, D. B., Jewur, S. S., Ruckenstein, E.: Catal. Rev.-Sci. Eng. **19** (2), 293 (1979)
101. Moiseev, I. I., Vargaftik, M. N., and Syrkin, Y. K.: Dokl. Akad. Nauk SSSR. **133**, 377 (1960)
102. Gates, B. C., Katzer, J. R., Schuit, G. C.: Chemistry of Catalytic Processes, New York: McGraw-Hill, 1979, p. 139
103. Bedell, K. R., Rainbird, H. A.: Hydrocarbon Process. **51** (Nov.), 141 (1972)
104. Nakamura, S., Yasui, T.: J. Catal. **17**, 366 (1970)
105. Nakamura, S., Yasui, T.: J. Catal. **23**, 315 (1971)
106. Arpe, von H-J, Hornig, L.: Erdöl Kohle Erdgas, Petrochem. **23**, 79 (1970)
107. Arzoumanidis, G. G., Rauch, F. C., Blank, G., Abstr. 163rd Meet. Amer. Chem. Soc., Inorg. Sect., paper 73

108. Bryant, D. R., McKeon, J. E., Ream, B. C.: J. Org. Chem. **33**, 4123 (1968)
109. Brownstein, A. N., List, H. L.: Hydrocarbon Process. 56 (Sept.), 159 (1977)
110. Takehira, K., Mimoun, H., DeRoch, I. S.: J. Catal. **58**, 155 (1979)
111. Hydrocarbon Process. **58** (Nov.), 167, 168, 169, 170 (1979)
112. Kilty, P. A., Sachtler, W. M. H.: Catal. Rev.-Sci. Eng. **10**, 1 (1974)
113. Murakami, Y., Tanaka, K.: Nippon Kagaku Kaishi, No. 11, 1603 (1977)
114. Verykios, X. E., Stein, F. P., Coughlin, R. W.: J. Catal. **66**, 147 (1980)
115. VanSanten, R. A., Moolhuysen, J., Sachtler, W. M. H.: J. Catal. **65**, 478 (1980)
116. Gans, M., Ozero, B. L.: Hydrocarbon Process. **55** (Mar.), 73 (1976)
117. Kiguchi, I., Kumazawa, T., Nakai, T.: Hydrocarbon Process. **55** (Mar.), 69 (1976)
118. French Patent 7,815,030 Dec. 22, 1978
119. W. German Offen 2,820,170 Dec. 7, 1978
120. Carra, S., Forzatti, P.: Catal. Rev.-Sci. Eng. **15**, 1 (1977)
121. Stein, F. P., Coughlin, R. W.: J. Catal. **66**, 147 (1980)
122. Hydrocarbon Process. **58** (Nov.), 189 (1979)
123. Hydrocarbon Process. **58** (Nov.), 188 (1979)
124. Wohlfahrl, K., Emig, G.: Hydrocarbon Process. **59** (June), 85 (1980)
125. Marlow, M.: Hydrocarbon Process. **59** (Nov.), 149 (1980)
126. Hucknall, D. J.: Selective Oxidation of Hydrocarbons, London: Academic Press, 1974
127. Varma, R. L., Saraf, D. N.: Ind. Eng. Chem. Prod. Res. Develop. **18** (1), 7 (1979)
128. Mars, P., van Krevelen, D. W.: Chem. Eng. Sci. **3** (special supplement), 41 (1954)
129. Hydrocarbon Process. **56** (Nov.), 194, 195, 196 (1977)
130. Hydrocarbon Process. **58** (Nov.), 207, 208, 209 (1979)
131. Wainwright, M. S., Foster, N. R.: Catal. Rev.-Sci. Eng. **19** (2), 211 (1979)
132. Wenderlein, H.: German Pat. 1,112,060 (1961)
133. Mizushima, F., Tanaka, T., Aigami, Y., Kakinoki, H., Suzuki, H.: Sekiyu Gakkai Shi. **7** (1), 15 (1964)
134. Mizushima, F., Tanaka, T., Aigami, Y., Kakinoki, H., Suzuki, H.: Sekiyu Gakkai Shi. **7** (11), 30 (1964)
135. Blechschmitt, K., Reuter, P., Wirth, F., Burger, G., Seubert, R.: U.S. Patent 4,036,783, July 19, 1977
136. Foster, N. R., Wainwright, M. S.: Process Chem. Eng. **31** (3), 28 (1978)
137. Schaal, G. E.: Hydrocarbon Process. **52** (Sept.), 218 (1973)
138. Hydrocarbon Process. **58** (Nov.), 123 (1979)
139. Voge, H. H., Adams, C. R.: Advan. Catal. **17**, 151 (1967)
140. Keulks, G. W.: J. Catal. **19**, 232 (1970)
141. Sachtler, W. M. H.: Catal. Rev.-Sci. Eng. **4**, 27 (1970)
142. Krenzke, L. D., Keulks, G. W.: J. Catal. **61**, 316 (1980)
143. Monnier, J. R., Keulks, G. W.: J. Catal. **68**, 51 (1981)
144. Voge, H. H., Wagner, C. D., Stevenson, D. P.: J. Catal. **2**, 58 (1963)
145. Adams, C. R., Jennings, T. J.: J. Catal. **3**, 549 (1964)
146. Adams, C. R., Jennings, T. J.: J. Catal. **2**, 63 (1963)
147. McCain, C. C., Godin, G. W.: Nature (London) **202**, 692 (1964)
148. Daniel, C., Keulks, G. W.: J. Catal. **24**, 529 (1972)
149. Miura, H., Otsubo, T., Shirasaki, T., Morikawa, Y.: J. Catal. **56**, 84 (1979)
150. Caporali, G.: Hydrocarbon Process. **51** (Nov.), 144 (1972)
151. Hydrocarbon Process. **58** (Nov.), 124 (1979)
152. Higgins, R., Hayden, P.: Catalysis Vol. 1, The Chemical Society, London, 1977
153. Sleight, A. W., Linn, W. J., Aykan, K.: Chemtech **8**, 235 (1978)
154. Burrington, J. D., Grasselli, R. K.: J. Catal. **59**, 79 (1979)
155. Burrington, J. D., Kartisek, C. I., Grasselli, R. K.: J. Catal. **63**, 235 (1980)
156. Paustian, J. E., Puzio, J. F., Stavropoulus, N., Sze, M. C.: Chemtech **11**, 174 (1981)
157. Das, A., Kar, A.: Ind. Eng. Chem. Proc. Res. Develop. **19**, 689 (1980)
158. Andersson, A., Lundin, S. T.: J. Catal. **65**, 9 (1980)
159. Gelbein, A. P., Sze, M. C., Whitehead, R. T.: Chemtech **3**, 479 (1973)
160. Sze, M. C., Gelbein, A. P.: Hydrocarbon Process. **55** (Feb.), 103 (1976)
161. Grasselli, R. K. et al.: J. Catal. **68**, 109 (1981)

162. Kirk-Othmer, Encyclopedia of Chemical Technology, New York: Wiley Interscience, 3rd Ed., vol. 7, 1979, p. 310
163. Weissermal, K., and Arpe, H-J.: Industrial Organic Chemistry, New York: Verlag Chemie, 1978, p. 42
164. Pan, B. Y. K.: J. Catal. **21**, 27 (1971)
165. Gillespie, G. R., Kenson, R. E.: Chemtech **1**, 627 (1971)
166. Schmidt, L. D., Luss, D.: J. Catal. **22**, 269 (1971)
167. Villadsen, J., Livberg, H.: Catal. Rev.-Sci. Eng. **17**, 203 (1978)
168. Allen, J. A., Clark, A. J.: Rev. Pure Appl. Chem. **21**, 145 (1971)
169. Kenney, C. N.: Catal. Rev.-Sci. Eng. **11**, 197 (1975)
170. Heckelsberg, L. F., Banks, R. L., Bailey, G. C.: Ind. Eng. Chem. Prod. Res. Develop. **8**, 259 (1969)
171. Bailey, G. C.: Catal. Rev.-Sci. Eng. **3**, 37 (1967)
172. Calderson, N., Chen, H. Y., Scott, K. W.: Tetrahedron Lett. (1967), 3327
173. Zuech, E. A.: Chem. Commun. (1968), 1182
174. Bassett, J. M., Leconte, M.: Chemtech **10**, 762 (1980)
175. Banks, R. L.: Catalyzed Disproportionation Reactions of Olefins; In catalysis in organic syntheses, ed. Jones, W. H., New York: Academic Press, 1980, p. 233
176. Banks, R. L.: J. Mol. Catal. **8**, 269 (1980)
177. Freitas, E. R., Gum, R.: Chem. Eng. Proc. **75** (Jan.), 73 (1975)
178. Hughes, W. B.: Chemtech **5**, 486 (1975)
179. Basset, J. M., and Leconte, M.: Chemtech **10**, 762 (1980)
180. Becker, M., Russell, J. L.: Chem. Eng. **80** (Apr. 2), 42 (1973)
181. Hydrocarbon Process. **58** (Nov.), 137 (1979)
182. Kirk-Othmer, Encyclopedia of Chemical Technology, New York: Wiley Interscience, 3rd Ed., vol. 2, p. 288

Chapter 2

Nature and Estimation of Functional Groups on Solid Surfaces

Hanns-Peter Boehm[1] *and Helmut Knözinger*[2]

1 Institut für Anorganische Chemie, Universität München, Meiserstraße 1, D-8000 München 2, FRG
2 Institut für Physikalische Chemie, Universität München, Sophienstraße 11, D-8000 München 2, FRG

Contents

1. Introduction

The lattice of a crystal is abruptly terminated at its surface. In consequence, the constituents of the lattice are unsaturated in their coordination, giving rise to the surface energy. Adsorption of substances which are foreign to the lattice may lower the energy of the system and reduce the coordinative "unsaturation". The interaction of the adsorptive with the adsorbent can very greatly in nature, *i.e.* from weak interaction by van der Waals forces, leading to physisorption, to strong chemical bonding or chemisorption. Chemisorption of molecules is frequently dissociative and often irreversible as well, that is the adsorptive cannot always be recovered in its original form. An example is the chemisorption of oxygen, O_2, on carbon giving rise to surface oxides [1] which yield CO and CO_2 on thermal decomposition.

It is evident that particularly stable surface compounds (also called surface complexes) are formed when the bonding of the hetero-atoms is similar in nature to the type of bonding within the solid. On covalently bonded solids such as diamond or graphite covalent bonding of the chemisorbed atoms or atom groups will be preferred; in the case of graphite with its strong structural anisotropy, covalent bonding without destruction of the structure is possible only at the edge of the carbon layers. On ionic crystals, foreign ions may be included and enriched in the surface even if their size would not allow substitution within the lattice. A special case is the metals. Many foreign atoms can be chemisorbed on metals; the bonding of non-metals to the surface is similar in nature to the bonding of ligands to metal atoms. In recent years great progress has been made in the preparation and study of metal-cluster complexes [2]. By imagining bigger and bigger metal clusters one can make in the mind a gradual transition to a metal with a chemisorbed layer. What has been explained for crystalline solids, holds analogously for amorphous solids. Also in amorphous substances, their constituent atoms show a preferred coordination. For instance in amorphous silica, the silicon atoms are tetrahedrally coordinated by oxygen, and oxygen is bonded to two silicon atoms in the same way as in crystalline SiO_2.

It must be kept in mind that the atoms (ions) in the surface of a solid do not necessarily retain the positions prescribed by the lattice structure. Reconstruction of atomically "clean" surfaces has been demonstrated by LEED in many cases. Obviously, this reconstruction lowers the surface energy by reducing the "unsaturation" of the surface. But also after chemisorption of foreign atoms a distortion of the surface layer can occur, and coordination numbers of surface atoms different from coordination within the lattice cannot be excluded.

The title of this chapter implies that it deals with surface atom groups and not with singly chemisorbed atoms. However, a clear distinction between surface functional groups and chemisorbed atoms or ions is not straightforward. Addition of H^+ to surface oxygen ions produces OH ions or groups which may react as an entity. Therefore, also chemisorbed atoms or ions will be included in the discussion if the resulting structure has the characteristics of a functional group. On the other hand, chemisorbed atoms or atom groups on metals will not be included since they are dealt with in detail in another chapter of this series. The chemistry of surface groups on certain solids has been described in detail several years ago [3]. The aim and scope of this chapter in the present series is not a comprehensive coverage of all possible surface groups on all known solids, but rather a systematic description of the general principles of their *identification and characterization*, their *generation*, and their *reactivity*.

It is obvious that these aspects are closely related to the surface properties of heterogeneous catalysts and catalyst precursors. Surface groups might themselves function as catalytically active species. They may determine, for instance, the metal-support interactions as in supported metal catalysts, and they may be involved in the bonding of active species on the surface. Examples of the latter type are the ion exchange adsorption of catalyst precursors or the function of surface groups as ligands for immobilized complexes and clusters. Because of the importance of the chemical properties of surface groups for the synthesis of catalysts, sections on surface group reaction chemistry and immobilization of complexes will also be included in this chapter.

2. Characterization, Identification, Estimation

A. Spectroscopic Methods

The identification and characterization of surface functional groups requires information on chemical composition and molecular structure of surface groups. A large variety of spectroscopic methods is available today which provide sufficient sensitivity and surface specificity for a meaningful detection of surface groups.

Vibrational spectroscopies [4–17] are certainly the most promising and most widely used methods for this purpose due to the very detailed information on molecular structure which is provided by vibrational spectra. Principally, photons, electrons and neutrons can be used as probes. A wide variety of special techniques has been developed for the photon vibrational spectroscopies, which have to be selected according to the surface and optical properties of the system to be studied. Because of their relative simplicity and wide applicability, infrared transmission absorption and laser Raman spectroscopy are most frequently used today, while diffuse reflectance, internal reflection and optoacoustic spectroscopy have found less application for vibrational studies. Double resonance excitation of fluorescence [18] may possibly find advantageous application for the vibrational characterization of surfaces and surface functional groups, though no example has yet been reported. This technique would provide the sharp lines characteristic of Raman or infrared spectroscopy but would have an increased sensitivity characteristic of fluorescence spectroscopy. If kinetic phenomena in the context of surface group reaction chemistry are to be studied by vibrational spectroscopy, time resolved infrared spectroscopy using rapid scanning Fourier transform (FT) spectrometers would be essential [19, 20].

Two electron vibrational spectroscopies have been developed. Electron energy loss spectroscopy (EELS) can hardly be applied to rough surfaces of polydisperse materials. Inelastic electron tunneling spectroscopy (IETS) on the other hand, is an interesting sophisticated technique which provides high resolution spectra within the entire vibrational frequency range. The experiment, however, requires very costly apparatus and is far less widely applicable than infrared transmission or Raman spectroscopy.

Inelastic neutron scattering is an attractive spectroscopy for vibrational analysis of surface species, namely hydrogen-containing surface groups, but it requires a neutron source with a sufficiently high flux.

The diffuse reflectance technique [21–24] is applied nearly exclusively when optical spectra in the visible and/or ultraviolet region of the electromagnetic spectrum are studied. These electronic spectra provide information on the molecular structure of surface groups or probe molecules interacting with surface functional groups as well as on electronically excited surface groups and perhaps vibrational modes in electronically excited states. Optical spectroscopy is moreover a powerful tool for the characterization of "heterogenized" transition metal complexes. Photoluminescence was less frequently used in the past, although it has proved rewarding for the characterization of surface electronic states and defect centers.

Magnetic resonance techniques are extremely valuable for molecular structure determination. However, there are certain limitations of the applicability of nuclear (proton) magnetic resonance (NMR) [16, 25–28] for the study of surface groups. Wide-line NMR has nevertheless successfully been applied for the characterization of surface hydroxyl groups and relaxation time measurements are undoubtedly most informative regarding the dynamic behavior of protons on surfaces. High resolution ^{13}C-NMR has been used for the elucidation of the molecular structure of surface groups and adsorbed molecules. Electron spin resonance (ESR) [16, 29–31] is the most sensitive traditional technique and provides important structural information of paramagnetic ions or atoms. The spin labeling technique may be applied in systems which are not paramagnetic themselves.

The electron spectroscopies [16, 32–34] of interest for surface functional group characterization are Auger electron and X-ray photoelectron spectroscopy (AES and XPS or ESCA). Both spectroscopies permit an elemental analysis within the topmost atomic layers of a solid surface, the only elements of chemical interest excluded from detection being hydrogen and lithium for AES and hydrogen for XPS. In favorable cases chemical shifts observed in photoelectron spectra also allow conclusions to be drawn on the chemical environment and/or the oxidation state of a particular element. Thus, very useful chemical and structural information can be deduced from XPS although quantitative measurements are difficult, particularly on porous polydisperse materials.

The surface spectroscopies which use ion probes, such as inelastic ion scattering spectroscopy (ISS) [34–37] and secondary ion mass spectrometry (SIMS) [33, 34] are destructive in nature. ISS has been applied only in very few cases for determination of the elemental composition within the topmost atomic layer of high-area solids of catalytic interest. No direct structural information can be obtained from ISS on such surfaces. SIMS on the other hand has been shown in recent years to allow conclusions to be drawn on structural details of surface groups from molecular fragments which are observed in the secondary ion mass spectra. This is possible at least under favorable so-called static conditions.

Finally, thermal desorption spectroscopy (TDS) [38–40] which was originally introduced as temperature programmed desorption (TPD), is a very useful technique which found frequent application due to its relative simplicity and breadth of information. Regarding surface functional groups, TDS would provide information on their thermal stability and on kinetic parameters of their possible decomposition or desorption, an analysis of the decomposition products can be made and quantitative analysis would determine the number of surface groups per unit surface area.

We will demonstrate the usefulness and applicability of those of the above mentioned spectroscopic techniques which are most relevant for the characterization, identification and possible estimation of surface functional groups by means of typical selected samples.

1. Vibrational Spectroscopies

a) Infrared transmission-absorption spectroscopy

Transmission infrared spectroscopy and its application to surface chemistry has been reviewed in the past [4–17]. The principle is well known from conventional infrared spectroscopy of solids in the transmission mode (thin pressed self-supporting wafers are usually being used for surface studies [13]). The applicability of the transmission technique is determined by the properties of the solid powder to be studied. Thus, samples which exhibit only weak bulk absorption, and the average particle size $\bar{d}$ of which is smaller than the wavelength of the infrared radiation in the region of interest will be optimally suited for the transmission mode. The particle size condition ($\lambda \gg \bar{d}$) which determines the wavelength range of suitably low scattering losses, is usually met in the mid and far infrared region in practice, whereas scattering losses become strongly involved in the near infrared region. On the other hand, most samples show strong bulk absorption in the low wavenumber region (roughly $< 1000\ cm^{-1}$). As a result, the accessible wavenumber range for transmission infrared spectroscopy will generally be limited for surface studies to the mid-infrared region ($4000 > \tilde{\nu} > 1000\ cm^{-1}$). However, when less than optimal spectroscopic conditions can be accepted, spectra may be obtained also in the near infrared [7]. An alternative for the reduction of scattering losses would be the immersion technique [41] in which the solid is immersed in a solvent having approximately the same refractive index (*e.g.* SiO_2 immersed in CCl_4).

Infrared transmission spectroscopy is a bulk rather than a surface specific technique. It is therefore necessary to prove for any detected species that it is a surface group. This can be realized in many cases by following changes in band position on exposure of the solid adsorbent to a suitable absorptive or by isotopic exchange experiments.

The sensitivity of the technique is dependent on the extinction coefficient of the surface groups which may vary from $5 \times 10^{-18}\ cm^2\ molecule^{-1}$ for the carbonyl stretching mode in CO ligands to between 10^{-20} and $10^{-19}\ cm^2\ molecule^{-1}$ for CH stretching modes in saturated hydrocarbon chains. The magnitude of the extinction coefficients renders a high surface-to-volume ratio desirable, the more so as the possible increase in sample thickness is limited by the concomitant increasing energy losses by absorption and scattering. Thus, the sensitivity of the technique strongly depends on the nature of the surface group and the physical properties of the solid. Assuming typical values of $100\,m^2\,g^{-1}$ for the adsorbent surface area, $20\,mg\,cm^{-2}$ for the weight of the irradiated geometric area, $10^{-19}\,cm^2$ molecule^{-1} for the extinction coefficient and 5% for a desirable absorption in order to obtain good quality spectra with standard infrared spectrometers, one estimates a lower limit of 0.02 for the surface coverage θ. This shows that transmission spectra can be obtained even at coverages below one tenth of monolayer. With the application of data acquisition techniques which are available today even for conventional dispersive spectrometers, the sensitivity of the technique can be increased further. Quantitative measurements of surface group densities should be possible, provided the Lambert-Beer law is applicable. It must be kept in mind, however, that this law is valid for optically homogeneous materials and deviations may occur for disperse substances.

A number of different transmission cells have been described in the literature. Typical designs of cells for *in situ* work at low and high temperatures are discussed in the reviews already mentined [8–15] and by Hobert *et al.* [42]. Gallei and Schadow [43] described a cell which permits work at pressures up to 3 MPa and in ultra-high vacuum at temperatures up to 870 K.

Transmission infrared measurements have largely been carried out with conventional dispersive instruments in the past. Long recording times are necessary to obtain sufficiently high signal-to-noise ratios. S/N, because of the generally low energy level (low transmittance of samples) even though high resolution is usually not required. Moreover, the scattering and absorption background of the solid adsorbent will be superimposed in the spectra. This can be compensated for by using an optically equivalent wafer in the reference beam or by computer elimination. The advantages of FT infrared spectrometers [44, 45] over dispersive spectrometers are twofold. Firstly, the energy throughput is rather high thanks to the relatively large circular entrance aperture. This, however, cannot always be utilized fully since it may be difficult to prepare sample wafers of equivalent size. The second, more important advantage of FT spectroscopy is the so-called Felgett or multiplex advantage, which allows a spectrum to be measured with a FT spectrometer in the same time but with considerably better S/N than with a conventional spectrometer, or in much shorter time with an equivalent S/N. This is obviously very helpful when spectra accumulation and signal averaging techniques have to be applied because of the lower scanning time needed. However, the cost of FT spectrometers is still considerably higher than that of the best conventional computerized spectrometers. An approach to *in situ* FT infrared spectroscopy on surfaces has recently been presented by Bouwman and Freriks [46].

i) Surface hydroxyl groups on oxide surfaces. Oxide surfaces are preferentially terminated by hydroxyl groups for energetic reasons [3]. They play a most important role as surface functional groups in catalysis and surface chemistry as either acidic or basic sites or as reactive surface groups. OH groups represent diatomic surface oscillators, which give rise to typical infrared spectra. They can therefore be considered as intrinsic surface probes, the vibration frequencies of which contain information on the coordination of the probe and hence on the local surface structure. Quantitative infrared spectroscopy permits the determination of the OH surface density and the effects of intermolecular interactions with suitable probe molecules provide information on the chemical properties of the surface OH groups, such as acidity and H-bond donor and/or acceptor strength.

Adsorbed molecular water may complicate the identification of surface OH groups since the OH stretching vibrations of both species occur in the same wavenumber range (3200–3800 cm^{-1}). However, the $\tilde{\nu}_2$ deformation vibration of the water molecule occurs between 1600 and 1650 cm^{-1}, whereas the deformation vibration of *e.g.* surface silanol groups was found near 870 cm^{-1} [47, 48]. A discrimination between surface OH groups and adsorbed molecular water becomes therefore possible on this basis, as shown for the Al_2O_3/H_2O [49, 50] and TiO_2/H_2O [51] systems. The absence of the deformation band near 1600–1650 cm^{-1}, however, does not necessarily indicate the complete desorption of molecular water, since the extinction coefficients of the deformation vibration is significantly lower than that of stretching vibrations.

The combination bands ($\tilde{\nu}_2 + \tilde{\nu}_3$) of molecular water and ($\tilde{\nu}_{OH} + \delta_{OH}$) of hydroxyl groups permit a distinction between both species. The ($\delta_2 + \delta_3$) band of water adsorbed on silica surfaces has been observed between 5100 and 5300 cm^{-1} (depending on the degree of hydration) while the ($\tilde{\nu}_{OH} + \delta_{OH}$) band of surface silanol groups is located at about 4550 cm^{-1} [47, 52–55]. As an example, Figure 1 shows the overtone and combination spectra (these are diffuse reflectance spectra, see section 2.A.1.b.ii) of a dehydrated aerosil sample for three dehydroxylation temperatures. The ($\tilde{\nu}_{OH} + \delta_{OH}$) band of the silanol groups occurs at 4550 cm^{-1} and the first overtone of the O—H stretching mode $\tilde{\nu}_{02}$ is found at 7285 cm^{-1} (the fundamental vibration $\tilde{\nu}_{01}$ is near 3750 cm^{-1}).

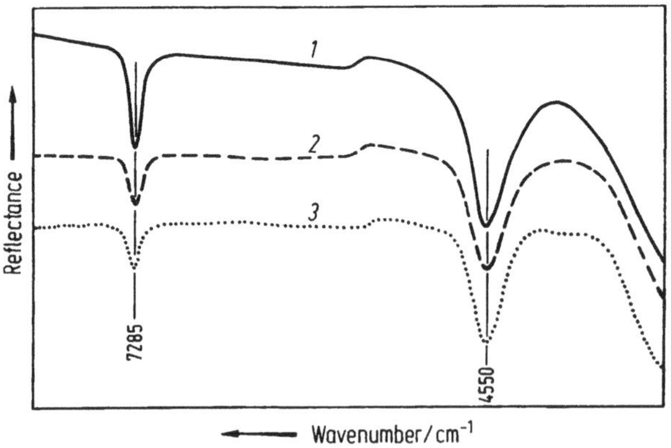

Figure 1. Diffuse reflectance spectra in the near infrared region of a SiO_2 surface. The specimen was dehydroxylated at 473 K (*1*), 773 K (*2*), and 973 K (*3*)

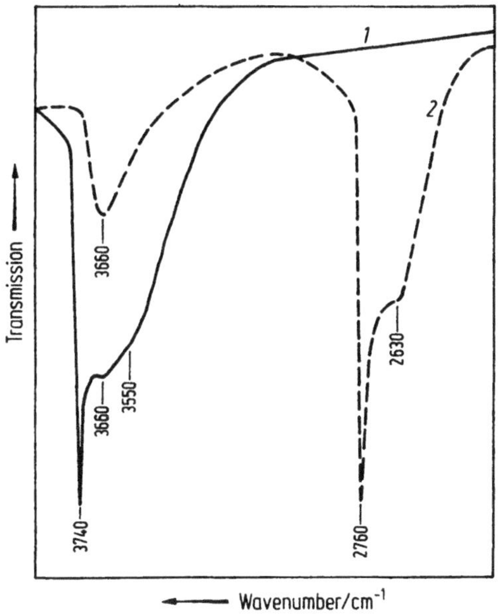

Figure 2. Infrared transmission spectra of hydroxyl (*1*) and deuteroxyl (*2*) groups on a SiO_2 surface (heat treatment 473 K)

A second problem is related to the discrimination between surface and inaccessible internal or bulk hydroxyl groups, which exist *e.g.* in silica particles. As shown in Figure 2, an Aerosil surface after dehydration at 473 K gives rise to a sharp band at 3740 cm^{-1}, a broader feature with maximum at 3660 cm^{-1} and a shoulder near 3550 cm^{-1}. The 3660 cm^{-1} band had been attributed to neighboring silanol groups which are perturbed by mutual H-bonding [56, 57]. More recently, however, this band was ascribed to inaccessible internal or bulk OH groups [58, 59], whereas the band at 3550 cm^{-1} is assumed to be due to mutually interacting surface silanol groups [58, 60, 61]. Deuterium-exchange with D_2O quantitatively shifts the 3740 cm^{-1} band to 2760 cm^{-1} and the shoulder near 3550 cm^{-1} to 2630 cm^{-1} (corresponding to a wavenumber shift by a factor of 0.74), while a broad band at 3660 cm^{-1} is observed in the OH stretching region which does not find an isotope-shifted analogue in the OD stretching region (see Fig. 2). The corresponding OH groups are thus inaccessible for deuterium-exchange and must be assigned to internal OH-groups, which cannot interact with adsorbed species [61].

As an example Figure 3 shows spectra in the OH-stretching region for a number of oxides and Table 1 summarizes reported vibrational frequencies of isolated OH and OD groups on the surfaces of various pure oxides.

The assignment of the OH stretching frequencies to certain types of OH species deserves detailed discussion. Peri [88] considered the (100) crystal plane of γ-Al_2O_3 exclusively as terminating the crystallites. He suggested the local environment of an OH group in the plane (varying number of O^{2-} neighbors depending on the degree of hydroxylation) as being responsible for the occurence of five distinct OH bands. The possibility of other

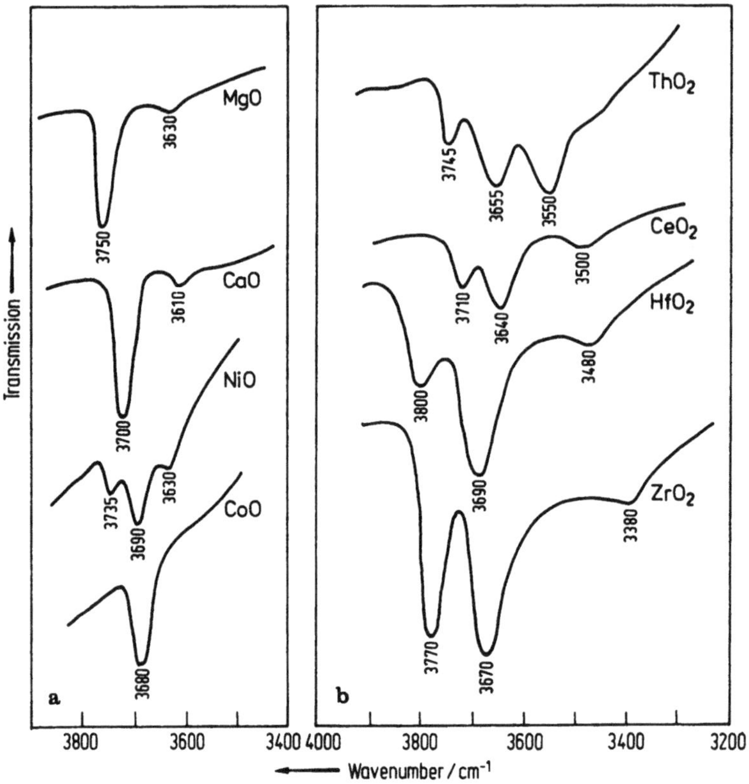

Figure 3. Infrared spectra of surface hydroxyl groups on various oxides. **a** MgO, CaO, NiO, CoO after evacuation at 723, 773, 593 and 473 K, respectively. **b** ZrO_2, HfO_2, CeO_2, ThO_2 after evacuation at 770 K. (Reproduced with permission from ref. [62])

terminating low index planes with different coordination of the surface OH groups was not considered in this model. On the contrary, most authors prefer a correlation between OH stretching frequencies of isolated OH groups and 3665 cm^{-1}, while Munuera *et al.* [89] prefer the (111) plane. The (110) various crystallographic planes existing in a given oxide. For instance, for TiO_2 (anatase) Primet *et al.* [79, 80] suggest two configurations of OH groups in the (001) plane to explain the two observed frequencies at 3715 and 3665 cm^{-1}, while Munuera *et al.* [89] prefer the (111) plane. The (110) plane was the favored crystal face for the assignment of OH frequencies on TiO_2 (rutile) [79, 81, 90]; Jones and Hockey [91], however, consider all three low index planes (110), (101) and (100). Dent and Kokes [92] suggested a model for the surface of ZnO which assumes the (0001) and $(000\bar{1})$ planes of the wurtzite structure to form the external crystal surface layer. Each of these two faces provides one possible configuration for surface OH groups, a singly coordinated OH in the (0001) plane and a triply bridging group in the $(000\bar{1})$ plane as shown in Figure 4. The observed narrow bands of

Table 1. Vibrational frequencies of free hydroxyl and deuteroxyl groups on pure oxide surfaces

Oxide	$\tilde{\nu}_{OH}/cm^{-1}$	$\tilde{\nu}_{OD}/cm^{-1}$	Ref.
BeO	3730, 3620	2745, 2660	[63]
	3735, 3630		[62]
MgO	3752, 3610		[64]
	3745		[65]
	3730, 3610	2760, 2670	[66], [67]
	3740		[68]
	3745		[63]
	3746, 3610		[69]
CaO	3707, 3695	2735, 2719	[70]
	3765, 3650		[71]
	3700, 3610		[62]
Al_2O_3	3680–3700; 3733; 3740–3745; 3760–3780; 3785–3800	2725–2733; 2755–2760; 2790–2803	[52], [72], [73]
SiO_2	3745–3750	2760	[52], [62]
GeO_2	3673	2702	[74]
SnO_2	3640	2640	[75]
TiO_2 (anatase)	3725, 3670, 3645		[77]
	3728, 3707, 3672, 3654, 3636		[78]
	3715, 3665		[79], [80]
TiO_2 (rutile)	3740, 3690, 3660	2755, 2730, 2700	[76]
	3685		[79]
	3700, 3670		[81]
α-Cr_2O_3		2700, 2675	[82]
ZnO	3670, 3620		[83]
	3680, 3620		[84]
ZrO_2	3770, 3670	2780, 2703	[85]
(monoclinic)	3760, 3660		[86]
(tetragonal)	3725, 3662		[86]
ThO_2	3742	2758	[87]

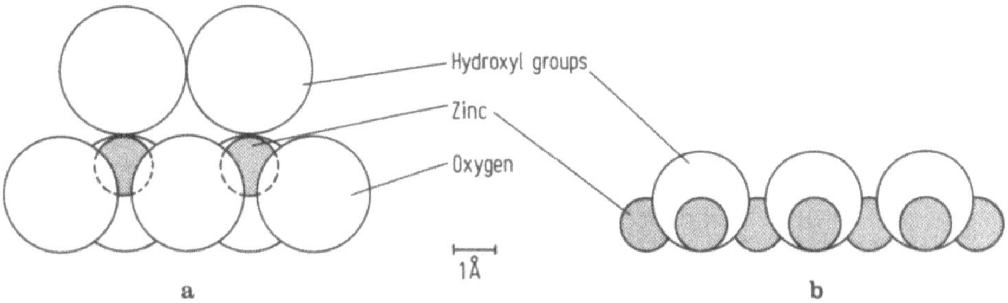

Figure 4. Hydroxyl groups on the surface of ZnO: **a** (0001) face; **b** (000$\bar{1}$) face. (Reproduced with permission from ref. [84])

isolated OH groups at 3660–3685 cm^{-1} and 3610–3630 cm^{-1} were therefore attributed to these two configurations [83, 84].

Tsyganenko and Filimonov [62] have generalized this approach to explain the frequently observed multiplicity of OH stretching bands by the assumption

Table 2. Vibrational frequencies of free surface hydroxyls on oxides with different crystal structures

Oxygen coord. number	Oxide	Frequency/cm^{-1}			
		Type I	Type II	Type III	Type IV
2	SiO_2	3750			
3	TiO_2	3725	3670		
	(anatase)		3650		
4	γ-Al_2O_3	3800	3740	3700	
	BeO	3735		3630	
	ZnO		3675	3622	
	ThO_2	3745		3655	
	CeO_2	3710		3640	
	HfO_2	3800		3690	
	ZrO_2	3770		3670	
6	MgO	3750			3630
	CaO	3700			3610
	NiO	3735		3690	3630
	CoO			3680	

Reproduced with permission from ref. [62].

that the surface hydroxyl oxygen atom may be bound to different numbers of lattice metal atoms. Coordination numbers of 1 to 4 for hydroxyl groups are found in many compounds [93]. For simplicity, the various OH groups having coordination numbers 1, 2, 3 etc. are referred to as type I, II, III, *etc.* hydroxyls

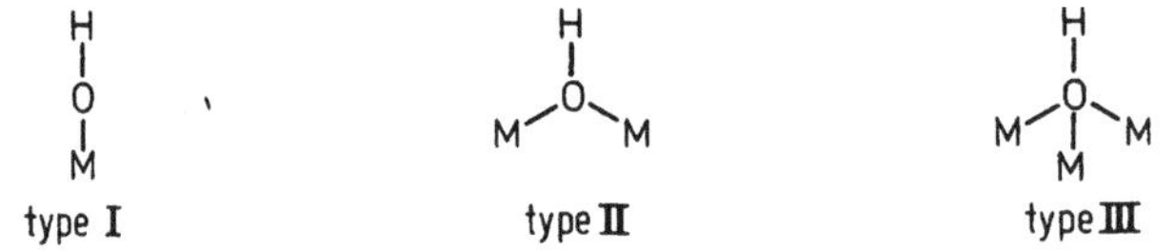

Table 2 summarizes the assignments given by Tsyganenko and Filimonov [62] on this basis. The first row in Table 2 gives the oxygen coordination number in the bulk lattice of the respective oxide. The maximum coordination number of a surface (hydroxyl) oxygen atom must necessarily be less by one unit.

The oxygen coordination number in SiO_2 is two; only one OH configuration type I is therefore expected which gives rise to a single sharp band at about 3750 cm^{-1} (Fig. 2) provided it is unperturbed. Further details regarding surface OH groups on silicas and more recent literature data will be discussed below.

In two crystalline forms of TiO_2, anatase and rutile, every oxygen atom is surrounded by three Ti atoms placed at the apices of an isosceles triangle. From this, Tsyganenko and Filimonov [62] infer that on the surface of anatase as well as rutile type I and II hydroxyls should be expected. Since the metal atoms bonded to the hydroxyl oxygen can occupy two fixed mutual positions corresponding to the side and base of the isosceles triangle, two kinds of type II hydroxyls may exist. After heat treatment at 643 K three intense bands have indeed been observed [94] (see Table 2) on the anatase surface at 3725, 3670 and 3650 cm^{-1}. These are assigned by Tsyganenko and Filimonov [62] accordingly to type I and the two kinds of type II hydroxyls, respectively.

BeO and ZnO crystallize in the wurtzite structure in which each oxygen atom is tetrahedrally surrounded by four metal atoms. As shown in Figure 4, singly bonded and triply bridging OH groups would be expected for the (0001) and (000$\bar{1}$) faces in this structure, which correspond to type I and III hydroxyls according to the nomenclature of Tsyganenko and Filimonov [62]. Thus, the two bands observed for BeO at 3735 and 3630 cm^{-1} are assigned to these free OH groups. In the case of ZnO, Tsyganenko and Filimonov [62] disagree with the previously mentioned assignment of Atherton *et al.* [83] and Mattmann *et al.* [84] of the reported two sharp bands of isolated OH groups at 3680 and 3620 cm^{-1} to type I and type III hydroxyls, respectively. They argue, that Zn^{2+} ions due to their relatively large ionic radius, may have a higher coordination number with respect to oxygen in the surface than in bulk ZnO. Therefore on the (0001) face the type I hydroxyls may be displaced and they may occupy positions between Zn atoms. Tsyganenko and Filimonov [62] thus assign the two observed bands of isolated OH groups on ZnO surfaces to type II and type III hydroxyls. It should be mentioned that the prism faces have never been included in this discussion.

ThO_2 and CeO_2 crystallize in CaF_2-structure with tetrahedral coordination of the oxygen atoms, while HfO_2 and ZrO_2 can be considered as having distorted CaF_2-lattices. One therefore expects type I, II and III hydroxyls on the surface of these oxides. Actually, two sharp bands of isolated OH groups are observed in the region between 3650 and 3800 cm^{-1} (Fig. 3a and Table 2) which can reasonably be assigned to type I and III hydroxyls. These are produced on crystal faces such as (111) which intersect one bond of the OM_4 tetrahedron

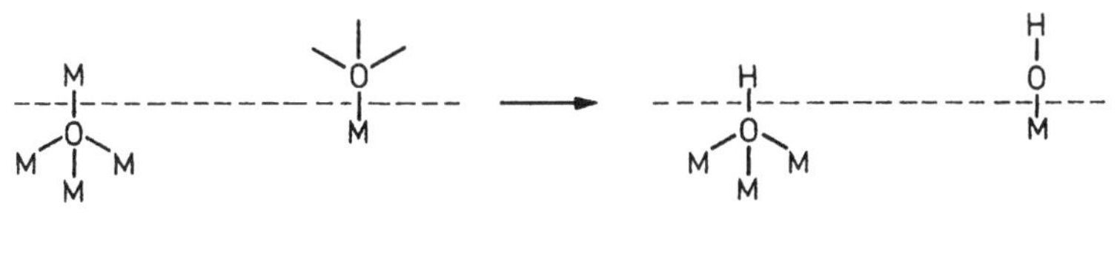

The additional low frequency bands observed for the four oxides (Fig. 3a) can be assigned to mutually H-bonded OH-groups [62].

MgO, CaO, NiO and CoO have the NaCl structure in which oxygen atoms are octahedrally coordinated to six metal atoms. Depending on the cleavage plane, OH groups of type I and V, II and IV, or type III exclusively can exist on the surfaces of these oxides. On MgO and CaO, the (100) faces usually predominate as cleavage planes which produces type I and V hydroxyls. The two sharp bands observed for MgO at 3750 and 3630 cm^{-1} and CaO at 3700 and 3610 cm^{-1} were therefore assigned to these two kinds

Table 3. Comparison of the OH(OD)-stretching fundamentals (cm^{-1}) for γ-, η-, and δ-Al_2O_3

	γ-Al_2O_3	γ-Al_2O_3	η-Al_2O_3	η-Al_2O_3	γ-Al_2O_3 + + δ-Al_2O_3	γ-Al_2O_3 + + δ-Al_2O_3
Reference	[95], [96]	[97]	[98]	[73]	[99]	[100]
$\tilde{\nu}_{OH}$	3700	3700	3710	3695	3670	3680
	3733	—	—	—	3685	—
	3744	3745	3740	3730	3727	3730
	3780	3760	—	3775	3775	—
	3800	3785	3785	3795	3790	3795
$\tilde{\nu}_{OD}$	2733	2725	2730			2715
	2759	2760	2755			2755
	2803	2790	2790			2800

of OH groups by Tsyganenko and Filimonov [62]. The single sharp low frequency band at 3680 cm^{-1} of isolated OH groups on CoO surfaces was tentatively attributed to a type III hydroxyl which may occur when the (111) face of the NaCl structure is predominantly exposed at the surface. On NiO surfaces two additional bands of lower intensity were observed at 3735

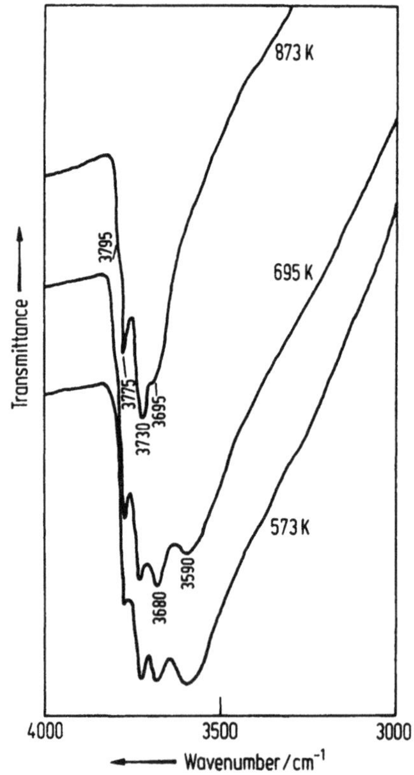

Figure 5. Infrared spectra of η-Al_2O_3 in the OH stretching region after various pretreatment temperatures. (Reproduced with permission from ref. [73])

and 3630 cm^{-1} besides the strongest type III band at 3690 cm^{-1}. These two bands are believed to belong to type I and V hydroxyls on (100) faces, respectively [62].

A special and most complicated case is that of alumina surfaces. The transitional phases η- and γ-Al_2O_3 have a defect spinel structure in which a certain number of cation positions remains unoccupied. Considering the (100) and (111) planes, Tsyganenko and Filimonov [62] concluded that only three types of OH-groups can occur which they identified as type I, II and III hydroxyls with OH stretching frequencies at 3800, 3740 and 3700 cm^{-1}. The observation of two additional, though weaker bands at 3780 and 3733 cm^{-1} [95, 96] (see Table 3 and Fig. 5) was accounted for by the fact that Al cations occupy octahedral and tetrahedral sites [101, 102]. A more detailed discussion which is based on essentially the same principles has been put forward by Knözinger and Ratnasamy [73]. The low index planes (100), (110) and (111) are considered as terminating the crystallites. Within an OH layer on top of the A-layer [73, 101, 102] of a (111)-plane two types of OH configurations can be distinguished, namely a terminal OH group which is coordinated to a single tetrahedral Al^{3+} cation

type Ia: 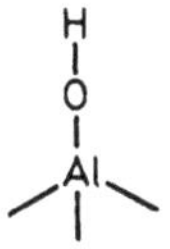

and a bridging OH group which links a tetrahedral and an octahedral cation

type IIa: 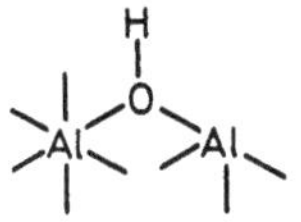

Similarly, in the OH layer on top of a B-layer of the (111)-plane two OH-configurations can again occur. Both are bridging groups. The type IIb configuration links two cations in octahedral positions,

type IIb: 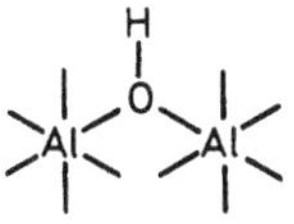

while the type III OH group is coordinated to three cations in octahedral interstices

type III: 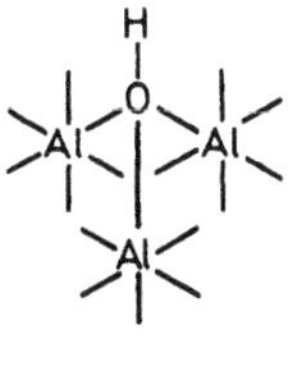

If possible vacant cation sites are also taken into account, a fifth configuration can exist as type I b

type I b: 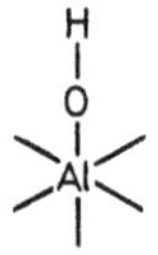

in which the OH group is coordinated to a single cation in an octahedral site. This configuration is obtained from configuration II a, if the tetrahedral cation is removed, or from configuration II b if one of the octahedral cations is removed. Similarly, type III would be transformed into type II b, when one of the cation positions would be unoccupied.

Table 4 summarizes the five possible OH configurations which can occur in ideal surface layers of aluminas. The coordination numbers of surface anions with respect to cations in octahedral (Al_{VI}) and tetrahedral (Al_{IV}) interstices are given in column 4 and the net charges at a surface oxygen atom and a surface OH group in the various configurations are compared in columns 5 and 6. These values were obtained as the sum of the negative charge of the anion and the sum of the strengths of the electrostatic bonds (= cation charge divided by cation coordination number) to the anion from the adjacent cations. Thus, the strength of the electrostatic bond [103] between an anion and an Al^{3+} cation in octahedral position is equal to +0.5, whereas that between an anion and an Al^{3+} in tetrahedral position would be +0.75. Because of energetic reasons the net charge in a stable ionic structure should be equal or nearly equal to zero [103]. Comparison of the values, given in Table 4, columns 5 and 6 readily shows that this requirement is better approximated or even completely fulfilled by OH groups than by oxygens as surface ligands. Oxide surfaces are therefore preferentially terminated by hydroxyl layers which can be formed via dissociative chemisorption of water. The same arguments have been applied by Boehm [3] when considering the TiO_2 surface.

The most important result of the above considerations is that one should expect a maximum of five different OH configurations on the surface of transitional aluminas with spinel structure, their actual occurence and relative abundance depending on the relative contributions of the different crystal faces. The OH groups in these various configurations bear slightly differing net charges. As a consequence, they should possess different properties, namely bond orders and force constants. The five observed frequencies of the stretching vibrations of isolated surface OH groups should therefore correlate with the estimated net charges of the five possible OH configurations. The corresponding band assignments are given in the last column of Table 4. The band with highest frequency (3800 cm^{-1}) is assigned to configuration I b which bears the most negative net charge (—0.5), while the band of lowest frequency (3700 cm^{-1}) is attributed to configuration III with a net charge of +0.5. The remaining three bands are assigned following decreasing frequency to the corresponding configurations with increasingly positive charge. This assignment is in agreement with the fact that the oscillator frequency ω_e of free OH^- species is $(3839 \pm 10)\ cm^{-1}$ and decreases by about $100\ cm^{-1}$ to $3735\ cm^{-1}$ for the neutral OH radical [104]. In these species a shift of the stretching frequency by $100\ cm^{-1}$ thus corresponds to a charge difference of one elemental charge. The same estimated charge and frequency regions are exhibited by the isolated OH group configurations on alumina surfaces.

The importance of the coordination of surface OH groups as a frequency determining factor had previously been pointed out by Zecchina [105], Borello *et al.* [106] and particularly by Morterra *et al.* [107]. They assigned the OH-stretching bands in the 3760–3800 cm^{-1} range to OH groups belonging to the coordination sphere of tetrahedrally coordinated Al^{3+} cations and those in the range 3700–3750 cm^{-1} to OH groups shared by an octahedrally and a tetrahedrally coordinated Al^{3+} cation. Peri [108] observed a single band at $3800\ cm^{-1}$ on $AlPO_4$, in which Al^{3+} occurs exclusively in tetrahedral coodination and Morterra *et al.* detected bands at 3730–3740 cm^{-1} on the surface of α-Al_2O_3 [107] and $MgAl_2O_4$ [109] with only octahedral Al^{3+} ions. These experimental results confirm the band assignments given in Table 4. Moreover, recent theoretical studies on the properties of surface

Table 4. Possible OH configurations on alumina surfaces (Reproduced with permission from ref. [73])

Crystal face	Layer	Configuration[a]		Coordination numbers of surface anion Al_{VI}	Coordination numbers of surface anion Al_{IV}	Net charge at O	Net charge at OH	$\tilde{\nu}_{OH}/cm^{-1}$
(111)	B		III	3	—	−0.5	+0.5	3700–3710
	B		IIb	2	—	−1.0	0	3740–3745
	A		IIa	1	1	−0.75	+0.25	3730–3735
	A		Ia	—	1	−1.25	−0.25	3760–3780
	A, B		Ib	1	—	−1.5	−0.5	3785–3800
(110)	C		IIb	2	—	−1.0	0	3740–3745
	C		Ia	—	1	−1.25	−0.25	3760–3780
	D		Ib	1	—	−1.5	−0.5	3795–3800
(100)			Ib	1	—	−1.5	−0.5	3785–3800

[a] Large circle oxygen atom or hydroxyl group; •: Al_{VI}; ∘: Al_{IV}

OH groups by means of the CNDO [110, 111] or LCAO—MO(Hückel) method [112] also support the simple empirical approach discussed above. It could particularly be shown, that the calculated bond orders correlate with the experimentally determined OH stretching frequencies [112]. The model calculations also show, that these properties are predominantly determined by the coordination number of the OH group and the nearest oxygen environment of the respective cations, while long range effects have only negligible influence [112]. These results confirm therefore the simple assumptions made for the band assignments in Table 4.

Although the considerations of Tsyganenko and Filimonov [62] suggested that only simple SiOH groups can exist on the surface of SiO_2, it is agreed upon that principally three different configurations of surface silanol groups may be distinguished [3, 52], i.e. isolated (a), geminal (b) and vicinal (c)

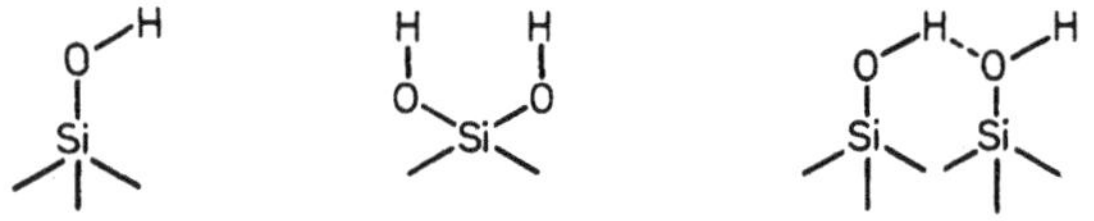

a b c

However, there is still no general agreement as to whether isolated and geminal silanol groups can be distinguished by their vibrational frequencies. The sharp band near 3750 cm^{-1} is undoubtedly due to the OH stretching vibration of unperturbed free silanol groups. Hair and Hertl [113] argued that this band could be resolved into three components. This was, however, disregarded by Hockey [114], who demonstrated that the splitting is an

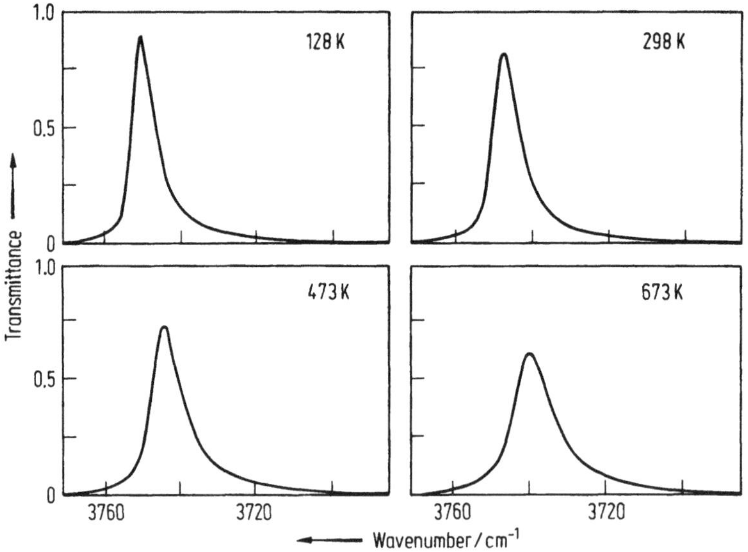

Figure 6. Spectra of isolated SiOH band at different temperatures. (Reproduced with permission from ref. [116])

Table 5. Experimental data for SiOH spectra shown in Figure 6 (Reproduced with permission from ref. [116])

Temp./K	Wave number/cm^{-1} (at max. intensity)	Peak intensity in absorbance	Band area/ arbitary scale
128	3751.4	0.97	6.56
398	3748.2	0.76	6.11
473	3744.7	0.57	6.05
673	3740.6	0.41	5.63

artefact due to residual atmospheric water vapor in the spectrometer. The discussion on this subject is still going on and different opinions have been put forward by various research groups. Thus, van Cauwelaert *et al.* [115] reported that the 3750 cm^{-1} band due to isolated surface hydroxyl groups has two distinct shoulders to high and low frequencies of the main band. The shoulders were more pronounced at higher temperatures and their exact position relative the main band depended on this temperature. Morrow and Cody [116] have reinvestigated the position and band shape of the 3750 cm^{-1} band of silica over a temperature range from 128 to 673 K. Figure 6 shows the band of isolated silanol groups at various temperatures and Table 5 summarizes the detailed spectroscopic data. The bands are asymmetric towards lower frequencies, but no distinct shoulders could be detected at any temperature. This asymmetry is believed to be due to a random intermolecular perturbation of surface silanol groups and the band splitting reported by van Cauwelaert *et al.* [115] was suggested to be caused by a slight imbalance by atmospheric water in the double-beam compensation [116]. The observed shift of the peak maximum towards higher frequency with decreasing temperature is theoretically quite involved and is not yet well understood, the concomitant intensification and narrowing on the other hand is not unexpected. No evidence for any band splitting of the first overtone of the silanol stretching vibration $\tilde{\nu}_{02}$ was found by Klier *et al.* [117]. Analogous to the band shape in the fundamental region, only a "tail" towards low frequency was observed, which was assumed to arise by various mechanisms such as by mutual perturbations of the vibrations of closely spaced OH groups, by vibrational broadening, by perturbations by the underlying lattice, by H-bonding of some silanols to others, and by surface heterogeneity in general. Van Cauwelaert *et al.* [118] later explained the discrepancies between their original work showing band splitting and the studies of Morrow and Cody [116] and Klier *et al.* [117] by the differences in the pressures exerted on the wafers during their preparation. The sensitivity of the position of the silanol stretching band to the sample preparation conditions was emphasized. Ryason and Russell [119], however, showed that the 3750 cm^{-1} band of isolated silanol groups was structureless and asymmetric as in the work of Morrow and Cody [116] for wafers prepared at pressures $\leqq 1.4 \times 10^9$ Pa, although they provide spectroscopic evidence that indicates surface heterogeneity. Van Rosmalen

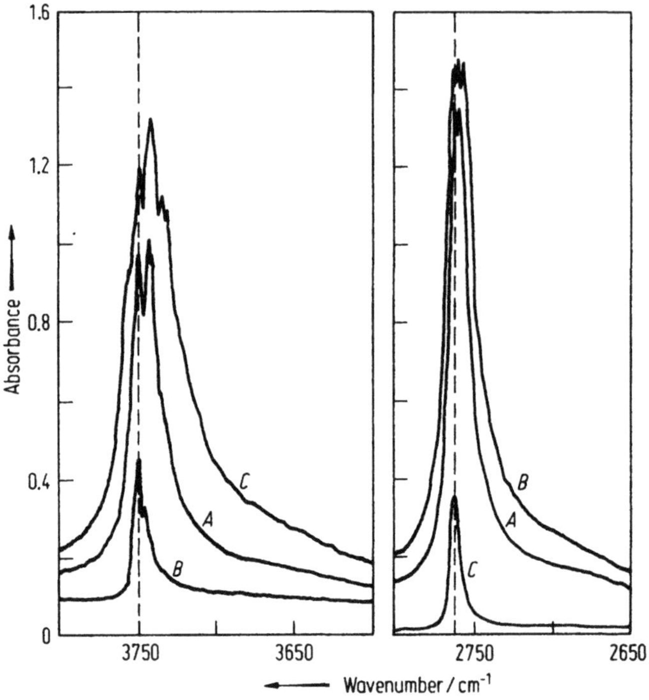

Figure 7. Spectra of a silica gel plate after degassing at 875 K during reaction with D_2O vapor: (*A*) after two exchange cycles; (*B*) after 20 exchange cycles; (*C*) after treatment of (*B*) with vapor phase H_2O. (Reproduced with permission from ref. [120])

and Mol [120] reported data recently, that seem to support the sensitivity of the band position and structure to preparation conditions. They carefully studied the OH stretching band of silanol groups on the surface of highly porous silica gels, which were prepared by the method described by Peri [121], as compared to that on silica powder or the widely used nonporous pyrogenic silica (Cabosil, Aerosil). After degassing at 875 K, they observe four absorption bands in the OH stretching region (Fig. 7) which were assigned to isolated single silanols (3749 cm^{-1}), a very weakly coupled ν_{sym}/ν_{asym} pair at 3742 cm^{-1} in geminal silanediol (structure b), and asymmetric H-bonded vicinal silanol pairs (absorbing at 3720 and 3500 to 3700 cm^{-1}). From their results, van Rosmalen and Mol [120] conclude that silica powders (Cabosil, Aerosil) having only one structureless band near 3750 cm^{-1} after vacuum treatment at the same temperature, and silica gel do not have identical surfaces. In agreement with this and previous work [116, 119], Ghiotti *et al.* [122] observed a structureless sharp OH stretching band on highly dehydroxylated aerosil silica (outgassed at 1073 K, 1.5 residual OH group per nm^2) at 3748 cm^{-1} (room temperature), which shifted to 3751 cm^{-1} at 123 K. For a highly hydroxylated aerosil silica (outgassed at 423 K, >6 residual OH groups per nm^2), however, the composite nature of the 3743 cm^{-1} band of free hydroxyl groups is revealed. Figure 8

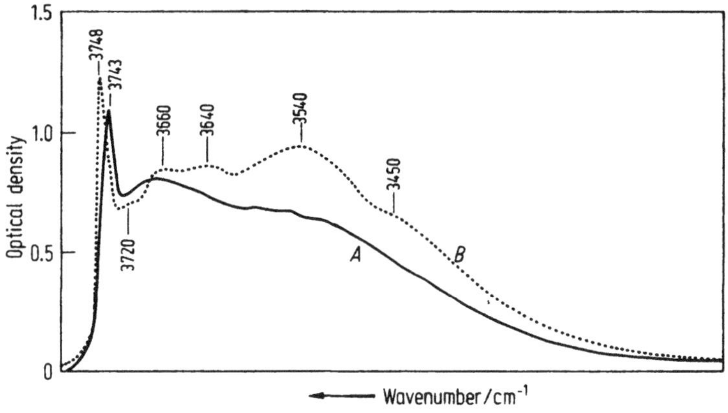

Figure 8. OH stretching spectra of hydroxylated Aerosil silica (6 OH nm^{-2}) outgassed at 423 K. (*A*) spectrum taken at room temperature; (*B*) spectrum taken at 123 K. (Reproduced with permission from ref. [122])

shows the room temperature spectrum of a hydroxylated sample in comparison with a spectrum taken at 123 K. The room temperature band at 3743 cm^{-1} has a larger band half width than that of the corresponding band of the highly dehydroxylated sample and seems to have a shoulder on the high frequency side. On cooling to 123 K, the band maximum shifts to 3748 cm^{-1}, and seems to exhibit a weak shoulder on the low frequency side. The low temperature spectrum of this hydroxylated sample shows additional bands at 3720, 3680, 3640, 3540 and 3450 cm^{-1}. In agreement with others [116], Ghiotti *et al.* [122] attribute the 3720 and 3540 cm^{-1} band pair to the OH stretching mode of silanols (one free and the other H-bonded) in vicinal pairs (structure c). The band pair at 3680 and 3450 cm^{-1} was assigned to the stretching modes in "free" and H-bonded groups of geminal H-bonded hydroxyls

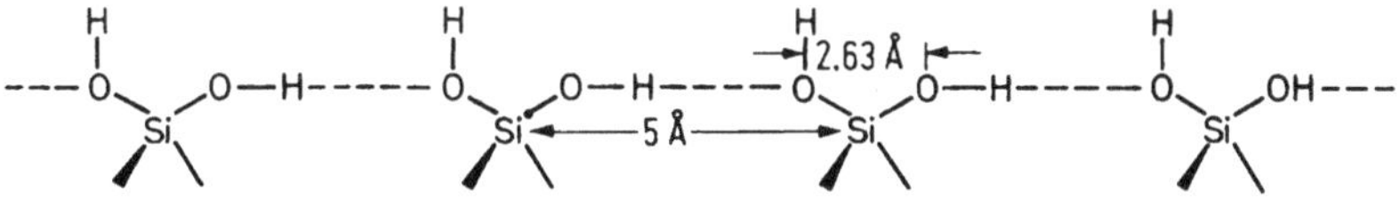

Chains of this type can occur on the (100) face of β-cristobalite, which has been considered as the terminating crystal plane on silica surfaces [123]. The 3640 cm^{-1} band is believed to contain contributions from internal OH groups (see Fig. 2), but possibly also from still another type of H-bonded neighboring OH groups. Fink and his coworkers [123a] have also reported spectroscopic evidence for the existence of germinal silanol groups.

In conclusion, a general agreement regarding possible band splitting of the 3750 cm^{-1} band of unperturbed surface silanols does still not seem to exist. It is, however, highly probable that the nature of the particular silica material, preparation conditions of the wafer and the degree of hydroxylation sensitively influence the shape and structure of this band.

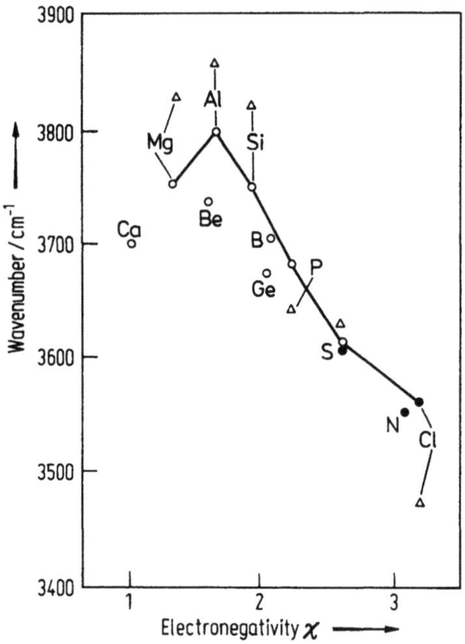

Figure 9. Stretching frequencies of unperturbed OH groups vs. electronegativity χ of the atom to which they form single coordination bonds: ○ surface OH groups of oxides [62]; and x-OH on modified silicas [110]; ● gaseous oxyacids [124–126]; △ theoretical values [110]. (Reproduced with permission from ref. [110])

Quantum chemical calculations have shown that the bond order of OH groups within O_3XOH clusters decreases with increasing electronegativity of the central atom X [112]. Figure 9 shows a correlation between experimental and calculated OH frequencies with the electronegativity of the atom X to which the OH forms a single bond. The predicted trend can clearly be seen for χ-values higher than 1.5. Analogous results have been obtained by several other workers [127–129]. Experimentally the formation of XOH groups on silica surfaces and the dependence of the OH stretching frequency on the nature of the atom X has been demonstrated by Fink *et al.* [130]. A series of spectra of these samples after outgassing at 873 to 973 K are shown in Figure 10. Sharp bands of unperturbed hydroxyls are observed at 3750 cm^{-1} ($\equiv$SiOH), 3707 cm^{-1} ($\equiv$BOH), 3682 cm^{-1} ($\equiv$GeOH), 3669 cm^{-1} ($\geqslant$POH), and 3665 cm^{-1} ($\geqslant$SnOH). This sequence entirely follows the predicted trends. Similar OH stretching frequencies have been observed for BOH [131–135] and POH [134, 136] groups on modified silica surfaces. In a study of "alumina-on-silica" a single band of $\geqslant$SiOH was detected at 3744 cm^{-1} after heat treatment at 773 K [137]. The failure to detect any AlOH bands was assumed to be due to the low Al concentration and mutual interactions of AlOH groups with other hydroxyl groups. Also *amorphous silica-aluminas* containing up to approximately 50 wt% Al_2O_3 did not give rise to OH stretching bands of hydroxyl groups associated with Al^{3+} ions [138–141]. After heat treatment at 773 K and higher temperatures only the stretching bands attributed to SiOH groups near 3750 cm^{-1} could be observed. It was therefore suggested [138] that AlOH groups were

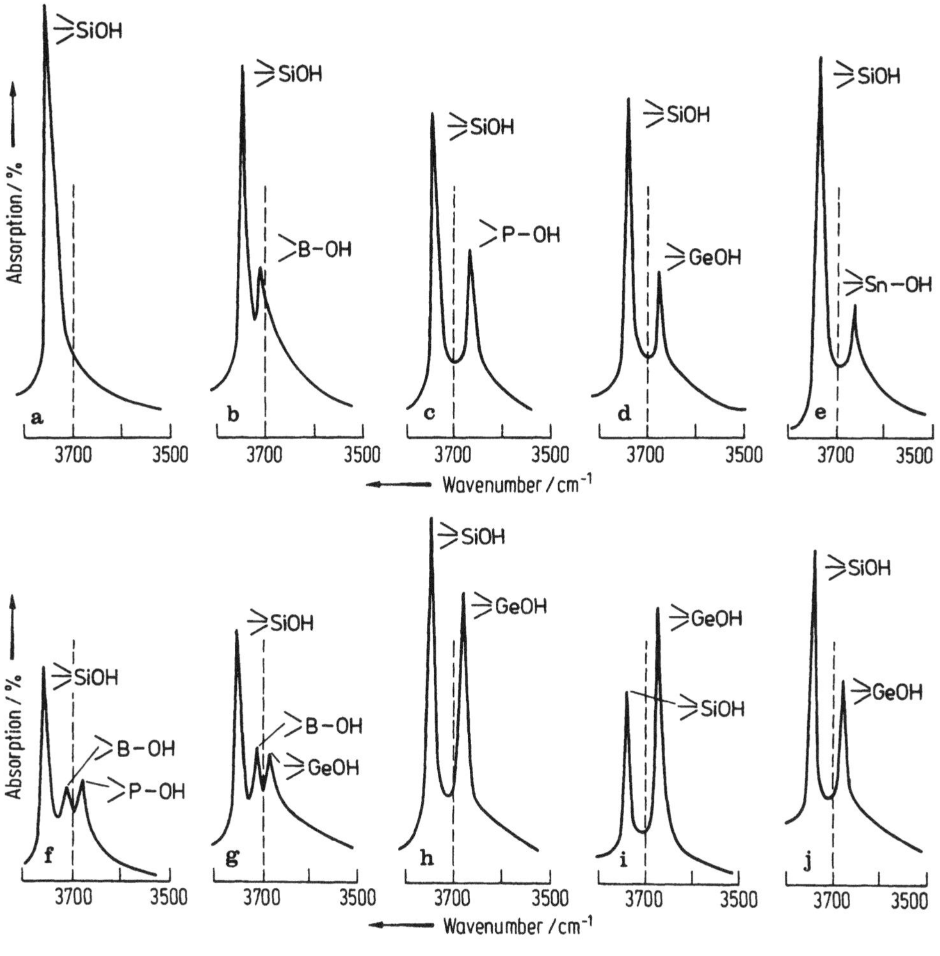

Figure 10. OH stretching bands of modified Aerosil silica surfaces: **a** Aerosil, outgassed at 973 K; **b** Aerosil, BCl_3 modified, hydrolyzed and outgassed at 973 °C; **c** Aerosil, PCl_3 modified, hydrolyzed and outgassed at 923 K; **d** Aerosil, $GeCl_4$ modified and outgassed at 923 K; **e** Aerosil, $SnCl_4$ modified and outgassed at 873 K; **f** Aerosil, PCl_3 and BCl_3 modified, hydrolyzed and outgassed 923 K; **g** Aerosil, $GeCl_4$ and BCl_3 modified and outgassed at 973 K; **h** Aerosil, $GeCl_4$ modified, hydrolyzed and outgassed at 973 K; 2 cycles; **i** same, 15 cycles; **j** same, after several hydration-dehydration cycles. (Reproduced with permission from ref. [130])

absent from the surface of silica-alumina gels (at least for materials containing less than 30 wt% Al_2O_3) and that hydroxyl groups combined with aluminium atoms are more readily removed than those associated with silicon atoms, a conclusion which is well supported by recent NMR studies (see section II.A.3.b.i). According to Peri [130], a band at 3650 cm^{-1} may or may not appear in spectra of silica-alumina, depending on the composition and extent of surface hydration. A model for the surface of silica-alumina catalysts which is based on the cristobalite (100) face of silica, was presented by Peri [142].

The *crystalline alumino-silicates* or *zeolites* consist of a network of SiO_4 and AlO_4 tetrahedra in which the charge of -1 of the AlO_4 tetrahedron is compensated in the structure by a cation to maintain electroneutrality. Among the large number of natural and synthetic zeolites, the ones having 8- and 12-membered oxygen rings have found major catalytic application, since their apertures are sufficiently large to allow penetration of molecules to be adsorbed. Thus, zeolites of practical interest are zeolites A, X, and Y, and synthetic mordenites, which have approximate SiO_2/Al_2O_3 ratios of 2, 2.5, 5 and 9–12, respectively. Consequently, these types of zeolites have most frequently been studied. Structural information of zeolite surfaces as obtained by infrared spectroscopy was reviewed by Ward [143]. Structural hydroxyl groups, namely in H-zeolites, may schematically be represented as follows

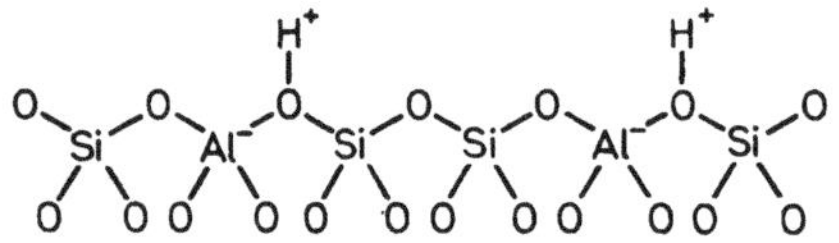

Sodium (X and Y) faujasites give rise to OH stretching bands at 3750 and 3690 cm^{-1} when they are dehydrated at temperatures below 523 K [144, 145]. The low frequency band near 3690 cm^{-1} occurs in all monovalent zeolite forms, the exact band position being dependent on the type of exchangeable cation (electric field strength) [146]. This band is attributed to molecular water which is coordinated to the cation. The 3750 cm^{-1} band is assigned as the stretching mode of silanol SiOH groups on outer surfaces of the crystallites where they are assumed to terminate faces at positions where bonding in the bulk would occur with adjacent tetrahedral Si or Al ions. This same band may, however, also be due to amorphous enclosures of SiO_2 [146]. Zeolites containing divalent charge compensating cations such as *e.g.* Ca^{2+}, give rise to four hydroxyl stretching bands at 3750, 3650 and 3540, and near 3600 cm^{-1}. The electric field strength of those cations is sufficiently high to lead to hydrolytic splitting of coordinated water molecules

The exact position of the low frequency band near 3600 cm^{-1} is dependent on the type of the cation and is associated with the O—H stretching mode of the MOH species [143]. The liberated proton is assumed to associate with a lattice O^{2-} ion to form two types of structural OH groups which produce the bands at 3650 and 3540 cm^{-1}. Adsorption studies with pyridine or ammonia showed that the 3650 cm^{-1} band was perturbed by adsorption interactions, while the 3540 cm^{-1} was not, at least at low coverages and temperatures. The OH group associated with this latter band is therefore inaccessible for the adsorbate molecules and must therefore be located within the sodalite cages or hexagonal prisms, while the group associated with the 3650 cm^{-1} band must protrude into the supercage [143, 144]. Dealuminated and ultrastable zeolites [147, 148] give rise to an additional band near 3600 cm^{-1}. OH nests which are formed at defect sites resulting from hydrolysis of framework aluminium [148]

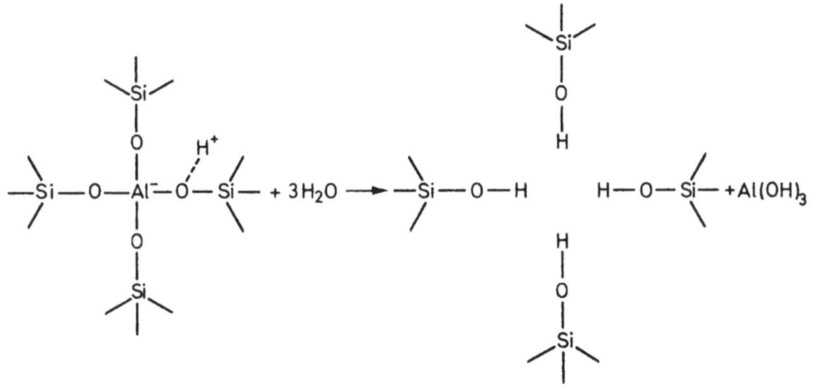

are considered responsible for this band [143]. However, hydroxoaluminium cations $Al(OH)_2^+$ may also be formed and contribute to the observed band [149].

Mordenites and clinoptilolites [146, 150] give rise to bands near 3600 cm^{-1} which are cation sensitive, besides the band at 3740 cm^{-1}.

A very important class of catalysts are the *supported molybdena catalysts*, namely $CoO—MoO_3—Al_2O_3$ hydrodesulfurization catalyst precursors. The OH stretching region of these materials has first been studied by Fransen *et al.* [151] and by Ratnasamy and Knözinger [152]. For molybdena on alumina with a molybdena loading close to monolayer capacity, a very broad, structureless band with maximum at approximately 3530 cm^{-1} was observed after calcination at 823 K [152], which was associated with the increased protonic acidity of these systems as compared to pure aluminas. Similarly, Topsøe [153] and Gil *et al.* [154] report on the appearance of a single broad band at 3680 and 3650 cm^{-1}, respectively, for molybdena on alumina. In contrast, Millman *et al.* [155] only observed a reduction in intensity of the original OH bands of the parent alumina and changes in their intensity ratios when the alumina hydroxyl groups were replaced by molybdena species, but no new band could be detected. This discrepancy is still not understood. It may be due to the nature of the support material and due to differences in catalyst preparation conditions. The OH stretching spectra of cobalt aluminates and of pure alumina showed identical band positions after heat treatment at >773 K [152, 153]. This suggests, that hydroxyl groups were either not coordinated to Co^{2+} or were more easily removed from Co^{2+} sites than from Al^{3+} sites. Copper and nickel aluminates behave analogously [156]. The $CoO—MoO_3—Al_2O_3$ catalysts on the other hand develop OH stretching spectra which differ from those of both molybdena on alumina and cobalt aluminate [152–154]. Two sharp bands are observed at 3790 and 3705 cm^{-1}, while the central bands of the alumina support at 3730 and 3740 cm^{-1} have disappeared [152]. An important influence of cobalt in controlling the interaction of molybdena species with the alumina support is apparent from these observations [152, 153]. Bands which could indiscriminately be attributed to MoOH groups have not been detected in any of these catalysts.

Quantitative measurements of hydroxyl group band intensities are generally complicated due to the overlap of the bands of individual hydroxyl groups. Although band deconvolutions of the five hydroxyl bands of alumina have been reported, these procedures are not straightforward. Quantitative determinations of hydroxyl group densities have therefore nearly exclusively been carried out for silica surfaces. The determination of OH group densities and the amount of molecularly adsorbed water from intensities of their fundamental vibrations in the infrared spectrum is rather difficult, since the OH stretching band is overlapped by the stretching band of molecular water and the bending mode of water near 1630 cm^{-1} is very weak and coincides with the Si—O overtone at 1640 cm^{-1}. The water combination band at 5265 cm^{-1} has therefore been preferred for the quantitative determination of the adsorbed amount of molecular water [157, 158]. The number of OH groups was then calculated from the difference of total water loss as determined by the ignition method and the spectroscopically determined amount of molecular water. Since these measurements were carried out by means of the transmission technique, the powders had to be suspended in CCl_4 to obtain sufficient transmittance. The method may suffer from the fact that the ignition method does also measure internal OH groups. Attempts have also been made to determine the extinction coefficient of the silanol stretching fundamental at 3750 cm^{-1}, the values obtained, however, differ greatly between 35 ± 1 liter mol^{-1} cm^{-2} [159] and 9.8×10^3 liter mol^{-1} cm^{-2} [160]. Deuterium exchange techniques (see also section II.B.1.a) are now preferred over other methods. Madely and Richmond [161] measured the OH group density indirectly by spectroscopically determining the D_2O/H_2O composition in liquid water, which was obtained after extensive D_2O exchange with silica surfaces and re-exchange with H_2O. The value reported was 4.4 ± 0.2 OH nm^{-2} which very closely corresponds to the generally accepted value for the OH saturation coverage [3, 52]. Davydov *et al.* [162] suggested that the isotopic composition of the water in the gas phase over an adsorbent is equal to that of the surface groups of the adsorbent, which bears "active" hydrogen, or in other words, the isotopic separation coefficient is assumed to be unity and moreover, equal for different adsorbents. In quantitative work for OH group determination an auxiliary adsorbent as a reference, usually aerosil because of its high transmittance, was therefore applied in addition to the sample which must not be infrared transparent in this approach. The content of hydroxyl groups involved in the deuterium exchange per unit surface area α_{OH} can be calculated from

$$\alpha_{OH} = \frac{2Km_{D_2O}}{mS} \qquad (1)$$

where

$$K = \left(\frac{[H]}{[D]}\right)_{sample} \approx \left(\frac{[H]}{[D]}\right)_{gas} \approx \left(\frac{[H]}{[D]}\right)_{ref} = \frac{D_{OD,\,init.} - D_{OD,\,equil.}}{D_{OD,\,equil.}} \tag{2}$$

m_{D_2O} and m are the amount of D_2O involved in the deuterium exchange and the adsorbent mass, respectively, and S is the surface area of the sample. The values [H]/[D] are the ratios of hydrogen and deuterium concentrations on the adsorbent surface under investigation, in the equilibrium gas phase and on the surface of the reference aerosil. $D_{OD} = \ln I_0/I$ are the optical densities of the 2760 cm^{-1} band of free deuteroxyl groups on the surface of the reference aerosil plate before (initial) and after (equil.) isotopic equilibration with the sample to be investigated. It is suggested that this technique could be used also for the study of other functional groups capable of deuterium exchange with D_2O, *e.g.* $—NH_2$, $>NH$, and —COOH.

ii) Properties of surface hydroxyl groups. Hydroxyl groups act as *H-bond donors* and they may develop *protonic acidity*. Their H-bond donor strength depends on the element, namely its electronegativity, to which they are coordinated [110, 112]. H-bonding interactions of adsorbate molecules with surface hydroxyl groups have been reviewed recently [52, 163]. The intensity of the OH stretching fundamental of free hydroxyl groups decreases on H-bonding, while intense and broad absorption bands appear at lower wave numbers. The wave number shift $\Delta\tilde{\nu}_{OH}$ to lower values on H-bonding is considered as a measure of the bond strength, *i.e.* the energy of specific interaction, and is determined by the acceptor strength of the H-bond acceptor and by the donor strength of the OH group as well. Although the possible differences between proton transfer (acidity) and H-bonding had been pointed out [164], a number of attempts have been undertaken to

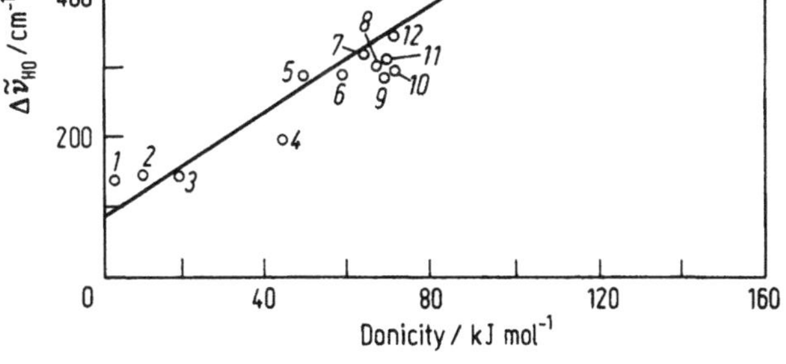

Figure 11. Correlation between wavenumber shift ($\Delta\tilde{\nu}_{OH}$) and donicity (DN) of adsorptives: (*1*) acetylchloride, (*2*) benzoylchloride, (*3*) nitrobenzene, (*4*) acetanhydride, (*5*) benzonitrile, (*6*) acetonitrile, (*7*) isobutylnitrile, (*8*) propionitrile, (*9*) methylacetate, (*10*) ethylacetate, (*11*) *n*-butylnitrile, (*12*) acetone, (*13*) diethylether, (*14*) tetrahydrofurane, (*15*) dimethylformamide, (*16*) diethylformamide, (*17*) dimethylacetamide, (*18*) diethylacetamide, (*19*) pyridine. (Reproduced with permission from ref. [175])

characterize the acidic properties of surface OH-groups by means of the wavenumber shifts $\Delta\tilde{\nu}_{OH}$.

H-bonding interactions have most extensively been studied on silica surfaces, since the single sharp band of the free silanols near 3750 cm^{-1} permits an accurate determination of $\Delta\tilde{\nu}_{OH}$ relative to this band. $\Delta\tilde{\nu}_{OH}$ values fall into the range of $100 \leqq \Delta\tilde{\nu}_{OH} \leqq 200$ cm^{-1} for π-electron donor molecules and may reach values as high as 1000 cm^{-1} for molecules containing lone-pair electrons [52]. Evidently the energy of interaction increases in this order. The wave number shifts for silica hydroxyl groups have been correlated with various molecular properties of the adsorbate, such as the macroscopic dielectric constant [165], the pK_a values in aqueous solution [166], the ionization potential [61, 167–170], the Hammett inductive parameter [174] and the Taft inductive parameter [61]. Linear correlations with the ionization potential did not show general validity but were only obtained for individual classes of molecules [52, 158–170]. Steric restrictions towards H-bonding of silanol groups with adsorbates having bulky substituent groups occur occasionally [170, 171]. Linear correlations were also reported between $\Delta\tilde{\nu}_{OH}$ and the specific H-bond interaction energy ΔH_{OH} [172–174], were ΔH_{OH} is defined as the difference

$$\Delta H_{OH} = \Delta H_{hydr.} - \Delta H_{dehydr.} \qquad (3)$$

between the differential heat of adsorption of the adsorbate on a hydroxylated and a strongly dehydroxylated (heat treated at 1173 K) surface at 50% coverage. Horill and Noller [175] have found a fairly general correlation for various classes of adsorbate molecules between $\Delta\tilde{\nu}_{OH}$ and the donicity DN (Fig. 11) as defined by Gutmann [176] by the negative enthalpy of formation of a donor acceptor complex between the electron-pair donor molecule (H-bond acceptor) and $SbCl_5$ in dilute 1,2-dichloroethane.

In solution, the wavenumber shifts of two H-bond donors can be correlated by plotting the relative wavenumber shift $\Delta\tilde{\nu}/\tilde{\nu}_0$ of one donor with respect to that for another [177]. Such graphs give linear relationships and chemically very different H-bond acceptors fit the same straight line provided the proton bearing atoms of the donors are identical [178]. Rouxhet and Sempels [141] succeeded in relating the wave number shifts of surface silanol groups analogously using *p*-fluorophenol, phenol and methanol as references as shown in Figure 12. The slope of these so-called BHW (Bellamy, Hallman, Williams [177]) plots could be successfully correlated with the pK_a values of the H-bond donors in aqueous solution (Fig. 13). The value for silanol groups of silica surfaces also fitted this same correlation when a pK_a value of 6.8 as determined by Schindler and Kamber [179] was applied. Using the same procedure, Hair and Hertl [134] had determined pK_a values of 15.5 for OH on magnesia, 8.8 for BOH on silica, 7.1 for OH on silica, and 0.4 for POH on silica. These examples demonstrate that it should be possible to estimate the pK_a values of hydroxyl groups not only on silica and related surfaces by this procedure. Tretyakov and Filimonov [180] have in fact reported the pK_a values of surface hydroxyls for a wide variety of oxides. Their data are summarized

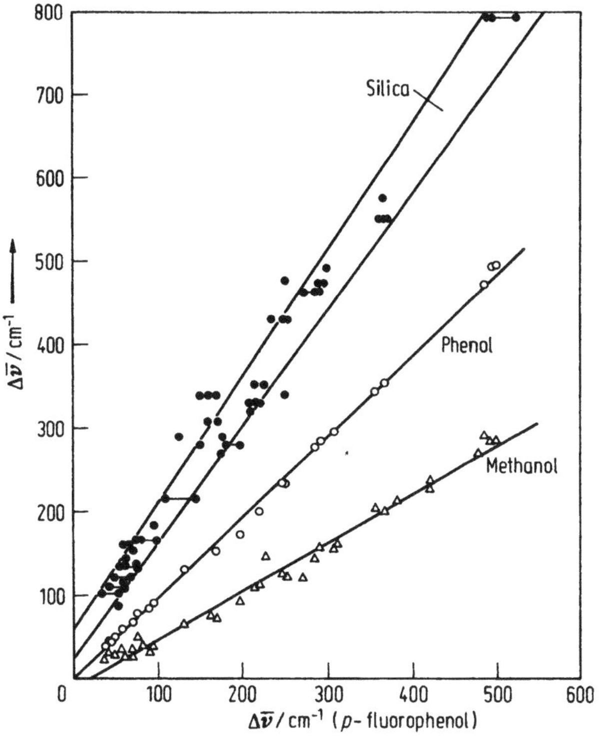

Figure 12. Plot of the hydroxyl wavenumber shift of various compounds as a function of the shift for *p*-fluorophenol upon interaction with the same H-bond accepting molecule. (Reproduced with permission from ref. [141])

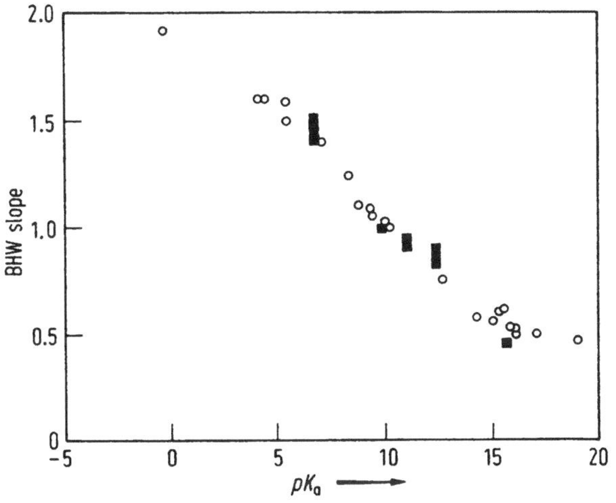

Figure 13. Correlation between BHW slope and the pK_a of various proton donors with OH as the active group. ●, phenol and *p*-fluorophenol (references for the BHW slope); ○, molecules with OH attached to a carbon atom (alcohols, phenols, organic acids); ■, water, acetoxime, trimethylsilanol, silica gel. (Reproduced with permission from ref. [141])

together with the observed wavenumber shifts of surface OH and OD groups in Table 6.

The weak bases benzene and acetonitrile have proved extremely valuable as probes for the H-bond donating power or acidity of surface hydroxyl groups in different environments. Figure 14 shows infrared spectra of the OH

Table 6. Wavenumber shifts $\Delta\tilde{\nu}_{OH}$ produced by benzene adsorption on various oxides and corresponding pK_a values (from ref. [180])

Oxide	$\tilde{\nu}_{OH}/cm^{-1}$	$\Delta\tilde{\nu}_{OH}/cm^{-1}$	pK_a
MgO	3752	35	≥18.5
γ-Al_2O_3	3710	110	8.5
Ga_2O_3	3670	90	10.5
SiO_2	3750	120	7.0
TiO_2 (anatase)	3725	190	
	3670, 3650	190	0.5–2.0
TiO_2 (rutile)	3735	130	6.5
ZrO_2	3770	90	10.5
	3670	65	
Ta_2O_5	3740	130	6.5

stretching region after benzene adsorption (coverage <0.2) onto modified silicas which bear BOH, POH or GeOH groups besides the SiOH groups [181]. The bands of OH groups which are perturbed by adsorbed benzene molecules are clearly discernible, the wavenumber shifts $\Delta\tilde{\nu}_{OH}$ being 100, 115, 119 and 245 cm^{-1} for BOH, SiOH, GeOH and POH, respectively. This sequence is equivalent to an acidity ranking, which indicates that the POH group is by far the most acidic group due to the high electronegativity of phosphorus (see also Fig. 9). The acidity of SiOH groups on silica surfaces containing P centers is also increased by the electronegativity difference between Si and P via inductive effects:

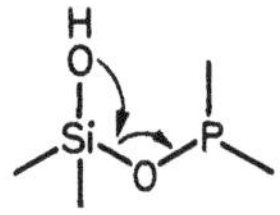

This phenomenon becomes evident from the appearance of a band at 3590 cm^{-1} in spectrum 2 of Figure 14, which corresponds to a wavenumber shift of 160 cm^{-1} relative to the 3750 cm^{-1} band of isolated SiOH groups. As mentioned previously (page 60), armorphous silica-aluminas did only show the typical sharp silanol band at 3750 cm^{-1} and acidity determinations gave a pK_a value which was close to 7, identical with that reported for pure silica [134, 179, 182], although the catalytic properties of these materials would suggest a much higher protonic acidity. Semples and Rouxhet [183], using benzene and acetonitrile as probe molecules could demonstrate that the sharp band at 3750 cm^{-1} of isolated OH-groups on the surface of silica-alumina must indeed contain contributions from at least two different types of OH-groups. On interaction with both benzene or acetonitrile, two broad bands develop at 3640 and 3440 cm^{-1}, or 3520 and 3200 cm^{-1}, respectively, as shown in Figure 15. This result provides clear evidence for the existence of hydroxyl groups having higher acidity than the silanol groups. These acidic OH groups gave rise to a BHW slope of about 3 with respect to phenol, which suggests a pK_a range of —4 to —8, the estimated values depending on the way the regression of the correlation in Figure 13

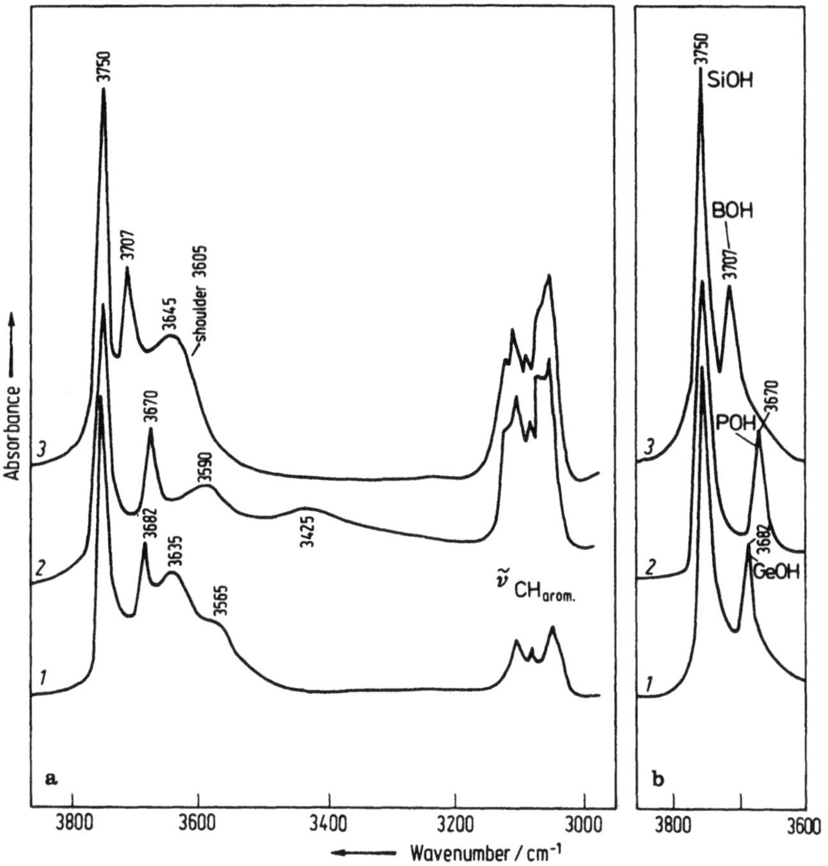

Figure 14. Infrared spectra, showing the effect of benzene interactions on the hydroxyl stretching bands on modified aerosil silica surfaces: (*1*) SiO_2 modified by $GeCl_4$, hydrolyzed, and dehydrated at 973 K; (*2*) SiO_2 modified by PCl_3, hydrolyzed, and dehydrated at 923 K; (*3*) SiO_2 modified by BCl_3, hydrolyzed, and dehydrated at 973 K. **a** spectra taken after benzene adsorption (coverage <0.1 for (*1*) and ~ 0.2 for (*2*) and (*3*). **b** dehydrated samples before adsorption. (Reproduced with permission from ref. [181])

is extrapolated towards negative pK_a values [141]. The acidity of fluorinated aluminas has been studied analogously [184]. In this system the fluoride ions preferentially replace the more basic (high frequency) hydroxyls, influence the stretching vibration of the remaining hydroxyls, and enhance the acidity of some of them.

Lercher and Noller [184a] have shown that the wavenumber shift for the carbonyl-stretching band of acetone observed upon adsorption on pure and mixed oxides correlated with Sandersons's intermediate electronegativity [184b] of the oxide.

The H-bonding interaction must necessarily modify the *potential curve of the OH oscillator* in that the anharmonicity constant should increase. This phenomenon can principally be studied by means of the overtone spectra of isolated and H-bonded surface hydroxyl groups, data, however,

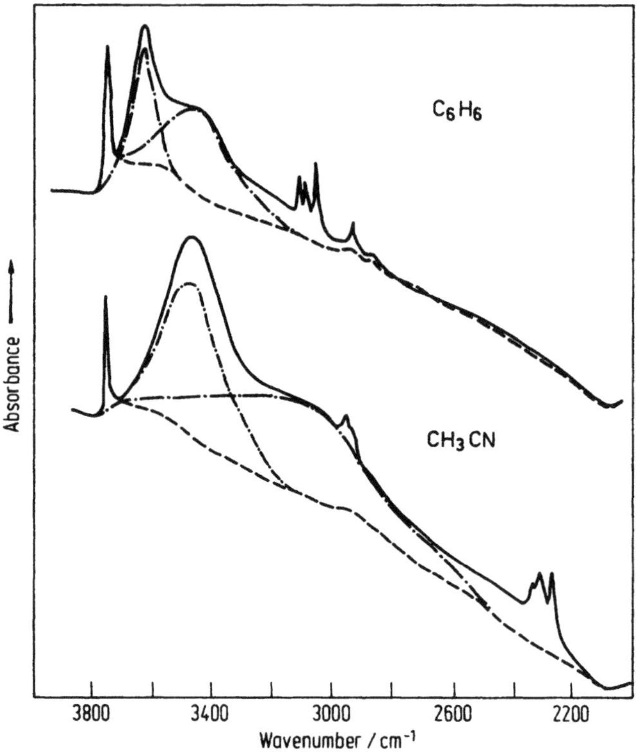

Figure 15. Infrared spectra of silica-alumina (75% SiO_2) outgassed overnight at 873 K: (– – –) before adsorption; (———) after adsorption of benzene or acetonitrile; (—·—·—) constructions showing the deconvolution of the band of hydroxyls perturbed by the adsorbate. (Reproduced with permission from ref. [183])

Table 7. Vibrational frequencies, anharmonicities, and dissociation energies of SiOH groups. (Reproduced with permission from ref. [185])

Adsorbed molecules	ω_{0-1}/cm^{-1}	ω_{0-2}/cm^{-1}	ω_e/cm^{-1}	x	D_0/kJ mol^{-1}
Isolated surface SiOH group	3749	7325	3921	2.3×10^{-2}	502
SiOH + C_6H_{12}	3720	7220	3943	3.0×10^{-2}	397
SiOH + CH_3COCH_3	3460	6400	3979	7.6×10^{-2}	146
SiOH + NH_3	2960	—	3670	12×10^{-2}	84

have only been published for silica surfaces. Table 7 shows data reported by Kazansky [185, 185a, b] for the wave numbers of the OH stretching fundamental and the first overtone of isolated silanol groups and silanol groups interacting with molecules of increasing H-bond acceptor strength. From these experimental data, the oscillator frequency ω_e and the anharmonicity constant x were estimated using a simple Morse function. The last column in Table 7 gives the calculated effective OH dissociation energy D_0. Adsorbed molecules have only a minor influence on the lower parts of the potential curve, since ω_e remains almost constant. They strongly perturb, however,

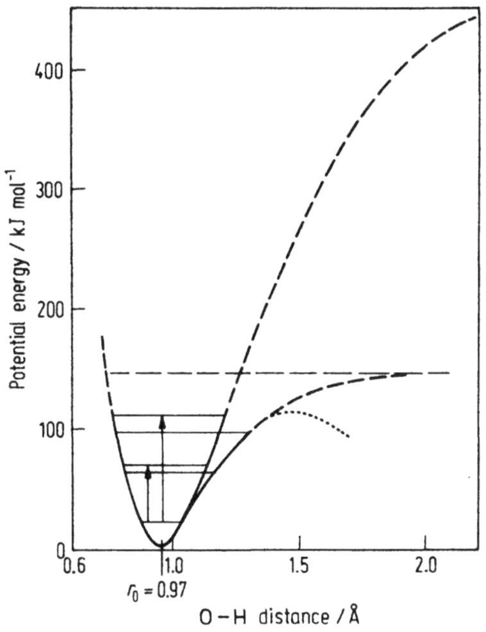

Figure 16. Potential energy of an isolated SiOH group interacting with adsorbed acetone. (Solid-line parts supported by infrared transitions; broken-lines: extrapolation according to Morse function; dotted curve: real shape of potential barrier for proton transfer). (Reproduced with permission from ref. [185])

the upper parts of the potential, as they increase the anharmonicity constant x and decrease the effective dissociation energy D_0. These effects become increasingly drastic as the basicity of the adsorbate molecule increases. Figure 16 shows the potential well of a silanol group and its modifications on H-bonding to acetone. The solid lines in Figure 16 of the potential curves are supported by experimentally observed infrared transitions, while the broken lines indicate regions of the potential which have been obtained by extrapolation on the basis of the Morse function. The dotted line indicates schematically the possible real shape of the potential barrier for proton transfer to the acceptor molecule. The barrier may lie between 100 and 140 kJ mol^{-1}. After subtraction of the zero point energy, this gives an estimate of activation energy for proton transfer of about 80 to 120 kJ mol^{-1} which seems quite reasonable for a weak acid silanol group. For stronger bases such as ammonia this barrier will still be lower (Table 7). The first OH overtone for silanol groups interacting with ammonia could not be detected which was explained by the assumption that the second vibrational level was located at energies higher than the barrier for proton transfer. Proton transfer from surface OH groups towards adsorbate molecules of lower basicity may occur at increased temperatures when the higher vibrational levels begin to be populated, provided the lifetime of the H-bonded complex is still long enough under these conditions. For a more detailed understanding of these proton transfer phenomena by means of the potential energy curve, the knowledge of the higher overtones of surface OH groups not only on silica surfaces would be highly desirable.

An analogous analysis has been carried out for zeolite hydroxyl groups [185c].

Protonated surface species may be formed when sufficiently basic acceptor molecules interact with acidic surface OH groups. Ammonia and pyridine are most frequently used for the study of acidic OH groups [52, 72, 163, 186–190], since their protonated forms NH_4^+ and PyH^+ can easily be detected by their characteristic vibrational modes. Thus, the ammonium ion NH_4^+ gives rise to the normal N—H stretching modes near 3230 and 3195 cm^{-1} and to the asymmetric NH_4^+ deformation mode at about 1430 cm^{-1}. The pyridinium ion PyH^+ typically absorbs at about 1485–1500, 1540, 1620 and 1640 cm^{-1}. These sets of vibrational modes permit an unequivocal distinction between the protonated molecule and simply H-bonded or coordinatively adsorbed species to be made. In certain cases, however, the individual protonated species escape detection via their molecular normal modes due to a very short lifetime of the protonated state. Instead a diffuse *continuous absorption* over the whole spectral range may be observed [52]. This had been explained by the formation of extremely highly polarizable H-bonds which are probably produced in dimeric species of the type

$$H_3N\text{-------}H^+\text{------}NH_3 \quad \text{or} \quad Py\text{-----}H^+\text{-----}Py$$

These bridges are characterized by a double minimum potential well which can be easily deformed due to the anisotropic and inhomogeneous environment of the surface species and thus leads to a quasi-continuum of vibrational levels [52]. Moreover, the proton can fluctuate between the two equilibrium positions within the H-bond fairly rapidly.

Experimental results are in qualitative agreement with the previously discussed acidity ranking for OH groups on various oxide surfaces in that ammonia and pyridine gave their protonated forms on amorphous silica-aluminas, only NH_4^+ was observed on silica and neither of the two protonated species could usually be detected on aluminas [72]. Proton transfer was reported between silanol groups on silica and amines in carbon tetrachloride [189].

Certain types of hydroxyl groups-depending on the element to which they are coordinated and their coordination number- may have *basic character* and even act as nucleophiles in surface chemical reactions. Suitable probe molecules for the detection of basic sites are not as frequent as those for acid sites. The use of pyrrole has recently been suggested by Scokart and Rouxhet [191], the wavenumber shift of the NH stretching mode being a measure of the basicity of the H-bond acceptor site on the surface. A base strength comparable to that of dimethylsulfoxide and pyridine was found for magnesia and aluminas, respectively. A distinction between OH groups and surface O^{2-} ions, however, is not possible.

The *nucleophilic character* of surface OH groups on aluminas shows up in a number of surface chemical reactions. Thus, CO_2 forms bicarbonate species, alcohols and ketones lead to carboxylates [72], nitriles to amides [72, 192], and pyridine may form a surface pyridone [72, 193]. All these reactions, the surface products of which have been identified by their infrared vibrational spectra, can be explained by a nucleophilic attack by the most basic (high frequency) OH group of the alumina surface, whereby coordination of the adsorbed molecule onto a neighboring Lewis acid site probably assists this reaction [193].

iii) Aprotonic surface sites. During dehydroxylation of oxide surfaces, water is produced by condensation of hydroxyl groups whereby coordinatively unsaturated surface oxygen and metal atoms/ions are being produced. These can function as Lewis basic and acidic sites, respectively, or else in concerted action as acid-base pair sites. Moreover, considerable strain may be introduced in the surface by the formation of these sites which can lead to new surface vibrational states not associated with adsorbed species. On silica surfaces, vibrational modes are observed on dehydroxylation in the temperature range between 300 and 870 K at 1020–975 and 860–800 cm^{-1} which are assigned as asymmetric and symmetric stretching vibrations of surface Si—O—Si links with angular distortions [48]. At still more severe dehydroxylation conditions (870–1020 K) two bands appear at 908 and 888 cm^{-1} [48, 194], which can be attributed to the normal modes of extremely strained localized and uncoupled Si—O oscillators. These could contain either non- or only very weakly bridging oxygen atoms [48] and should be strongly dipolar. Evidence for the existence of aprotonic sites of this type on the surface of partially dehydroxylated silica surfaces has also come from adsorption studies of a variety of probe molecules [53, 195–206], which suggest the importance of such sites for the chemical modification and functionalization of silica surfaces (II.A.1.a.iv.). Morterra [207] reported on the observation of surface Al—O modes near the high-frequency part (1100—1000 cm^{-1}) of the Al—O lattice modes of alumina which are responsible for the so-called cut-off. Surface vibrational modes have also been reported for a few other oxides [208].

The coordinatively unsaturated lattice ions (oxygen ions as Lewis bases and metal ions as Lewis acids) are not considered as surface functional groups in the present context. Since they may, however, play an important role for surface functionalization and since Lewis acid sites must be recognized for an unequivocal distinction from protonic sites, we will briefly deal with their characterization. For the detection of Lewis acid sites by means of infrared spectroscopy, suitable probe molecules have to be used. The effect of the coordination bond between an electron pair donor molecule and the Lewis acid (electron pair acceptor) site is taken as evidence for the nature and strength of the surface acid site. The use of ammonia, amines, pyridines, ketones, nitriles, diazines, nitrogen dioxide and carbon dioxide [72, 163] and carbon monoxide [73] as probe molecules has recently been reviewed. Only a few additional comments on the subject will therefore be made here. As already mentioned above, ammonia and pyridine are most frequently used as probes for the distinction between Brönsted and Lewis acid sites. Morterra *et al.* [109, 209] have recently shown that the vibrational modes of coordinated pyridine are also sensitive to the coordination number of the metal center. They could discriminate between Al^{3+} in octahedral and tetrahedral sites on the alumina surface.

Molecular hydrogen has been used to detect acid-base pair sites on which dihydrogen can dissociate heterolytically. On ZnO [210, 211] and

Table 8. Characteristic wavenumbers of different probe molecules adsorbed on silica, zinc oxide, and alumina. (Reproduced with permission from ref. [213])

	Pyridine $\tilde{\nu}_{8a}$: 1582	$\Delta\tilde{\nu}$	Ammonia δ_s: 988	$\Delta\tilde{\nu}$	$(CH_3)_3CCN$ $\tilde{\nu}_{C\equiv N}$: 2238	$\Delta\tilde{\nu}$	CD_3OCD_2H [a] $\tilde{\nu}_{CH}$··: 2886	$\Delta\tilde{\nu}$
SiO_2	1596[b]	+14	~1050	(+62)	2251[b]	+13	2914[b]	+28
ZnO	1605	+23	1207	+219	2275	+37	2923[c]	+37
			1182	+194				
Al_2O_3	1614[d]	+32	1225	+237	2299	+61	2956[c]	+70

[a] The wavenumbers reported for the ether are relative to the most intense $\tilde{\nu}_{CH}$ band. They correspond to following amounts adsorbed: 5 µmol g^{-1} (ZnO); 40 µmol g^{-1} (Al_2O_3); and 270 µmol g^{-1} (SiO_2).

[b] After desorption at room temperature at 10^{-2} Pa.

[c] Band shifts to lower wavenumbers on further addition of ether.

[d] Accompanied by a shoulder towards higher wavenumbers.

MgO [212], Zn—H and Mg—H groups, respectively, and new OH groups have been identified by their characteristic stretching vibrations.

Lavalley and Caillot [213] have recently suggested the use of deuterated dimethylether CD_3OCD_2H as a probe molecule. The CH stretching vibrations in CH_3OCH_3 are known to be sensitive to the ether oxygen environment and to shift towards higher frequencies when the lone pair electrons become engaged in coordination bonds. The spectra in this frequency range, however, are very complex due to Fermi resonance phenomena. This difficulty is removed in the penta-deuterated compound CD_3OCD_2H. The free molecule (in CCl_4) shows to CH stretching bands: $\tilde{\nu}_{CH\cdot\cdot} = 2886\ cm^{-1}$ and $\tilde{\nu}_{CH} = 2966\ cm^{-1}$, which are due to two different conformations with the C—H oscillator located trans and gauche to a lone electron pair, respectively. In the CD_3OCD_2H—$AlCl_3$ adduct the low frequency band $\tilde{\nu}_{CH\cdot\cdot}$ is split into two bands at 2968 and 2985 cm^{-1}, while the $\tilde{\nu}_{CH}$ mode occurs at 3040 cm^{-1}. The splitting of the $\tilde{\nu}_{CH\cdot\cdot}$ band is due to the fact that the two electron pairs in the coordination compound are no longer equivalent. The low frequency component of the $\tilde{\nu}_{CH\cdot\cdot}$ bands is being used by Lavalley and Caillot [213] for the detection of surface acid sites, since its position is very sensitive to coordination. Frequency shifts $\Delta\tilde{\nu}$ relative to the $\tilde{\nu}_{CH\cdot\cdot}$ mode of the free molecule at 2886 cm^{-1} should thus be a measure of the coordination bond strength or acid strength of the acid site. The $\tilde{\nu}_{CH\cdot\cdot}$ band position is also influenced by H-bonding. An easy distinction between H-bonded and coordinatively bonded molecules, however, is possible by means of their desorbability. Table 8 shows a comparison of $\Delta\tilde{\nu}$ values obtained for the $\tilde{\nu}_{CH\cdot\cdot}$ mode on adsorption of CD_3OCD_2H on SiO_2, ZnO and Al_2O_3 with frequency shifts of coordination sensitive normal modes of other probe molecules, namely pyridine, ammonia and *tert.* butyl cyanide. Obviously, the shifts of the $\tilde{\nu}_{CH\cdot\cdot}$ band increase as the Lewis acid strength of the oxide increases. Thus, the CD_3OCD_2H molecule is a very sensitive probe for surface Lewis acid sites, which has the advantage of a much lower basicity as compared to most conventional probe molecules.

Moreover, it coordinates to Lewis acid sites exlusively, so that their surface density can easily be determined by gravimetric techniques.

Knözinger and Stolz [214] first demonstrated the importance of steric effects when 2,6-disubstituted pyridines were used as probe molecules. 4-methyl pyridine was more strongly bound to Lewis acid sites on alumina than pyridine due to its higher basicity. 2,4,6-trimethyl pyridine despite its still higher basicity was quantitatively displaced by pyridine, since the coordination bond is significantly weakened by the steric hindrance exhibited by the methyl groups in 2- and 6-position. Benesi [215] suggested 2,6-dimethyl-pyridine as a selective probe for protonic sites in the presence of Lewis acid sites, although it has been shown that steric hindrance with this molecule does not completely prevent its coordination onto Lewis acid sites [214]. Selective, but not exclusive interaction of 2,6-dimethyl pyridine with protonic sites was also detected by infrared spectroscopy during adsorption on CaY, HY and NaY zeolites [216] and on fluorided aluminas [217]. As at least local reorganizations of the surface may occur under the action of temperature and adsorbate, surface cations may possibly be displaced from their position in a coordination site towards the surface and thus facilitate the coordination bond formation. 2,6-di-*tert*. butyl pyridine is so strongly hindered that a coordination onto Lewis acid sites is definitely impossible [217–219]. On the basis of infrared studies, however, it was infered that the protonation of this molecule by Brönsted sites should be possible [217, 218]. On the other hand, 2,6-di-*tert* butyl pyridine does not form N···H—O H-bonds with silanol groups but rather interacts with the silanol groups via the π-electron system [52, 171] as evidenced by a frequency shift of OH stretching vibrations of 100 cm^{-1} which is typical for OH···π interactions. The same type of interaction was also observed on alumina [219], so that the possible protonation of 2,6-di-*tert* butyl pyridine still seems to be open to question. However, even if protonation were possible, the adsorption of this probe molecule would not necessarily be selective, since a strongly chemisorbed species was detected on alumina surfaces in addition to the weakly interacting H-bonded form [218, 219]. This strongly held species was identified as an N-oxide ($\tilde{\nu}_{N-O} = 1275\ cm^{-1}$), which is probably formed on exceptionally exposed and electron deficient oxygen ions [219].

The relative numbers of Brönsted and Lewis acid sites can be estimated in principal, provided the ratio of the extinction coefficients is known for two bands typical for a protonated and coordinated probe molecule, respectively. Attempts have been made to evaluate this ratio for pyridine normal modes, using the 19b mode, which occurs near 1450 cm^{-1} for coordinated pyridine species and near 1545 cm^{-1} for the pyridinium ion [137, 217, 220–222]. The most reliable value is probably that given by Hughes and White [221]

$$\frac{\varepsilon\,(1450)}{\varepsilon\,(1545)} = 1.08 \pm 0.09 \tag{4}$$

This value was recently verified by Matulewicz *et al.* [217] who found a ratio of 0.9 ± 0.1. Using the apparent integrated absorption intensities reported by Hughes and White [221], Ward and Hansford [223] estimated the limits of detectability of Brönsted acidity to be of the order of magnitude of 10^{-2} meq g^{-1} for silica-aluminas with BET surface areas between 350 to 500 m^2 g^{-1}. Bielanski and Datka [224] reported an integrated extinction coefficient of (1.06 ± 0.07) cm μmol^{-1} for the 1545 cm^{-1} band of the pyridinium ion adsorbed on decationized Y zeolites. Jacobs and Heylen [216] and Matulewicz *et al.* [217] have also estimated the relative extinction coefficients for the corresponding bands of coordinated (1579 cm^{-1}) and protonated (1642 cm^{-1}) 2,6-dimethyl pyridine, which is again very close to unity as for pyridine. The relative number of Brönsted and Lewis acid sites can thus be estimated to a good approximation from the relative intensities (peak heights) of the respective bands of the pyridine or dimethyl pyridine species.

iv) Functionalized oxide surfaces. Surface functional groups on oxide surfaces can be introduced by chemisorption or chemical reaction of a wide variety of reagents with surface hydroxyl groups and/or acid-/base pair sites which involve coordinatively unsaturated surface metal and oxygen ions. The introduction of certain foreign elements such as F or Cl may significantly alter the acidic and basic character of the solid surface (see II.A.1.a.ii). These foreign elements, as far as they substitute surface

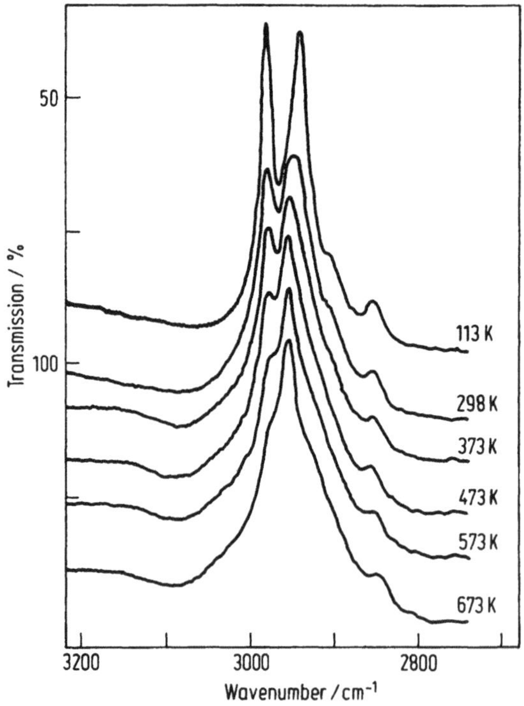

Figure 17. Infrared spectra ($\tilde{\nu}_{CH}$) of CHD_2OH chemisorbed on silica. The percent transmission scale refers to the 113 K spectrum. (Reproduced with permission from ref. [225])

atoms of the substrate have to be considered as modifiers of intrinsic surface properties rather than as functional groups.

Reaction of alcohols on oxide surfaces leads to the formation of *surface alcoholate* species usually through dissociative chemisorption on acid-base pair sites with the simultaneous formation of new hydroxyl groups. The mechanisms which are operative on silica surfaces, *i.e.* the esterification with silanol groups and the reaction with "strained" siloxane bridges are well known and have been reported frequently [52, 206]. The most detailed vibrational analysis of surface methoxyl groups on silica was given by Morrow [225]. For a surface $\geq SiOCHD_2$ species two bands could be detected at 2979 and 2940 cm^{-1} in the low temperature spectra in Figure 17, which are assigned to CH stretching vibrations of the methoxyl species in two different configurations. In the surface species the SiOC atoms should lie in a local plane of symmetry. There are two possible positions off this plane for the H atom in trans orientation relative to the oxygen lone pair (*as*-$SiOCHD_2$) and one position in the plane (*s*-$SiOCHD_2$). The two observed bands at 2979 cm^{-1} and 2940 cm^{-1} are assigned to the *s*-and *as*-$SiOCHD_2$ species (species A), respectively. In addition, a weaker band is observed at 2954 cm^{-1} at low temperature in the spectra of Figure 17. As the temperature is raised, this band grows in intensity and eventually becomes the dominant spectral feature. Therefore, the 2954 cm^{-1} band can be assigned to a new chemisorbed species B which, in view of the reversibility of the spectral changes with changing temperature, must exist in equilibrium with the above mentioned low-temperature species A. Species B must be symmetrical since it gives rise to only a single CH stretching band. It has been suggested [225] that the infrared bands of species A are probably hot bands arising from transitions from the first vibrationally excited state of the SiOC bending mode such that a rapid flipping of the SiOC bond can occur in this excited state. Classically the CH bonds would see an "average" interaction with the rigid framework and all CH bond strengths would become equal.

Azrak and Angell [226] have shown by infrared spectroscopy, that surface functional groups such as $-(CH_2)_xSH$ and $-(CH_2)_xNH_2$ can be created on silica surfaces by esterification of substituted alcohols with silanol groups. The reaction of mercaptoethanol with silica at 520 K leads to chemically bonded $-(CH_2)_2SH$ groups ($\tilde{\nu}_{SH} = 2580\ cm^{-1}$) which were thermally stable at 520 K and resisted hydrolysis at 370 K. Analogously, reaction of silanol groups with 3-aminopropanol gave surface $-(CH_2)_3NH_2$ groups ($\tilde{\nu}_{NH_2} = 3330$ and $3390\ cm^{-1}$) which were stable on heating at 520 K.

Conventionally, functional groups of this type are created on silica surfaces by reaction of *alkoxysilanes* $(RO)_xR'_{x-3}Si(CH_2)_yX$ with silanol groups [227], where the functional grouping may be *e.g.* $X = -CH{=}CH_2$, $-PPh_2$, $-NH_2$, $>NH$, $-SH$, $-C{\equiv}N$, $-NHCO$ 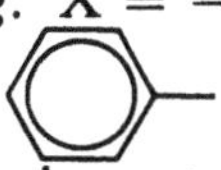 $-N_2^+Cl^-$ *etc.* Surface modified materials of this type are of primary importance for the *immobilization of complexes* [228] and *molecular clusters* [229] and of *enzymes* [230].

Glass surfaces can be modified in the same way. Figure 18 shows transmission infrared spectra of a modified silica surface bearing cyclohexylaminopropyl groups [231]. Spectrum A corresponds to the modified silica after drying at 350 K. Residual silanol groups are still present on the surface. H-bonding interactions occur between these silanols and amine groups and among amine groups as indicated by the band near 3450 cm^{-1}, the broad absorption extending to below 3000 cm^{-1} and the weak NH stretching band at 3300 cm^{-1}. The large width of the CH_2 stretching bands at 2855 and 2930 cm^{-1} suggest strong perturbations of the skeleton possibly due to a dense packing on the support surface. A complete reorganization of the surface ligands occurs when the amine functions interact with Cu^{2+} ions to form immobilized Cu^{2+} aminoaquo complexes (spectrum B in Fig. 18). In this situation, the CH stretching bands sharpen significantly, H-bonding interactions are reduced as indicated by the much lower absorption below 3200 cm^{-1} and by the appearance of a band at 3740 cm^{-1} of unperturbed surface silanols and at 3455 cm^{-1} of probably unperturbed NH groups. The two bands at 3370 and 3322 cm^{-1} have to be assigned as NH stretching modes of amine groups which are engaged in coordination bonds with Cu^{2+} in two different configurations.

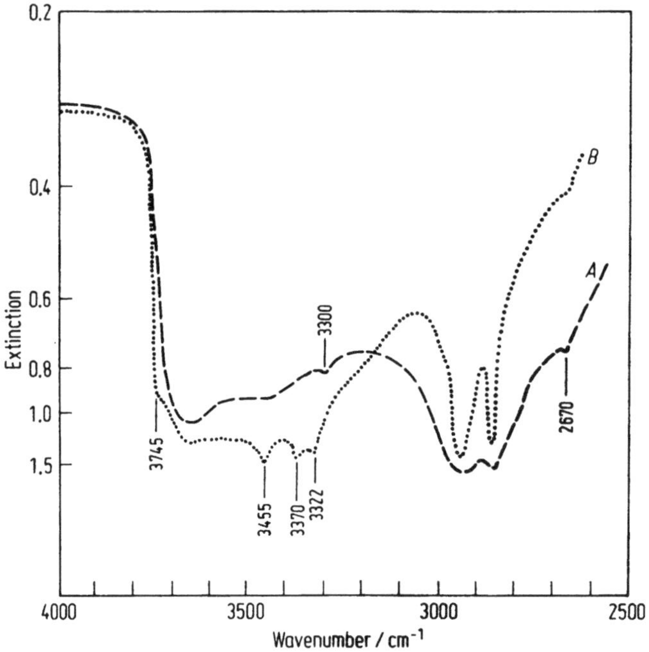

Figure 18. Infrared spectra of silica modified with cyclohexylaminopropyltrimethoxysilane (*A*) and of (*A*) after formation of complexes with $CuCl_2$ (*B*)

An infrared control of the products obtained by surface modifications is extremely important since side reactions may lead to rather ill-defined systems (see section II.A.4). Thus, the nitrile groups of nitrilopropyltetraethoxysilane were partially transformed into amine groups during modification [231]. The molecular structures of multiple amino coupling agents (*e.g.* N-2-aminoethyl-3-aminopropyltrimethoxysilane and trimethoxysilylpropyldiethylenetriamine) have been studied by FT infrared spectroscopy by Chiang and Koenig [232]. Moreover, infrared evidence had suggested [231] possible reactions between the amine groups with strained siloxane bridges (see section III.A.4) during modification of silica with (2-aminoethyl)-aminopropylsilane. This reaction could only be avoided by pretreatment of the surface with methanol which preferrentially eliminated these siloxane bridges.

The dissociative chemisorption with N—H bond rupture had been observed for CH_3NH_2 and $(CH_3)_2NH$ according to [203]

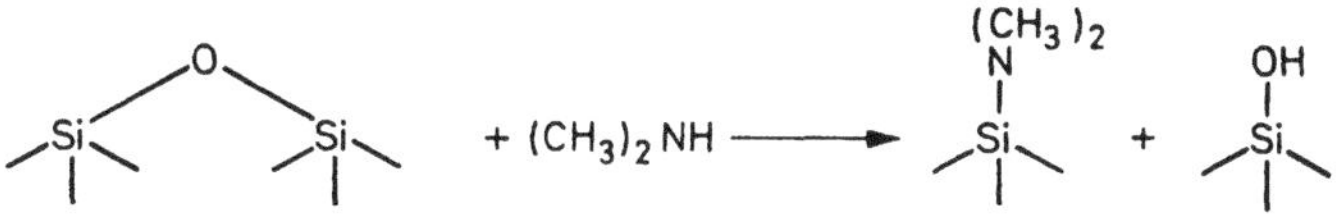

Ammonia tends to react on silica [121, 201, 203, 204, 233, 234] and other oxide [235–240] surfaces in an analogous way and can be used to generate $—NH_2$ and $>NH$ groups. Observed normal modes of these two functional groups are summarized in Table 9 for various oxides. The mechanism for the formation of these groups on silica may be described [201, 203, 233, 234] as follows

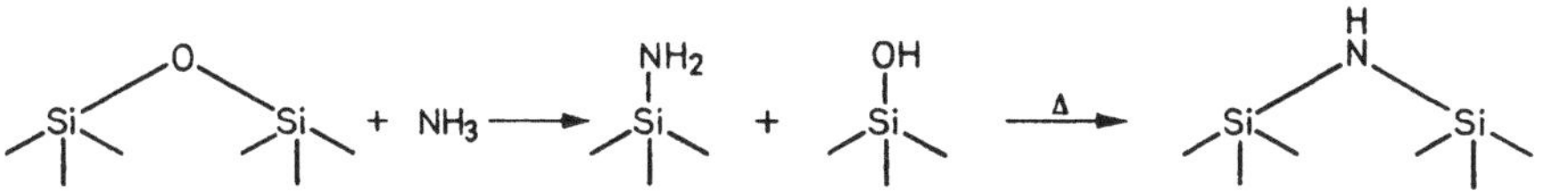

and

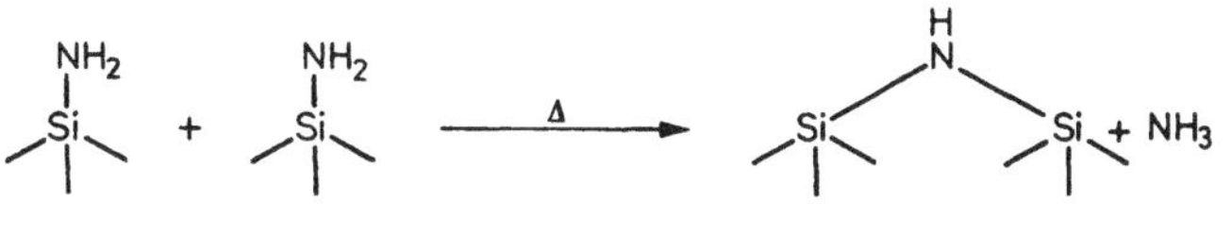

The imide groups may be hydrolyzed in the presence of water vapor [233]

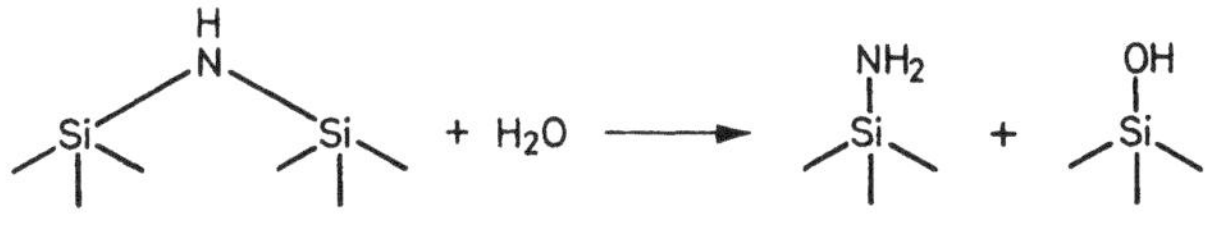

Dihydrogen has also been shown to react with strained siloxane bridges to form SiH_2 (2301 and 2285 cm^{-1}) and SiH (2294 and 800 cm^{-1}) groups at temperatures above 770 K [234].

Halogen containing molecules (such as chloromethylsilanes, BF_3, BCl_3, $AlCl_3$, $GeCl_4$, $TiCl_4$ etc.) and B_2H_6 have frequently been used to determine the number and configuration of surface hydroxyl groups, especially on silica [3, 227, 242–246].

The analysis of conversion data is complex since different surface compounds can be formed with the same reagent depending on the configuration of the hydroxyl groups. Geminal and vicinal hydroxyls *e.g.* cannot be distinguished on the basis of the stoichiometry of their reaction with MCl_3.

Table 9. Normal modes of surface $—NH_2$ and $>NH$ groups (cm^{-1})

Oxide	$\tilde{\nu}_{as, NH_2}$	$\tilde{\nu}_{s, NH_2}$	δ_{NH_2}	$\tilde{\nu}_{Si-N}$	$\tilde{\nu}_{N-H}$	Ref.
SiO_2	3540	3452	1550	932	—	[201], [203]
	3535	3455	1555	—	3390	[233], [234]
	3526	3446	—	—	—	[121]
	3526	3446	1555	—	—	[235]
Al_2O_3	3386	3335	1510	—	—	[236]
	3520	3440	1560	—	(3340)?	[238], [239]
TiO_2 (rutile)	3400	3350	—	—	—	[240]
CaO	3321	3264	1540	—	—	[235]
MgO	3380	3315	1555	—	—	[235]
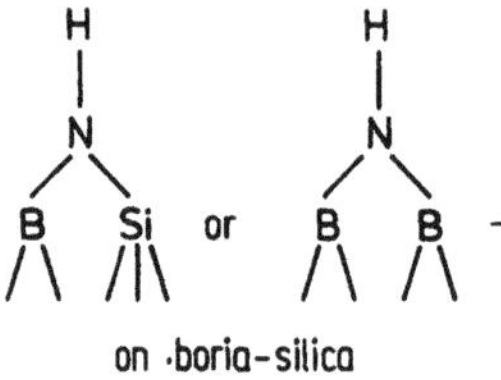	—	—	—	—	3450	[241]
Ge-NH_2 on silica	3490	3410	1560	—	—	[205]

Moreover, infrared spectroscopic evidence had suggested that reaction with strained siloxane bridges may occur to some extent [202, 203]

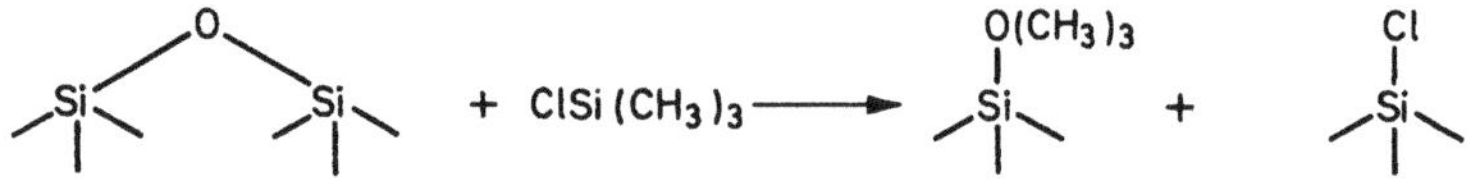

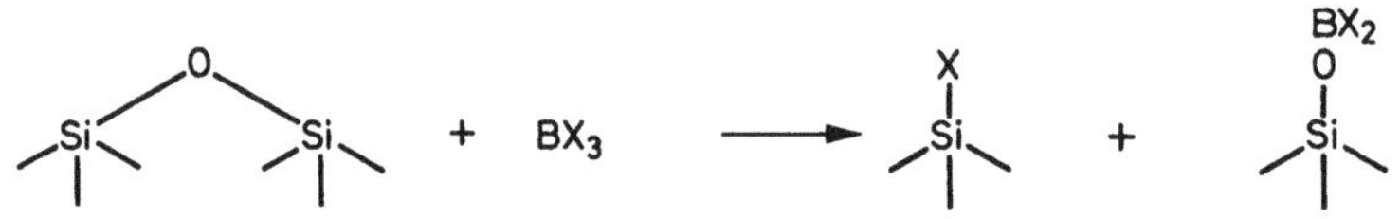

An infrared (or Raman) control of these surface reactions is desirable for various reasons. Firstly, the vibrational spectra permit a distinction of the different surface compounds formed and, secondly, the configuration of MOH groups which are produced on hydrolysis of the halogen surface compounds formed from MX_n reagents can be predicted from the configurations of the latter

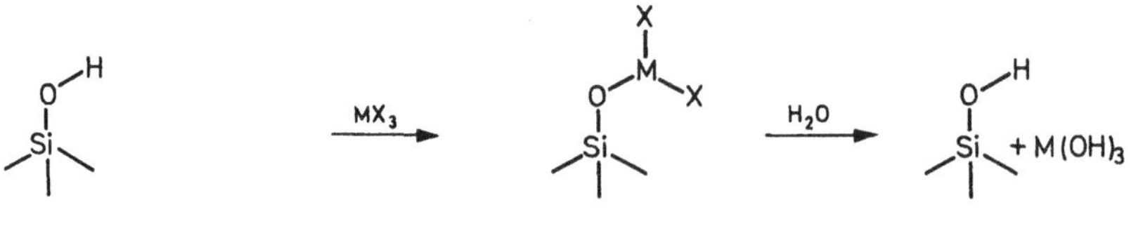

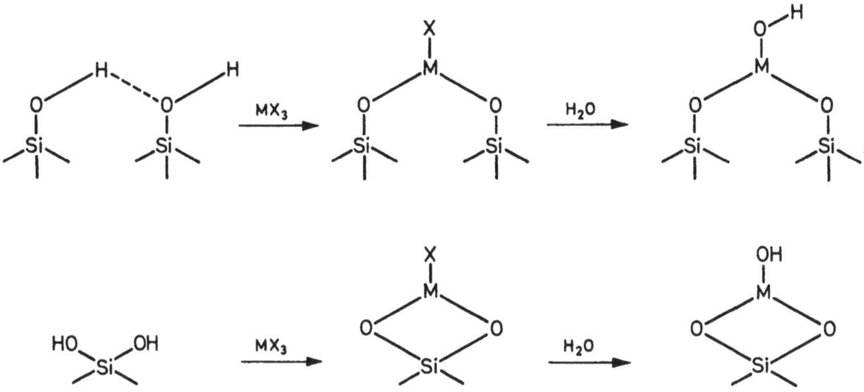

A serious problem with these systems is the fact that most normal modes of the heavier inorganic halides occur in or near the far infrared, where the adsorbent absorbs very strongly and the spectrometer sensitivity (of conventional dispersive instruments) is relatively low. Using a special sample preparation technique, Bermudez [247] succeeded in overcoming these problems when studying the SiO_2/BCl_3 system. A characteristic spectrum is shown in Figure 19 and the assignments of the observed fundamental vibrations of the unhydrolyzed and hydrolyzed surface species are summarized in Table 10.

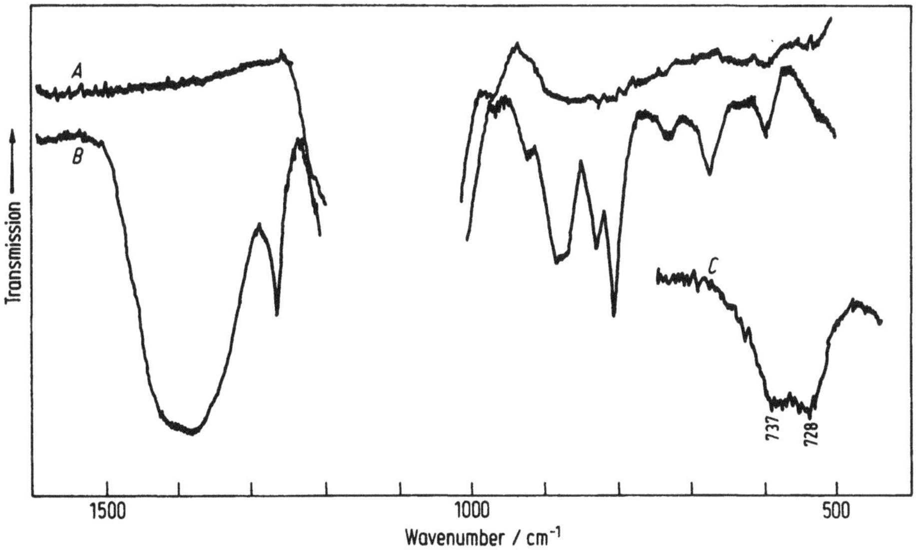

Figure 19. Spectrum of BCl_3 on silica gel: (*A*) background after 5 hr evacuation at 298 K; (*B*) after chemisorption of BCl_3 and 12 hr evacuation at 298 K; (*C*) weak overlapping bands at 730 cm^{-1} re-scanned with fivefold expansion of transmission and frequency scales. Blanc section indicates the region where instrument response is poor. (Reproduced with permission from ref. [247])

Table 10. Tentative assignment of absorption bands in the vibrational spectra of unhydrolyzed and hydrolyzed products from chemisorption of BCl_3 on SiO_2 (from ref. [247])

		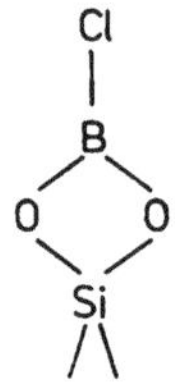		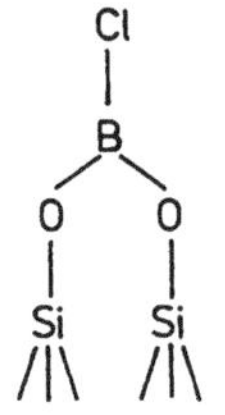	Assignment
1410					B—O stretch
			1380		antisym.
					B—O stretch
		1265			antisym.
					Si—O stretch
927					antisym.
					B—Cl stretch
		889		}	sym. B—O
				880 }	stretch
831					SiOB bending
		808			"ring defor-
					mation" or
				737	torsion
	728		677		BCl stretch
				596	
		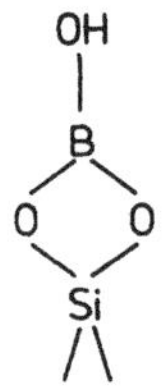		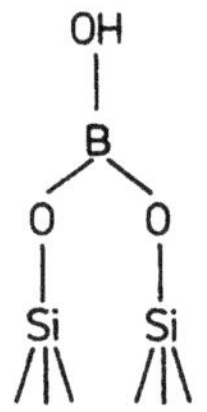	
		3710		3695	OH stretch
			1400		antisym.
					BO stretch
		1267			antisym.
					SiO stretch
		883			sym. BO stretch
				850	
		807			"ring defor-
			731		mation" or
					torsion

Normally, the fundamental vibrations of the reaction products of chloromethylsilanes on oxide surfaces have only been identified in the CH stretching and deformation region. A recent study of this type by Harrison and Thornton [248] describes the interaction of trimethylchlorosilane and dimethyldichlorosilane with SnO_2.

The formation of surface functional groups from the reaction of *ethyl isocyanate* with silica surfaces was studied by Eley *et al.* [249]. These authors demonstrated the formation of an urethane by reaction of silanol groups with the adsorbate

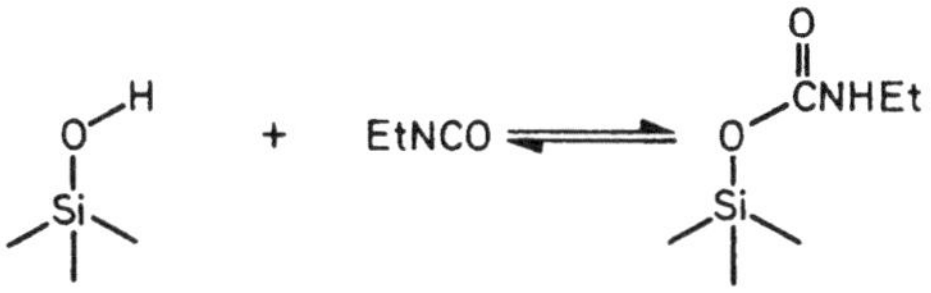

Also in this system, surface siloxane bridges may be involved. A dissociative chemisorption of ethylisocyanate at high temperatures leads to the formation of surface isocyanate species

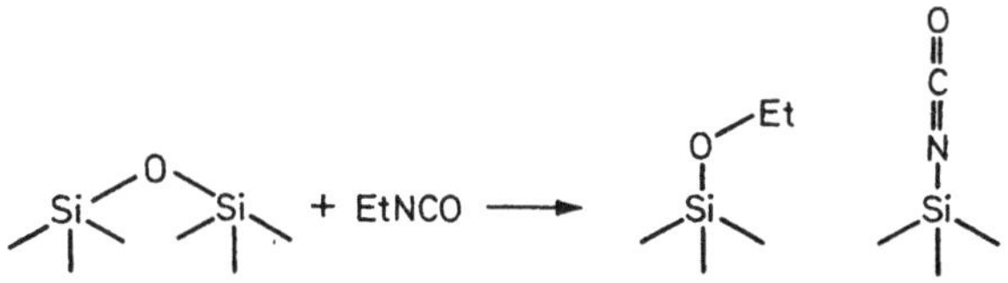

Surface coordination compounds can perhaps be considered as surface functional groups. Various coordination compounds, such as carbonyls and π-allyl complexes of group IV b, VI b and VIII b elements can be reacted with surface hydroxyl groups of oxides to generate surface coordination compounds [250]. Most of these systems are active catalysts for various hydrocarbon reactions as *e.g.* olefin and diene polymerization, metathesis, and olefin hydrogenation [250]. Vibrational spectroscopy would permit a structural characterization of these compounds and of their oxidation and reduction products.

Yermakov [250] has shown a convenient route for the generation of various surface organometallic compounds. This route consists of the reaction of π-allyl complexes of transition elements with reactive surface groups on oxides. Thus, molybdenum complexes may be bound onto hydroxylated silica by the reactions

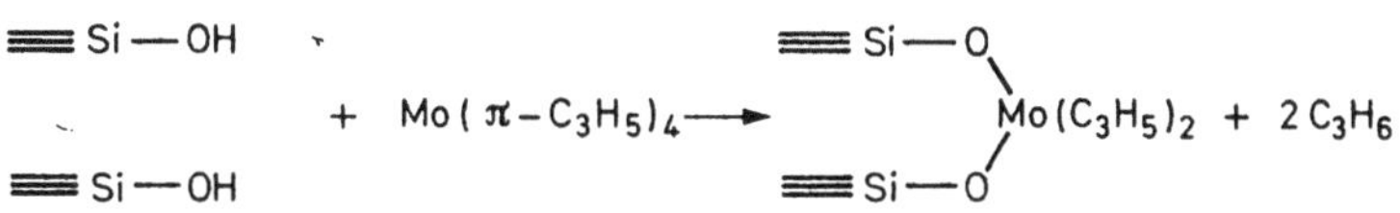

or

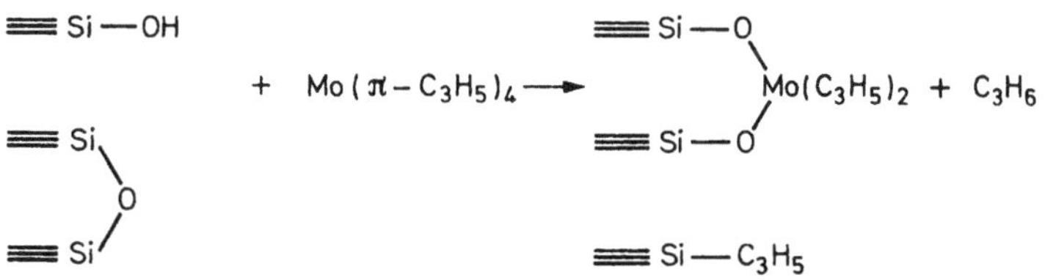

Figure 20 shows spectra of a silica-supported molybdenum complex. The band at 3740 cm^{-1} of isolated silanol groups is strongly reduced after reaction

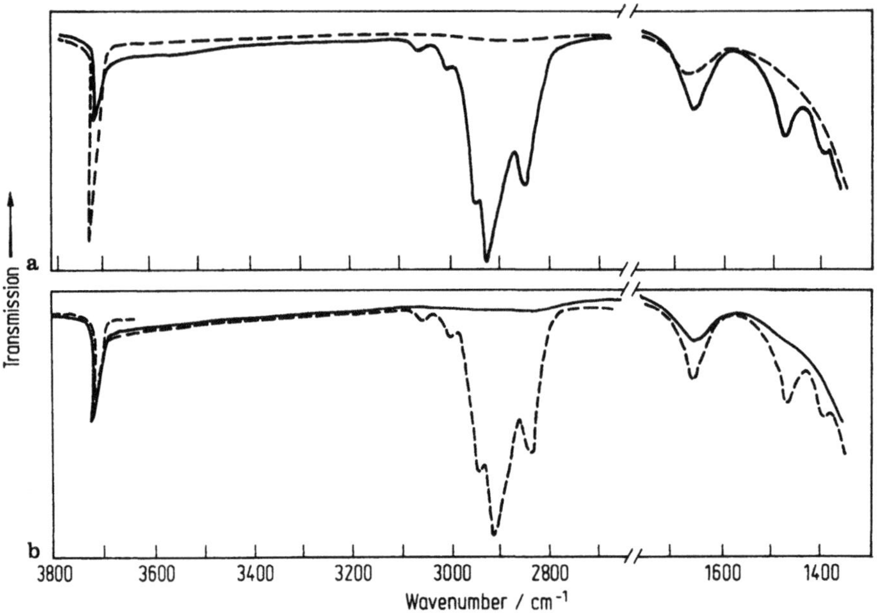

Figure 20. IR spectra of the catalyst obtained by the interaction of tetrakis-π-allylmolybdenum with silica. **a** (– – –) SiO_2 dehydrated at 973 K; (——) SiO_2 after adsorption of $Mo(\pi\text{-}C_3H_5)_4$. **b** (– – –) SiO_2 after adsorption of $Mo(\pi\text{-}C_3H_5)_4$; (——) the same sample after treatment with hydrogen at 673 K. (Reproduced with permission from ref. [250])

with $Mo(\pi\text{-}C_3H_5)_4$. C—H stretching vibrations appear in the region between 2800 and 3000 cm^{-1} and the band at 1650 cm^{-1} is attributed to a C=C stretching mode. The spectrum does not show an absorption around 1500 cm^{-1} which would correspond to the C⋯C⋯C vibration in a π-allyl group. It was therefore suggested that the allyl ligand might be bound to the molybdenum ion by σ rather than by a π-allyl bond in the surface complex. Surface organometallic compounds of this type may undergo various transformations depending on temperature and gas phase. For instance, an allylic surface zirconium compound can be transformed into a zirconium hydride compound on treatment in H_2 atmosphere [251].

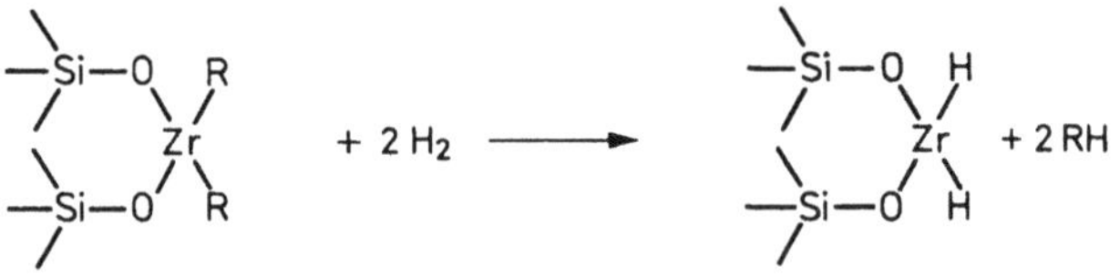

The ZrH_2 species can be identified by its characteristic Zr—H stretching vibrations which appear at 1620 and 2190 cm^{-1}.

A second route for the generation of surface complexes of group VIb metals is the adsorption and thermal decomposition of the respective carbonyls, *e.g.* $Cr(CO)_6$, $Mo(CO)_6$ and $W(CO)_6$. The decomposition of

Table 11. Comparison of carbonyl stretching frequencies for subcarbonyls of molybdenum, tungsten and chromium. (Reproduced with permission from ref. [254])

Species	Frequencies ($\pm 5\ cm^{-1}$)		
	M = Mo	M = W	M = Cr
$M(CO)_6$	2120, 2020, 1990	2125, 2000	2120, 2000
$M(CO)_5$	2080, — — [a]	— [a] 1750	2080, 1880
$M(CO)_4$	2040, 1960, 1750	— —	2000, 1760
$M_2(CO)_8$	2040, 1960, 1640	2000, 1960, 1590	— — [b]
$M(CO)_3$	2000, 1900, 1680	1990, 1880, 1640	— — [b]
$M_2(CO)_6$	2000, 1900, 1590	1990, 1880, 1545	— — [b]

[a] Not resolved
[b] Not observed

these carbonyls on alumina supports leads through a number of subcarbonyl species [252] which may be represented in the following scheme [253]

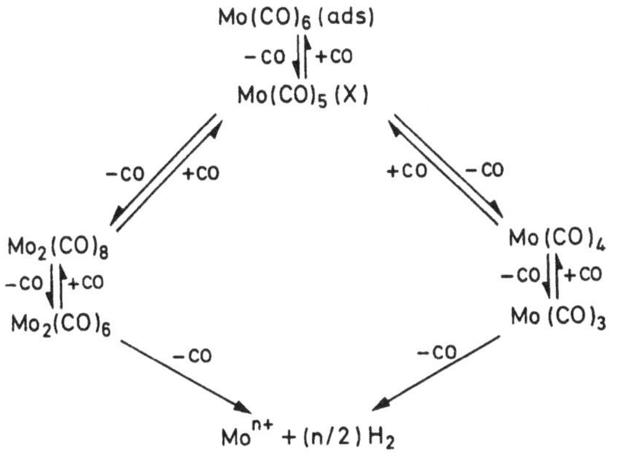

The carbonyl stretching bands of the various molybdenum subcarbonyl species [253] are summarized in Table 11 and compared with those of the subcarbonyl species observed during adsorption and decomposition of $W(CO)_6$ and $Cr(CO)_6$ [254]. The decomposition sequences of $Mo(CO)_6$ and $W(CO)_6$ are very similar. $Cr(CO)_6$, however, does not form the proposed bimolecular species and complere decarbonylation of the adsorbed chromium complex is achieved more readily.

A very important class of catalysts for olefin polymerization (Philips process) consists of silica supported chromium species (Cr/silica polymerization catalyst). These catalysts can be prepared by impregnation of silica surfaces with chromic acid CrO_3, calcination and reduction with CO. The active form of the catalyst most likely consists of Cr^{2+} species [255–257] although Beck and Lunsford [258] are in favor of Cr^{3+} species. The oxidized catalyst precursor CrO_3/SiO_2 gives rise to a broad absorption band centered at 925 cm^{-1}, which was interpreted as an indication of the presence of dichromate species on the support surface [259]

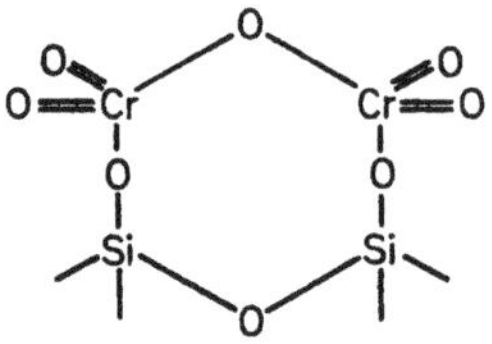

This species, however, does not exist exlusively in these catalyst. At low loadings and low calcination temperatures a chromate species seems to predominate which may form di- (or poly) chromates at increasing temperatures [260–262].

The reduced catalyst can be characterized by the infrared spectra of probe molecules, such as CO and NO. Since the oxidized high-temperature precursor state was described by the dichromate structure, the formation of low-valent Cr ion pairs on the support surface seems to be plausible. On adsorption of CO at a coverage corresponding to a 1:1 ratio for Cr:CO three carbonyl stretching vibrations at 2191, 2186 and 2181 cm^{-1} were observed [263], which can be assigned to three types of linear Cr—CO complexes, namely a Cr^{3+}—CO complex (2191 cm^{-1}) and two different Cr^{2+}—CO complexes (2186 and 2181 cm^{-1}). At higher coverages (Cr:CO > 1), CO also adsorbs in a bridging form which gives rise to a band at 2095 cm^{-1}. The corresponding surface species must be due to the presence of Cr ion pairs and were described as follows [263]

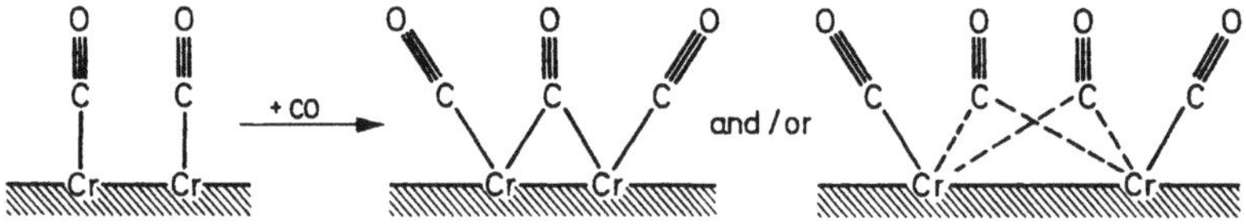

Interestingly, NO did not discriminate between the two types of Cr^{2+} species detected by CO adsorption [264]. NO adsorption led to the formation of dinitrosyl complexes $Cr^{2+}(NO)_2$ which exhibit a characteristic band pair at 1747 and 1865 cm^{-1} of the symmetric and antisymmetric NO stretching modes [258, 264], as shown in Figure 21. The additional weak band at 1810 cm^{-1} was assigned to a linear Cr^{2+}—NO complex. CO adsorption did apparently not detect the corresponding "shielded" Cr^{2+}ions. The Cr^{3+}ions responsible for the Cr^{3+}—CO complexes also formed dinitrosyl complexes characterized by a band pair at 1755 and 1880 cm^{-1}. Simultaneous adsorption of CO and NO gave information on the coordination of the Cr species [265]. While Cr(NO) and $Cr^{3+}(NO)_2$ only undergo ligand displacement reactions, the $Cr^{2+}(NO)_2$ species can coordinate a further CO ligand forming $Cr^{2+}(NO)_2CO$ complexes. From this result it was infered [265] that Cr^{2+} ions should be trigonal originally and should be able to accomodate up to three ligands to complete an octahedral coordination sphere.

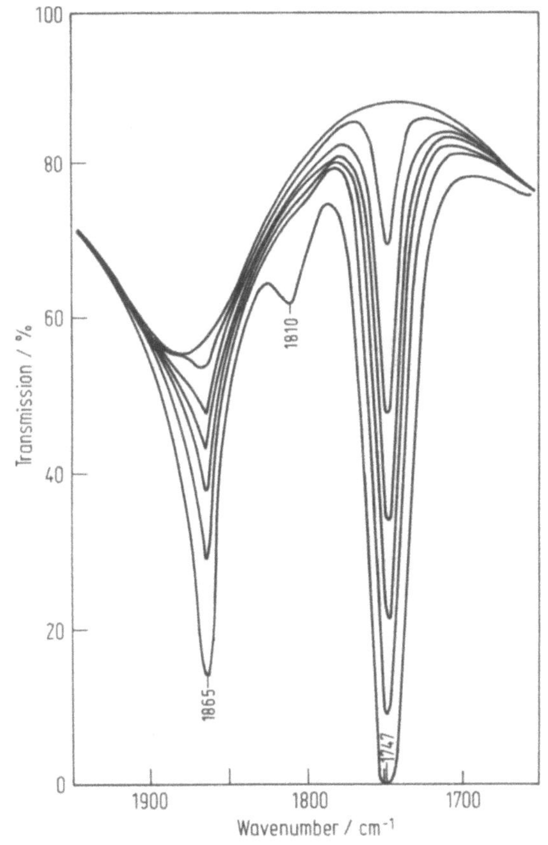

Figure 21. Infrared spectra of the adsorption of small doses of NO on a standard reduced 0.5% CrO_3/SiO_2 sample (transmittance vs. wavelength in cm^{-1}). The final pressure is 9.3 Pa, i.e. the vapor pressure at 77 K. (Reproduced with permission from ref. [264])

Kazusaka and Howe [266] observed analogous Cr^{2+} and Cr^{3+} dinitrosyl species on catalysts which are prepared by adsorption and decomposition of $Cr(CO)_6$ on Al_2O_3. The same surface species had previously been reported for reduced CrO_3/Al_2O_3 catalysts [267]. Adsorption of NO on $Mo(CO)_6/Al_2O_3$-derived catalysts gave typical bands of Mo^{4+} dinitrosyl complexes [266] which were also detected on reduced conventional MoO_3/Al_2O_3 catalysts [268].

The results discussed above, may serve as examples for the usefulness of CO and NO as probe molecules in infrared transmission spectroscopy. These probes may provide information on the coordination and valence state of surface coordination compounds and about the existence of pairs or clusters of metal atoms or ions.

b) Infrared reflection spectroscopies

The transmission technique fails when opaque and/or strongly scattering ($\lambda \leqq \bar{d}$) materials are to be studied. A variety of infrared reflection methods have been developed which make use of the specular reflection or diffuse reflectance depending on the nature of the solid adsorbent and its optical properties [269]. The *reflection-absorption technique* is applied for vibrational spectroscopy [15, 16] on single crystal surfaces and *ellipsometry* [269] is a widely used technique which permits the measurement of the optical constants of materials with flat surfaces. Ellipsometry provides high accuracy and sensitivity to thin films or to adsorbed materials at low coverage. Most important for the detection of surface functional groups among the reflection techniques are the *internal reflection spectroscopy* (a modification of the well-known attenuated total reflection (ATR) method) and the *diffuse reflectance technique*. The application of various modulation techniques [270] may be advantageous because of their improved signal-to-noise ratios. Wavelength modulation in particular produces ac signals at the detector which are proportional to the first derivative of the reflectance. The spectra are therefore free of any baseline drifts and spectral structures are more sensitively detected.

i) Internal reflection spectroscopy (IRS). In IRS a monochromatic light beam is totally reflected (angle of incidence greater than the critical angle θ_c) at the interface between a non-absorbing crystal (having a highly polished surface) with refractive index n_1 and an optically rarer medium with refractive index $n_2 < n_1$. A so-called evanescent wave is produced in the optically rarer medium, the amplitude of which decays exponentially with increasing distance from the interface, the range of penetration being around $\frac{\lambda}{7}$ to $\frac{\lambda}{10}$. If the optically rarer medium is absorbing, the totally reflected beam will be attenuated. Transmission-like spectra can thus be obtained. Several review articles were devoted to the IRS technique [271–274].

The most attractive features of the IRS technique can be summarized as follows:

(i) strongly absorbing or opaque materials which do not permit the application of the transmission technique, can be studied;

(ii) disperse materials and powders having high scattering coefficients provide good quality internal reflection spectra although they cannot be brought in good optical contact with the geometrically flat crystal surface. In cases where the penetration depth of the evanescent wave into the sample medium is small compared to the dimensions of the particles, scattering does not seem to cause problems [275];

(iii) the sample thickness does not play a role provided it is larger than the penetration depth of the evanescent wave;

(iv) the intensity of "absorption bands" can be increased by multiple reflections.

IRS has certainly found its most attractive application for the detection of surface functional groups on *carbons*. Good qualitity infrared transmission spectra of oxygen surface compounds on carbons have been reported by Zawadzki [276]. However, these transmission spectra could only be obtained with extremely thin, specially prepared carbon films. In general, the high extinction coefficient and the high refractive index of carbons render transmission studies impossible. A carbon film of 3.7 μ thickness would transmit only about 1 % of the incident energy at a wavelength of 5 μ [275].

Mattson and Mark [277] recorded the internal reflection spectra for graphite, carbon black and activated carbon. They suggested that the predominant oxygen containing surface functional groups on activated carbon are ketone carbonyl groups. The relative abundance of all oxygen-containing functional groups was found to be dependent upon the type of carbon and its pretreatment.

The electric field amplitudes at the totally reflecting interface are different for the three spatial directions and depend sensitively on the polarization of the incident beam and on the angle of incidence. Absorption measurements as a function of polarization and angle of incidence provide therefore the possibility to determine the average orientation of surface functional groups on crystallographically well-defined surfaces of infrared transparent crystals. Haller and Rice [278] have studied the internal reflection spectra of hydroxyl groups on the (0001), ($11\bar{2}3$) and ($41\bar{5}0$) faces of an α-Al_2O_3 crystal. The experimental absorbance values were consistent with randomly oriented OH dipoles. This and the width of the bands observed were explained in terms of proton mobility on the surface of the crystals at room temperature. Similar studies on MgO and TiO_2 single crystal faces have been reported by the same authors [279].

It should be mentioned that IRS can also successfully be used to obtain the vibrational spectra of surface groups on working electrodes [272, 280].

ii) Diffuse reflectance spectroscopy (DRS). Diffuse reflectance techniques have been reviewed by Kortüm [21] and by Wendlandt and Hecht [22] and they are discussed in several other articles [16, 23, 24]. These techniques can be applied for samples in the form of powders or pellets which are usually terminated by two parallel surfaces. The front surface is illuminated by diffuse light or a parallel beam under a fixed angle of incidence [21]. If isotropic scattering is assumed, the reflectance R (defined as the ratio of backward and forward fluxes at the illuminated front surface) can be expressed in terms of phenomenologically defined absorption and scattering coefficients K and S, respectively, according to the Schuster-Kubelka-Munk theory [21–24]

$$R = \frac{1 + R_g[b \coth Y - a]}{a + b \coth Y - R_g}; \tag{5}$$

where $Y = Sbd$, R_g is the reflectance of the background, d = the thickness of the sample layer, $a = 1 + K/S$, and $b = \sqrt{(a^2 - 1)}$. For infinitely thick samples ($Y \to \infty$; order of magnitude about 1 mm in the UV-VIS region, about 10 mm in the infrared fundamental region [21]), this equation reduces to the simple so-called Kubelka-Munk function

$$\mathrm{F}(R_\infty) = \frac{(1 - R_\infty)^2}{2R_\infty} = \frac{K}{S}; \tag{6}$$

(R_∞ = reflectance at infinite sample thickness). This expression determines directly the ratio $\frac{K}{S}$, which is equivalent to the true absorption spectrum provided the scattering coefficient is independent of the wavelength. Normally, this condition is met in the ultraviolet-visible and near infrared region. In the mid-infrared (fundamental) region, however, the scattering coefficient will become wavelength dependent when the wavelength is larger than the mean particle size ($\lambda > \bar{d}$). On the other hand, samples having large particle sizes which cannot be studied by transmission spectroscopy, are well suited for DRS.

Normally, in the ultraviolet-visible and near-infrared region the diffuse scattered radiation from the sample is collected in an integrating sphere to reach the detector. In this spectral range large detector areas (about 1 cm^2) can be allowed and high sensitivity detectors are available. Infrared detectors on the other hand have comparably low sensitivity, and small detector areas are required to reduce the thermal noise from the environment. Kortüm and Delfs [281] have therefore designed an equipment in which the sample is placed behind the exit slit of a monochromator in one focus of an ellipsoidal mirror, which collects the diffusely scattered radiation from the sample in the second focus where the detector is located. Similar devices have been described later [282, 283]. In conventional dispersive instruments the sample cannot normally be placed behind the exit slit of the monochromator without significant technical modifications. FT spectrometers, however, permit an easy adaptation of the ellipsoidal mirror device. Apparatus for FT diffuse reflectance spectroscopy have recently been described in the literature [284–286].

The DRS technique has frequently been used for studies in the near-infrared region. Figure 1, as an example, shows the overtones and combination vibrations of surface OH groups on SiO_2. These spectra have been recorded in the conventional way with an integrating sphere.

Reported diffuse reflectance spectra in the infrared fundamental region

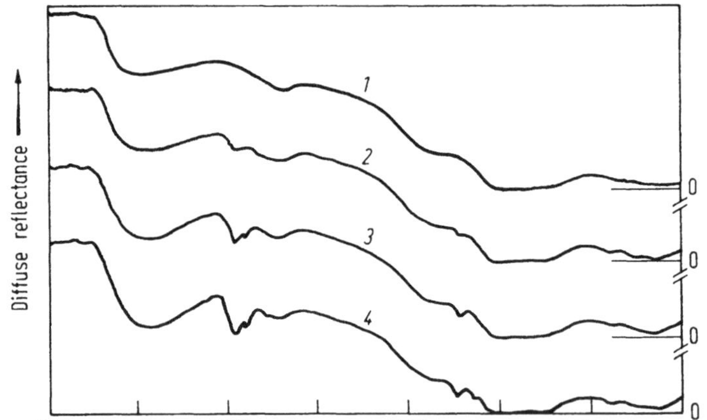

Figure 22. FT diffuse reflectance spectra of pure glass fibers (spectrum *1*) and fibers covered with γ-methylacryloxypropyl-trimethoxysilane (spectra *2–4*, coverage increases from spectrum *2* through *4*). (Reproduced with permission from ref. [286])

are scarce. Kortüm and Delfs [281] have studied the adsorption of ethylene on a SiO_2—Al_2O_3 catalyst and of hydrogen cyanide on different metal oxide surfaces. Their results demonstrate the potential utility of the DRS technique for the study of surface functional groups on solids which cannot be used for the transmission technique due to their scattering characteristics. It may be foreseen that the technique will be more widely applied in its FT version. An example of FT-diffuse reflectance spectra is shown in Figure 22, representing the vibrational spectra in the fundamental region of glass fibers covered with γ-methylacryl-oxypropyl-trimethoxysilane. All spectra show a background drift towards lower wavenumbers caused by the wavelength dependence of the scattering coefficient. The intense absorption of the Si—O vibration of glass gives rise to low diffuse reflectance values below 1500 cm^{-1} superimposed by some contributions of regular reflections near 1050 cm^{-1}. Discrete bands can be recognized near 2900 cm^{-1} (C—H stretching modes), 1720 cm^{-1} (C=O stretching mode), and at 1640 cm^{-1} (C=C stretching mode). The silane concentration of the sample giving spectrum 2 was 3.68 mg per gram glass fiber, which had a specific surface area of 0.01 $m^2\ g^{-1}$. The absolute quantity of the detected silane was <0.1 µmol. It is evident that vibrational spectra of such low surface area materials with their special geometric shape could hardly be obtained with any of the other vibrational techniques.

c) Raman Spectroscopy

The application of Raman spectroscopy to surface studies has been reviewed in the past [16, 287–291]. As compared to other vibrational spectroscopic techniques Raman spectroscopy can widely be used for investigations of surface species on oxides, on supported and bulk metals (including the phenomenon of the so-called "surface enhanced Raman scattering" [291]), on supported oxides and at the water-solid interface. The information depth of Raman spectroscopy depends strongly on the optical properties of the sample and should typically be of the order of magnitude of 2 nm. Raman spectroscopy provides several unique possibilities:

(i) the interference of gas phase spectra is negligibly weak. *In situ* Raman spectra of working catalysts at elevated pressures and temperatures can therefore be obtained;

(ii) simple glass or quartz cells can be used;

(iii) the Raman scattering probabilities of the solid adsorbent or support are frequently very low, especially of most oxides, so that Raman spectra of surface species can be recorded in the low wavenumber region down to approximately 50 cm^{-1}.

The basic experimental requirements for conventional laser Raman spectroscopy including data manipulation and signal averaging are described in the review articles mentioned above [287, 288, 290]. Suitable *in situ* Raman cells for surface investigations have been described by Brown *et al.* [292], Cheng *et al.* [293], and by Jeziorowski *et al.* [294].

Some major problems which may be encountered in laser Raman spectroscopy are:

(i) sample sensitivity to heating effects of the laser beam including desorption of surface species. This phenomenon may become very severe when coloured samples are studied;
(ii) low sensitivity of the technique;
(iii) background fluorescence, a problem which is sometimes so severe on oxide surfaces that weak Raman signals remain undetectable.

The heating effects of the laser beam can be reduced by simply applying low laser power levels ($\leqq 100$ mW). Alternatively, cylindrical lens focus techniques have been applied in order to spread the laser beam into a line image on the sample. As discussed recently by Freeman *et al.* [295], this approach may additionally provide increased signal intensity. Finally, sample rotation will decrease the energy flux onto the sample.

As a consequence of the inherent low sensitivity of Raman spectroscopy, adsorption systems and surfaces containing strong Raman scatterers have found preferred interest in the past [287–291]. Typically, the detection limits will be of the order of magnitude of 5–10% of a monolayer for samples having specific surface areas of 100–200 $m^2\,g^{-1}$. An optimization of spectrometer sensitivities is therefore urgently desired. The intensity of light reaching the detector depends on a variety of parameters, of which only the spectral slit width, the scanning rate and the laser power can be changed when commercial spectrometers are used. All other parameters are normally invariably preselected and determined by the instruments design. However, the illumination and viewing of the sample (lens aperture, focal length and demagnification) will strongly influence the instruments sensitivity and a given commercial instrument does not necessarily provide the optimal situation for surface studies [295]. One may therefore wish to change illumination and viewing of the sample in ciritical cases.

Most Raman spectrometers work in the scanning mode using photomultipliers or photon counting detector systems. The signal-to-noise ratio S/N in the final spectrum is proportional to the time during which data are accumulated within the band pass of the monochromator. The use of "multiplex" (position sensitive) detection where the spectral information of the entire spectral range is available simultaneously at the multichannel detector, improves the instrument sensitivity considerably ($>$x 1000) over that of conventional scanning instruments [295]. The value of this increased optical efficiency was demonstrated by Freemann *et al.* [295] and by Heaviside *et al.* [296], when they studied the adsorption of pyridine on γ-Al_2O_3. Figure 23 compares the Raman spectra of chemisorbed pyridine on γ-Al_2O_3, which were recorded with photon counting (PC) and multiplex detection by means of an optical multichannel analyzer (OMA). It is clearly evident from this comparison, that multiplex detection yields significantly higher S/N ratios at considerably lower data accumulation times. This detection method will therefore permit Raman spectroscopy of weak Raman scatterers even at low surface concentrations.

The fluorescence background often encountered in Raman spectra of oxide surfaces [16, 287, 288, 291] must partly be attributed to contaminations by hydrocarbons, which can be eliminated by heating the samples in an oxygen

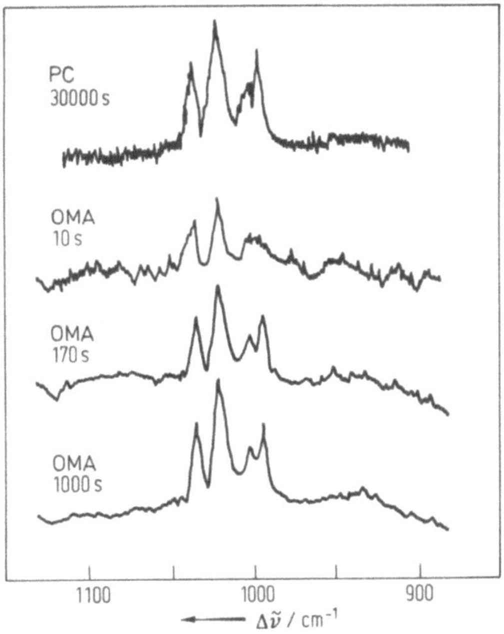

Figure 23. A comparison of Raman spectra of pyridine adsorbed on γ-Al_2O_3 taken with total data accumulation times noted in the figure. PC = photon counting; OMA = optical multichannel analyzer. (Reproduced with permission from ref. [295])

atmosphere at temperatures >700 K. Saperstein and Rein [297] reported that the fluorescence background from a 4A zeolite could effectively be reduced by washing in 0.2 N NaOH and calcining at 673 K. Considerable contributions to the fluorescence background from hydroxylated oxide surfaces are due to laser induced electronic excitations of surface hydroxide ions [298]. The fluorescence can be reduced to an extremely low level by thermal dehydroxylation [298] or by ion exchanging the OH^- ions by *e.g.* molybdate ions [299]. Thus, fluorescence seems to be an intrinsic property of oxide surfaces which bear basic hydroxide ions (see also section II.A.2.b.ii) and it can only be overcome by experimental techniques. On the other hand, information on the nature and properties of surface OH^- groups may perhaps be extracted from the fluorescence background [298].

In some cases a simple change of the exciting frequency may reduce the fluorescence background in the spectral range of interest [300]. Alternatively, time resolved Raman techniques may be applied. These methods, however, are experimentally complicated and depend on the fluorescence decay times of the samples; they can therefore hardly be developed as routine techniques for surface studies.

Frequency modulation (FM) Raman spectroscopy [270, 301] is a relatively simple technique, which produces the final spectrum in form of the first derivative of the Raman intensity. The background is therefore effectively elimintated and the S/N ratio is improved as compared to that of a conventional Raman spectrum recorded under comparable conditions. Brückner *et al.* [302] have recently described an experimental set-up for FM Raman spectroscopy. An example is shown in Figure 24, which compares the

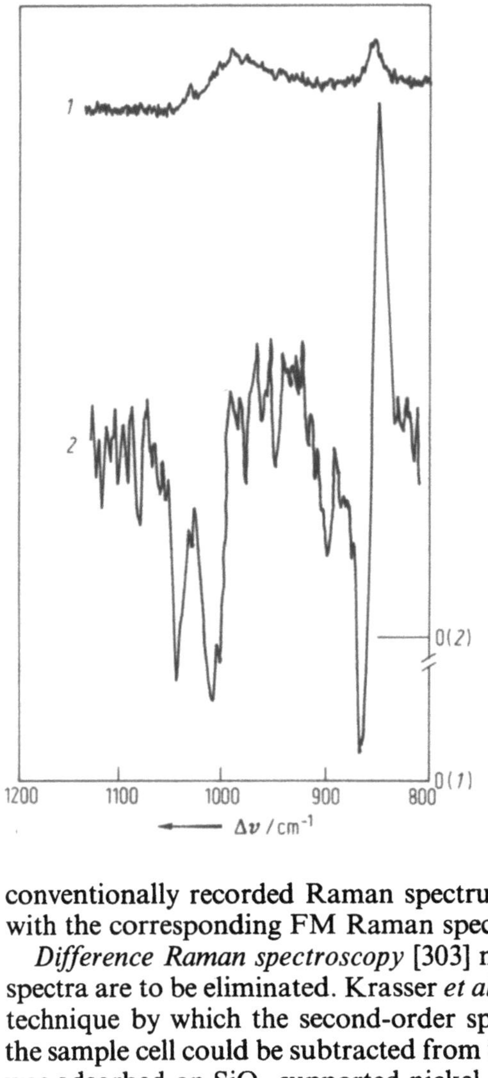

Figure 24. Comparison of a conventional (1) and frequency modulation (2) Raman spectrum of a 3% NiO–12% MoO_3 catalyst supported on γ-Al_2O_3

conventionally recorded Raman spectrum of a MoO_3—NiO/Al_2O_3 catalyst with the corresponding FM Raman spectrum.

Difference Raman spectroscopy [303] may be of interest when background spectra are to be eliminated. Krasser *et al.* [304] described a difference Raman technique by which the second-order spectrum of fused quartz windows of the sample cell could be subtracted from the Raman spectra of benzene which was adsorbed on SiO_2-supported nickel.

i) Surface hydroxyl groups. Murray and Greytak [305] observed the O—H stretching mode of surface silanol groups on SiO_2 at 3750 cm^{-1}. The Si—O stretch of surface silanols was associated with a band at 980 cm^{-1}, while the corresponding wagging mode is Raman inactive. The 980 cm^{-1} band shifted to 998 cm^{-1} on adsorption of NH_3. This wavenumber shift was explained as being due to the increased polarity of the Si—O bond resulting from hydrogen bond formation with NH_3. An enhacement in the Raman spectrum below 100 cm^{-1} which increased with NH_3 coverage and disappeared on evacuation, was also noticed. This enhancement was attributed to the formation of extremely polarizable hydrogen bonds, which are also responsible for the continuous absorption in the infrared spectra of the SiO_2/NH_3 system (see section II.A.1.a.ii). Analogous enhancements have been observed in the very low frequency Raman spectra of aqueous acid and salt solutions [306].

ii) Probe molecules. The Raman spectra of NH_3 adsorbed on SiO_2 surfaces [305] are in agreement with the infrared transmission spectra of this system (see section II.A.1.a.ii). The hydrogen bonded NH_3 species gave rise to Raman bands at 3414, 3324 and 3233 cm^{-1}. In addition, a more strongly bound species with a Raman band at 3285 cm^{-1} was detected, which is attributed to a coordinatively chemisorbed NH_3 molecule.

Typical Raman spectra of pyridine chemisorbed on γ-Al_2O_3 are shown in Figure 23. The bands observed in Figure 23 are to be attributed to the in-plane symmetric (A_1) ring deformation modes $\tilde{\nu}_1$ of pyridine of various adsorption states. This band is most useful for probing surface properties, since it shows predictable wavenumber shifts when pyridine interacts with various surface sites [16, 287–291]. The bands at 991 and 1031 cm^{-1} are due to physically adsorbed pyridine, that at about 1000 cm^{-1} is assigned to hydrogen bonded pyridine. Coordinatively adsorbed pyridine which characterizes Lewis acid sites, gives rise to the band at 1025 cm^{-1}. The results obtained with pyridine as a probe molecule on various oxide surfaces using Raman spectroscopy did not provide new information as compared to that obtained

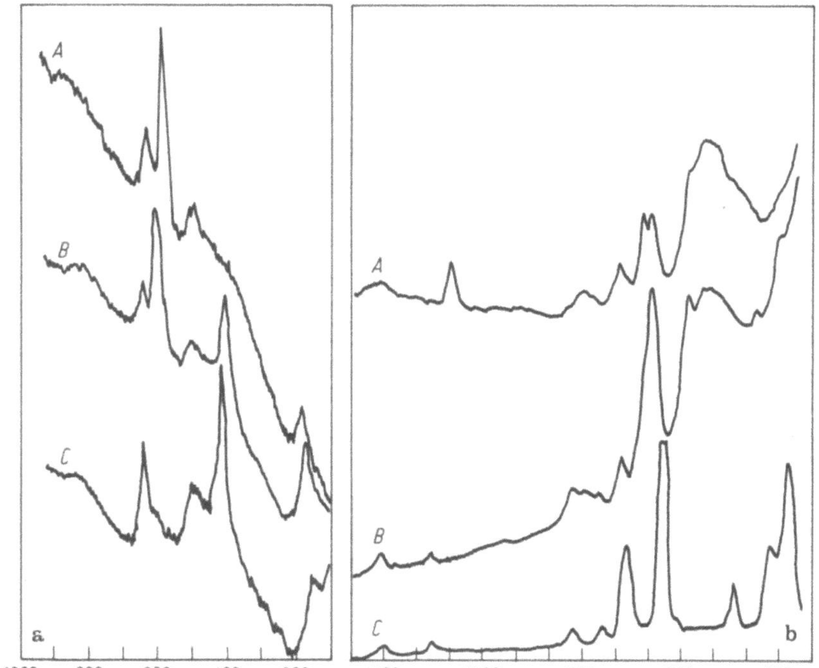

Figure 25. **a** Raman spectra of chemisorbed. (*A*) $GeClMe_3$, (*B*) $GeCl_2Me_2$, and (*C*) $GeCl_3Me$. **b** Raman spectra of (*A*) chemisorbed $AlMe_3$; (*B*) chemisorbed hexamethyldisilazane, and (*C*) liquid hexamethyldisilazane. (Reproduced with permission from ref. [307])

by infrared transmission spectroscopy (see section II.A.1.a.ii and iii). However, the normal mode $\tilde{\nu}_1$ near 1000 cm^{-1} would not be detectable by infrared transmission because of the strong lattice absorptions in this region.

iii) Functionalized oxide surfaces. The difficulties connected with the determination of the configuration of surface hydroxyl groups on SiO_2 have been discussed (see section II.A.1.a.iv). Raman spectroscopy should provide low-frequency spectral data of the reaction products between hydrogen-sequestering agents and silanol groups. This information would be required for a complete vibrational analysis and would not normally be accessible by infrared transmission. Morrow and Hardin [307] have studied by Raman spectroscopy the vibrational spectra of various hydrogen-sequestering agents (methylated and nonmethylated silicon and germanium chlorides, $TiCl_4$, $Al(CH_3)_3$, and $HN[Si(CH_3)_3]_2$ (hexamethyldisilazane)), which react with surface hydroxyl groups of silica. Their results strongly emphasize the successful application of Raman spectroscopy for investigations of the vibrational spectra of surface functional groups in the low-frequency region (< 1200 cm^{-1}). Some typical spectra are shown in Figure 25 and band assignments for M—C, SiOM and M—Cl modes are summarized in Table 12. The chemisorbed species formed from the germanium compounds were

$$\geq SiOGe(CH_3)_3,\ \geq SiOGeCl(CH_3)_2,\ \text{and}\ \geq SiOGeCl_2CH_3$$

(spectra A, B and C, respectively, in Fig. 25a). It would have been very interesting to compare Raman spectra of chemisorbed BCl_3 (and other boron compounds) with the corresponding infrared spectra reported by Bermudez [247] and shown in Figure 19. Surprinsingly, however, Raman bands of boron surface groups could not be detected [291]. This was also the case for the low-frequency modes of chemisorbed methoxy-groups, although the C—H stretching bands were observed [308]. These results were tentatively explained by Morrow [307, 308] by the assumption that the low-frequency modes would strongly couple to the lattice or phonon modes of the solid adsorbent when the mass of the atom M attached to the surface via a —SiOM

Table 12. Assignments of MC, MCl, SiOM stretching modes of chemisorbed species. (Reproduced with permission from ref. [307])

Reagent	$\tilde{\nu}_{as,MC}/cm^{-1}$	$\tilde{\nu}_{s,MC}/cm^{-1}$	$\tilde{\nu}_{s,SiOM}/cm^{-1}$	$\tilde{\nu}_{s,MCl}/cm^{-1}$
$TiCl_4$			490	400
$GeCl_4$			480	410
$GeCl_3Me$		630	490(440)[a]	410
$GeCl_2Me_2$	640	595	490	400
$GeClMe_3$	620	573	490	
$SiClMe_3$	690	600		
HMDS[b]	690	600	490	

[a] 440 cm^{-1} assigned to $\tilde{\nu}_{as,MCl_2}$
[b] hexamethyl disilazane

linkage is less than that of silicon. The band would then be smeared out and become essentially undetectable in the background at low frequency. It was therefore suggested to use probe molecules which have metal atoms which are heavier than silicon for Raman studies of this type [307].

M—Cl modes have also been observed if metal chlorides impregnated on oxide supports in the first step of the preparation of supported metal catalysts [309]. It may therefore be possible to follow the generation of supported metal catalysts by Raman spectroscopy.

iv) Supported oxide catalysts. Supported oxides — with alumina as the preferred support — are an important class of catalysts or catalyst precursors (*e.g.* molybdates, tungstates, rhenates or vanadates supported on alumina). Raman spectroscopy has been used in recent years with great success for a structural characterization of these systems. This type of spectroscopy is unique in this respect since all normal modes of the supported oxides fall into the wavenumber range below 1000 cm^{-1}, which is

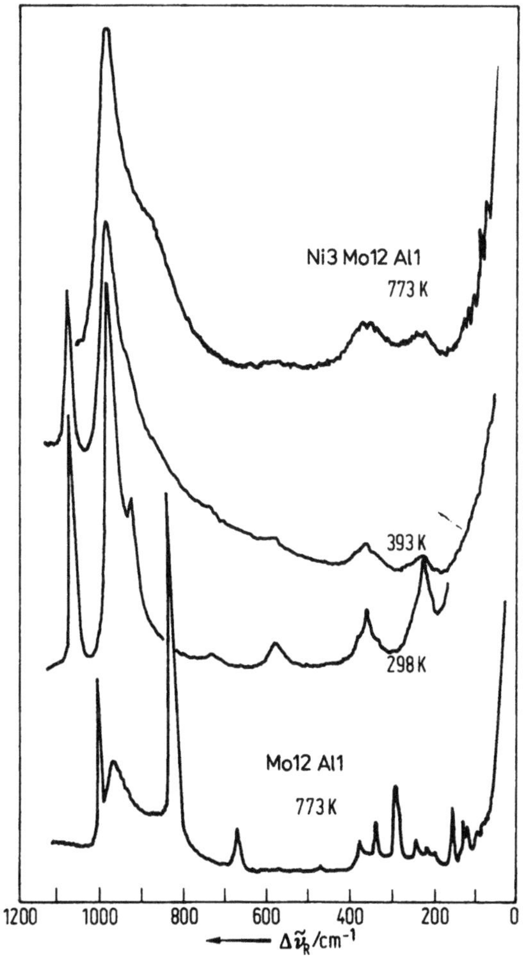

Figure 26. Raman spectra of nickel-molybdate catalysts. (Reproduced with permission from ref. [313])

mostly inaccessible for infrared spectroscopy. Complete Raman spectra of these materials in the wavenumber range to below 100 cm^{-1} have been reported [310–314].

As an example, the Raman spectra obtained during the various steps of formation of a nickel promoted molybdena-alumina catalyst are shown in Figure 26 (see also Figure 24). The series of sharp lines observed for sample Mo12Al1 (12 wt% MoO_3 supported on γ-Al_2O_3 by impregnation at *pH* 1 and calcination 773 K) are characteristic for the formation of bulk MoO_3, while the broad features at 960 and 850 cm^{-1} are typical for an interaction species. This latter surface species was described as a two-dimensional surface polymolybdate with Mo^{6+} in octahedral coordination [310, 311, 315–319]. The complete set of Raman bands of this species can be assigned as follows: 310–370 cm^{-1} (M=O bending), 900–1000 cm^{-1} (Mo=O stretching), 200–250 cm^{-1} (Mo—O—Mo deformation), 400–600 cm^{-1} (symmetric Mo—O—Mo stretching and 700–850 cm^{-1} (antisymmetric Mo—O—Mo stretching). These bands are clearly seen in the spectra of Figure 26 after a second impregnation with $Ni(NO_3)_2$ (sample Ni3Mo12Al1) to give 3 wt · % supported NiO after heat treatments at 298, 393 and 773 K (the band at 1049 cm^{-1} is due to the NO_3^- ion).

The MoO_3 content of 12 wt · % corresponds to a theoretical molybdate monolayer on the γ-Al_2O_3 support used. At decreasing loading in the sub-monolayer range, the surface concentration of the polymolybdate species is reduced in favor of a distorted monomeric tetrahedral MoO_4^{2-} species [310, 316].

Supported molybdate catalysts promoted by Ni^{2+} [313, 314] and Co^{2+} [312, 315–318] have been studied by Raman spectroscopy and Sombret *et al.* [320] have applied the *Raman microscope* technique to obtain information about the spatial distribution of surface species and compounds. Other supported oxide catalysts have equally successfully been characterized by their Raman spectra in the low frequency region, namely tungsten oxide on γ-Al_2O_3 [321, 322], rhenium oxide on alumina [323], and vanadium oxide on alumina [324, 325].

d) Infrared photoacoustic spectroscopy

In photoacoustic spectroscopy [326, 327], PAS, a sample is subjected to illumination with chopped radiation. This may produce temperature pulsations in the surface layers of the specimen, the amplitudes of which can be made to depend on the absorption spectrum of the sample. The temperature fluctuations can be measured photoacoustically through the sound waves which will be generated in a gas (usually He) atmosphere contacting the sample or in a piezoelectric substrate. Any sample whose optical absorption length is similar or larger than the thermal diffusion length is well suited for PAS. This requirement is easily fulfilled for infrared chromophores. The most typical characteristics of PAS are as follows:

(i) the samples for PAS need not be thin, flat, smooth or powdered, they can be used in their "natural" state, and minimum sample preparation is needed. This is especially attractive in cases where sample preparation such as grinding, pressing *etc.* may alter the properties of a given material;

(ii) optically opaque samples can be investigated;

(iii) the spectral information is simply determined by the absorption of the sample. It does not depend on the totally or diffuse reflected components of the radiation (as in IRS and DRS, respectively), which may be weak and difficult to interpret;

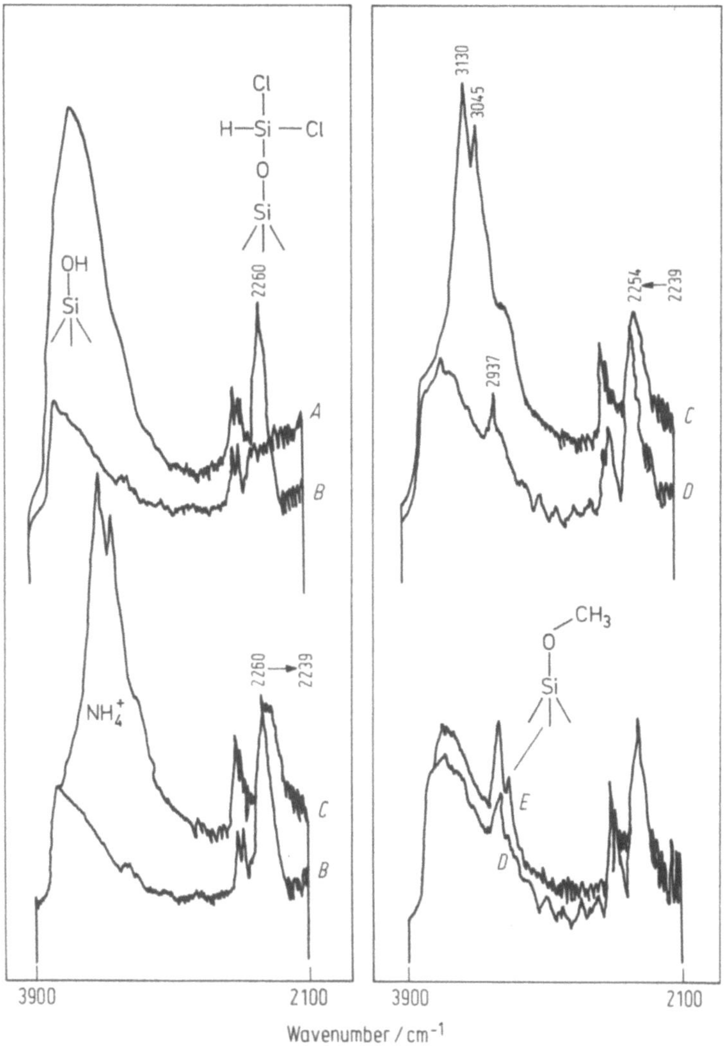

Figure 27. Surface reactions on silica, after 773 K degassing, observed via infrared photoacoustic spectroscopy (*A*) after 773 K degassing, (*B*) after reaction with $HSiCl_3$, leading to the formation of ~0.9 monolayer of $\equiv Si—O—SiHCl_2$ species. (*C*) after exposure to NH_3, leading to the formation of some NH_4Cl and ammonia coordinated to surface silane. (*D*) after degassing, to remove sorbed NH_3, (*E*) after reaction with methanol, leading to the formation of some surface $\equiv Si—OCH_3$ species. LiF prism. (Reproduced with permission from ref. [328])

(iv) the thermal diffusion length can be varied by using different light modulation frequencies. This provides the possibility to deduce depth profiles and to discriminate between surface and bulk contributions to the experimental PA spectrum;

(v) low acoustical background may be achieved, which is independent of the light irradiation intensity. This enables one to improve the signal to noise level linearly with the incident light intensity.

Low and Parodi [328] have modified a conventional dispersive infrared spectrometer for its application for computer-controlled IR-PAS. They also describe photoacoustic cells for investigations of chemisorption phenomena on disperse high-area adsorbents. From their results it can be concluded that the technique has submonolayer sensitivity and good-quality spectra of surface groups in the mid-infrared region can be obtained. This has particularly been demonstrated for partially dehydroxylated silicas and for silicas on which the hydroxyl groups were modified by NH_4^+, CH_3O-, and $HSiCl_2O$-groups, *etc.*; the IR-PA spectra of these systems permit an immediate comparison with their IR-transmission spectra. One may have to deal with silica adsorbents having large particle sizes and high scattering losses so that transmission is poor (catalyst supports, chromatographic silicas, pellets *etc.*). If one would not like to subject these samples to any severe treatments for one reason or another, PAS might be superior to transmission spectroscopy.

Figure 27 shows some IR-PA spectra of surface groups on silica which were reported by Low and Parodi [328]. The authors have also reported the PA spectrum of anthracite. Riseman et al. [328a] have studied adsorbed pyridine by IR-PAS.

Since the sensitivity of the technique is not very high, dispersive spectrometers need unfavorably long measurement times. Fourier transform infrared photoacoustic spectroscopy (FT-IR-PAS) has therefore been advanced recently [329, 330, 330a]. Rockley and Devlin [331] have obtained high-quality infrared spectra of a variety of fresh and aged coals, which demonstrated the formation of carbonyl and hydroxyl surface groups on oxidation.

A comparison is made in Figure 28 of the FT-diffuse reflectance spectrum of a low-aromatic coal with the FT-PA spectrum of the same material. The PA spectrum has clearly a much higher spectral contrast and flatter baseline than

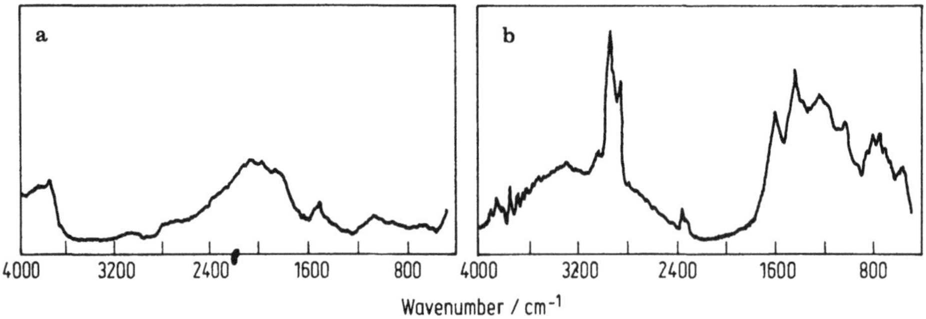

Figure 28. Low aromatic coal spectra: **a** diffuse reflectance spectrum; **b** photoacoustic spectrum. (Reproduced with permission from ref. [329])

the diffuse reflectance spectrum of the same sample. Thus, IR-PAS namely in its FT version is already a technique for study of surface groups which can well compete with IRS and DRS. It will be especially advantageous for opaque samples having large particle sizes, which would cause problems due to poor optical contact in IRS and of extremely low reflectivity in DRS. *Photoacoustic reflection-absorption spectroscopy* (PARAS) [332] and the contact-free spectroscopy by *photothermal radiometry* (PTR, pulsations in the thermal reradiation produced by modulated sample irradiation are measured) [332] might be further developed as useful techniques for certain applications in surface functional group spectroscopy.

e) Electron vibrational spectroscopies

High resolution *electron energy loss spectroscopy* (EELS) [16, 333] is a powerful technique for the detection of vibrations of atoms and molecules chemisorbed on metal surfaces. EELS provides a very high sensitivity (depending on the surface dipole moment as few as 10^{10} molecules may be detected on an area of only 1 mm^2), its spectral resolution, however, is restricted to the order of magnitude of about 60 cm^{-1}, while infrared and Raman spectroscopy easily achieve resolutions of 2 cm^{-1} and better. Although EELS has the great advantage of being capable to open up the entire vibrational region from below 200 cm^{-1} up to 4000 cm^{-1}, its application remains restricted to the study of geometrically flat surfaces since the method uses the specular reflection of electrons. However, attempts have recently been reported [334] to study model supported metal catalysts which were fabricated by evaporating a small quantity of rhodium onto an oxidized aluminium substrate. This lends some support to the possibility of detecting surface functional groups by EELS on model systems consisting of thin insulating oxide layers grown on metal supports.

The second electron vibrational spectroscopy is *inelastic electron tunneling spectroscopy* (IETS). Excellent reviews dealing with theoretical and experimental aspects of the technique have been published by Weinberg [335] and Walmsley [336]. The IETS sample is a metal-insulator-metal tunnel junction. The current produced by electrons tunneling through the insulating barrier between the metal electrodes when a bias voltage is applied, is measured. Since additional channels open up for the tunneling electrons which traverse the insulating barrier inelastically by exciting vibrational modes within the barrier, the electron current across the barrier will exhibit a small abrupt increase at bias voltages which correspond to the vibrational excitation energy of the respective oscillator. It has been shown, that the second derivative of the current d^2I/dV^2 as a function of the applied voltage V corresponds to an optical absorption spectrum such as infrared or Raman spectra [337]. The most attractive features of the IETS technique are the following:

(i) the entire vibrational spectral region from 200–4000 cm^{-1} is accessible;
(ii) the sensitivity is comparable to that of EELS;
(iii) resolutions better than 10 cm^{-1}, in some cases near 1 cm^{-1} can be obtained;

(iv) for molecular adsorbates and surface functional groups infrared-active, Raman-active and optically forbidden transitions are allowed and observed in IETS with intensities of nearly the same magnitude [337]. Together with the accessibility of the low frequency region, IETS therefore permits the detection of a complete vibrational spectrum in a single experiment.

Besides these strong potentials of IETS the technique also suffers from some special features which may restrict its general application as a routine technique:

(i) the measurements must be run at temperatures usually $\leqq 4.2$ K to obtain the high resolutions mentioned above;

(ii) the need for the metal-insulator-metal tunnel junction requires a special sample preparation, which usually consists in growing a thin (typically 2–3 nm thick) oxide layer on the surface of the corresponding metal film. The oxide layer is then covered by the second metal electrode. Normally, chemisorption on and sample treatment of the oxide layer has to be carried out before the second metal is deposited. However, Jaklevic and Gaerttner [338] have recently reported on an infusion technique which permits the external doping of a complete tunnel junction with organic molecules.

The tunnel junction which was most frequently studied in the past is the Al—Al_2O_3—Pb junction, where the Al_2O_3 layer is used as adsorbent. Another reported tunnel junction for chemisorption studies is Mg—MgO—Pb [339]. The surface groups that have been studied by IETS (preferrentially on Al_2O_3) comprise *e.g.* OH^- [338, 339], CN^- [338, 340], NCS^- and OCN^- [340], RO^- [342, 344], $RCOO^-$ [337–339, 342, 343], iron cyanide complexes [341] and O_ySiH groups on alumina [345].

Zirconium tetraborohydride supported on alumina was investigated by Evans and Weinberg [346] by IETS and this is perhaps the most interesting system in the present context dealing with surface functional groups. $Zr(BH_4)_4$ on Al_2O_3 has considerable practical importance since it is known to be effective in catalyzing polymerization reactions of olefins. Figure 29 shows the IET spectrum of $Zr(BH_4)_4$ adsorbed on Al_2O_3 at 300 K over the spectral range from 240–4000 cm^{-1}, and Table 13 gives the band assignments. From a detailed analysis of their IET spectra, the authors infer that during $Zr(BH_4)_4$ adsorption one or more of the BH_4 ligands are displaced in order to accomodate zirconium-surface bonds. As a consequence, the remaining BH_4 ligands seem to adopt a bidentate bonding to the Zr central atom through bridging hydrogen atoms rather than a tridentate bonding which is typical for $Zr(BH_4)_4$. The multiple bands observed in the terminal B—H (2407, 2437 and 2490 cm^{-1}) and bridging B—H (2142, 2177, 2230 and 2252 cm^{-1}) region suggested the formation of more than one type of surface complex, probably having distinct compositions. This picture is consistent with the detection of two bands at 693 and 910 cm^{-1}, which were assigned as Zr—O modes. In addition, bands in the B—O region (1378 and 1457 cm^{-1}) suggested the formation of $\mathrm{O{-}B}{<}^{\mathrm{H}}_{\mathrm{H}}$ species through reaction of the

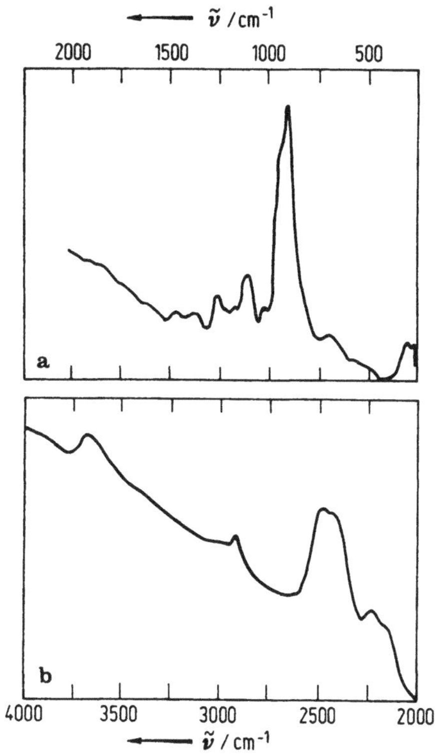

Figure 29. IET spectrum for $Zr(BH_4)_4$ adsorbed on Al_2O_3 at 300 K over the spectral ranges **a** 240–2000 and **b** 2000–4000 cm^{-1}. Substrate features which have been documented previously include an Al phonon in the underlying Al film (299 cm^{-1}), the Al—O stretch of bulk Al_2O_3 (945 cm^{-1}) and its harmonic overtone (1863 cm^{-1}), the C—H stretch of contaminant hydrocarbons (2930 cm^{-1}), and the O—H stretch of surface hydroxyl groups (3675 cm^{-1}). (Reproduced with permission from ref. [346])

displaced BH_4 ligands with the alumina surface. The interaction of these catalysts with H_2O, D_2O and D_2 [346] and with ethylene, propylene and acetylene [347] as studied by IETS has also been reported by Evans and Weinberg. The study by IETS of supported homogeneous cluster compounds has recently been reviewed by Weinberg [347a].

f) Inelastic neutron scattering

Inelastic neutron scattering spectroscopy (INS) is a promising technique for investigations of adsorbed molecules and surface groups, although it has the disadvantage of the need to be performed at a central facility. The origin of neutron scattering is the interaction between the neutron and the nucleus of the scattering atom. Dipole or polarizability selection rules are not to be obeyed in INS and therefore all normal modes of vibration are active. The intensity of scattering is weighted by the nuclear incoherent inelastic cross section of the scattering nucleus and the square of the amplitude of vibration of the nucleus in the particular mode [348] (see also literature cited in ref. [348]). As a consequence, INS is particularly sensitive to H containing species since both these parameters are greater for protons than for any other nucleus. Due to the high penetration power of neutrons, INS can be applied to any kind of solid material, including metals, but for the same reason it is obviously not surface sensitive and possible bulk contributions to neutron scattering spectra have to be eliminated when surface phenomena are to be

Table 13. Band Positions (cm^{-1}) and Assignments for $Zr(BH_4)_4$ Supported on Al_2O_3. (Reproduced with permission from ref. [346])

After vacuum treatment at 300 K	After vacuum treatment at 475 K	Assignments
264	264	BH_4—Zr—BH_4 bend or Zr—BH_4 torsion
299	299	Al phonon .
323	323	metal-oxide or Zr—BH_4 torsion
480–580	480–580	Zr—BH_4 or Zr—O modes
693	685	Zr—O stretch
910	910	Zr—O stretch
945	945	bulk Al—O stretch
1030		CH bend (contamination)
1106	1106	BH deformation
1130	1121	BH deformation
1173	1165	BH deformation
1220	1214	BH deformation
1260	1252	BH deformation and B—O stretch
1378	1378	B—O modes
1457	1457	B—O modes
1870	1873	harmonic of 945 cm^{-1}
2142	2137–2258 (weak, broad)	bridging B—H stretch
2177		bridging B—H stretch
2230		bridging B—H stretch
2252		bridging B—H stretch
2407		terminal B—H stretch
2437	2455 (broad)	terminal B—H stretch
2490		terminal B—H stretch
2930	2930	C—H stretch (contamination)
3675	3675	O—H stretch

studied. The recent literature on INS as applied to investigations of adsorbed molecules and surfaces has been reviewed by Hall and Wright [349] and by Wright [350]. Extremely interesting results have been obtained *e.g.*, for chemisorbed hydrogen on metals [349, 351], and the adsorption of water on oxide surfaces has been studied with regard to their librational and translational modes [349, 352–354]. However, though certainly desirable and most informative, INS studies on surface functional groups in the sense of the present chapter have not yet been performed to the authors' knowledge. Only occasionally is the observation of the surface hydroxyl bending vibration on oxide surfaces between 700 and 900 cm^{-1} mentioned in connection with studies on the water adsorption [353].

2. Optical spectroscopies

Spectra of surfaces, adsorbed molecules and surface functional groups in the visible and ultraviolet region of the electromagnetic spectrum (approximately 200–800 nm or 50000–12500 cm^{-1}) can be obtained as absorption spectra by the *diffuse reflectance* technique or by means of the *photo-acoustic* method. The principles of these spectroscopies have been briefly discussed in section II.A.1.b.ii and II.A.1.d., respectively, and are described in detail in various articles and books [16, 22–24, 326, 327]. In addition, *luminescence spectroscopy* can provide valuable information about the nature and electronic state of surface functional groups.

a) Absorption spectroscopy

i) Probe molecules. Indicator dyes are used to test acidic and basic properties of surfaces (see section II.B.2.). Colour changes occuring on adsorption of a dye due to proton addition can principally be followed visually. However, in view of the uncertanties of a visual judgment of the colour shown by an adsorbed indicator molecule (particularly when the adsorbent is coloured itself), spectrophotometric methods for qualitative and quantitative studies are to be preferred [355]. Figure 30 compares the solution spectra of benzeneazodiphenylamine ($pK_a = +1.5$) and its acid form with that obtained for the dye after adsorption on a silica-alumina catalyst containing 12% Al_2O_3. The adsorbed dye is entirely in its acid form. No significant shift of the wavelength of the principal band as compared to that in an acid solution is observed. The spectra of the physisorbed dyes in their base form are usually obtained when the surface acidity is not sufficient for protonation to occur. An interesting feature of the spectra of physisorbed dyes is a rather large red-shift of the principal bands in the adsorbed state as compared to the dye in solution [355].

Triphenylmethane can be converted into the corresponding carbenium ion via a hydride abstraction by an electrophilic site on oxide surfaces; catalyst protons need not be involved. The triphenylcarbenium ion formation is characterized by a double-peaked absorption band in the 410–435 nm region of the spectrum [356]. Extinction coefficients of these bands are of the order of magnitude of 10^4 liter mol^{-1} cm^{-1}, which permits a sensitive detection of the amounts of carbenium ions formed. However, the use of these molecules for the titration of acid sites on surfaces appears to be

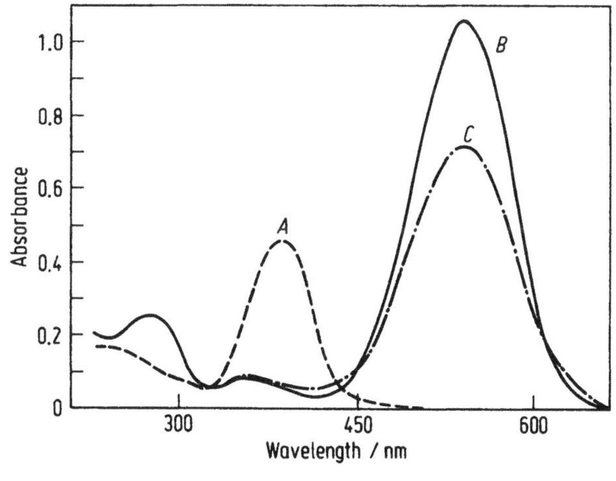

Figure 30. Spectrum of benzeneazodiphenylamine. (*A*) in isooctane solution; (*B*) in ethanolic HCl; (*C*) adsorbed on a silica-alumina catalyst (12% Al_2O_3). (Reproduced with permission from ref. [355])

complicated by side reactions and by the fact that the carbenium ion formation is photocatalyzed [356].

ii) Functionalized surfaces. The electronic spectra of surface groups, particularly the ligand field spectra of surface complexes of transition metals provide information on the local environment (symmetry) and the oxidation state of the particular transition metal ion. Numerous examples for this application of electronic spectra can be found in the literature [16, 24, 269]; only one example is therefore discussed, namely the CrO_3/SiO_2 system which is also dealt with in sections II.A.1.a.iv and II.B.3.b. It had been discussed, that under certain conditions the predominant surface structures in the oxidized state are dichromates which produce mainly Cr^{2+} species and a small amount (approximately 5–15% of total Cr content) of Cr^{3+} species on reduction with CO [259]. As an example, Figure 31 shows the diffuse reflectance spectra of a 0.5% CrO_3/SiO_2 sample, as reported by Zecchina *et al.* [259]. Very similar spectra were described by Krauss and coworkers [256, 357] and by Fubini *et al.* [261]. The strong bands at 21,700, 28,500 and 39,000 cm^{-1} with a shoulder at 26,000 cm^{-1} of the O_2 activated sample can be attributed to the surface dichromates, while the weak band near 14,000 cm^{-1} may indicate the existence of small quantities of Cr^{5+} species. After reduction in CO, bands at 12,000 and 7,500 cm^{-1} with shoulders near 15,000 and 10,000 cm^{-1} can be assigned to *d—d* transitions of Cr^{2+} complexes. The complexity of these features, however, suggests the presence of more than two Cr^{2+} species, a conclusion which is corroborated by the appearance of two strong ligand charge transfer (LCT) bands at 31,000 and 38,500 cm^{-1}. It was therefore suggested that Cr^{2+} ions occur in at least two coordinatively unsaturated environments in the reduced catalysts.

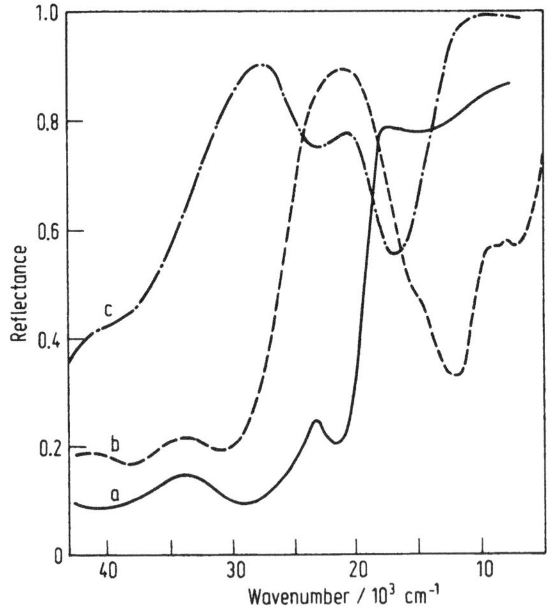

Figure 31. UV-visible reflectance spectra of a 0.5% CrO_3/SiO_2 sample after various treatments: (*a*) after activation in O_2 at 873 K; (*b*) after reduction in CO at 623 K; (*c*) after adsorption of NH_3. (Reproduced with permission from ref. [259])

Fubini *et al.* [261] detected four types of Cr^{2+} species which exhibited different reactivity. The adsorption of ammonia, which particularly shifted the LCT bands to higher frequencies is in accordance with this conclusion. On the basis of such spectral data, Hierl and Krauss [357] proposed the following reaction scheme

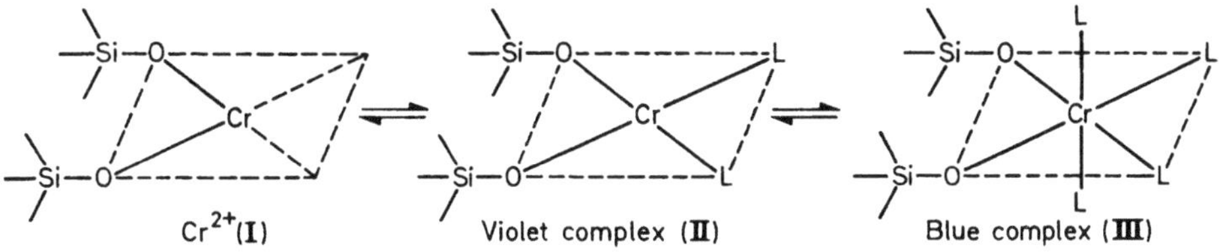

which is in agreement with the general features of the coordination chemistry of Cr^{2+}.

b) Luminescence spectroscopy

i) Probe molecules. π-donor molecules interact with Lewis acid sites on oxide surfaces and may form electron-donor-acceptor (EDA) complexes. As shown by Oelkrug and Radjaipour [358] these EDA complexes can give rise to luminescence spectra. Examples are shown in Figure 32 of the excitation and emission spectra of anthracene adsorbed on γ-Al_2O_3 and β-Ga_2O_3. The absorption spectra are similar to those of the uncomplexed anthracene molecule in that the number of bands is identical. The excitation bands 24,200, 27,700 and 36,000 cm^{-1} can be correlated to the L_a band (center at *ca.* 28,000 cm^{-1}), the L_b band (superimposed by the L_a band) and the B_b band.

It could be shown that the fluorescence bands (Fig. 32b) were due to the EDA complexes, since the transition energies of a variety of π-donor molecules adsorbed on γ-Al_2O_3 followed the approximate equation

$$h\nu_{CT} = I_D - E_A + C \tag{7}$$

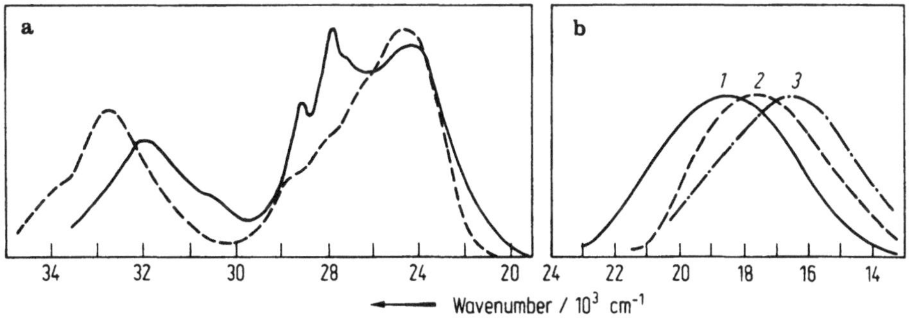

Figure 32. **a** Excitation spectra of anthracene chemisorbed from high-vacuum on γ-Al_2O_3 (– – –) and β-Ga_2O_3 (———). **b** Fluorescence spectra of anthracene chemisorbed on (*1*) γ-Al_2O_3, activated at 723 K; (*2*) β-Ga_2O_3, activated at 723 K; and (*3*) TiO_2 (anatase), activated at 473 K. (Reproduced with permission from ref. [358])

or Mulliken's charge transfer theory [358, 359]. I_D in the above equation is the ionization potential of the donor molecule, E_A the electron affinity of the acceptor surface and C a constant which depends on the donor-acceptor distance. Fluorescence spectra of chemisorbed aromatic hydrocarbons could therefore be expected to probe the Lewis acidity of oxide surfaces. However, the order of the transition energies of the fluorescence maxima of chemisorbed anthracene on the surfaces of γ-Al_2O_3, β-Ga_2O_3 and TiO_2 (anatase) is contrary to the order which would follow from the Lewis acidity of the surfaces [358]. This may be due to other factors than only the electron affinity E_A influencing the fluorescence transition energy. In particular, the surface structure may be important and modify the quantity C for different surfaces. These questions have not yet been elucidated.

Kessler and Wilkinson [359] have recently reported on a diffuse reflectance flash photolysis technique, by which they were able to measure triplet-triplet absorption decay times and phosphorescence lifetime. This technique is capable of direct observation of primary photophysical and photochemical processes of adsorbed species and surface groups on highly reflecting surfaces of catalytic interest.

ii) Surface hydroxyl groups. Highly dehydroxylated, high area oxide surfaces, namely alkaline earth oxides, give rise to strong emission bands which are assigned to $^1\psi(MeO) \leftarrow {}^3\psi(Me^+O^-)$ charge transfer transitions on sites in low coordination compatible with high index planes [360, 361]. Similar phenomena have also been observed for thoria surfaces [362]. This type of fluorescence is effectively quenched by admission of O_2.

However, another type of fluorescence is observed on these same surfaces in their hydroxylated state [362, 363]. For instance, on a hydrated MgO surface, a luminescence at *ca.* 470 nm is excited by 365 nm light. As shown in Figure 33, the intensity of this luminescence is strongly dependent on the degree of hydration of the surface and is therefore attributed to surface hydroxide ions. The increase of the luminescence intensity during heat

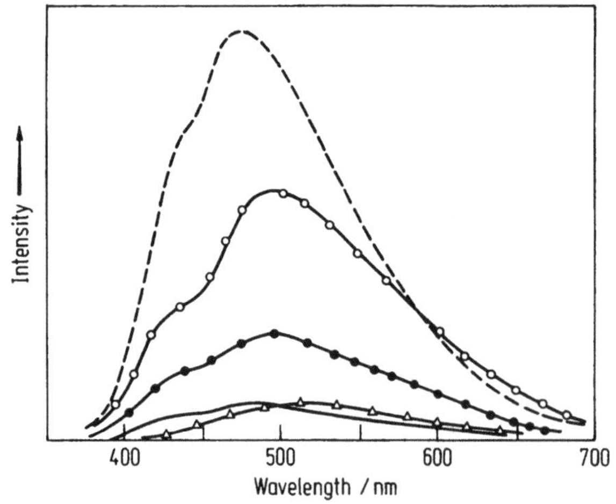

Figure 33. Emission spectra of hydrated MgO, under 365 nm excitation in vacuo at 300 K, outgassed at successively higher temperatures: (———) outgassed at 300 K; ● outgassed at 473 K; (– – –) outgassed at 573 K; ○ outgassed at 973 K; △ outgassed at 1073 K. (Reproduced with permission from ref. [363])

treatments at temperatures up to 573 K is due to the increased surface population by OH^- ions via dissociation of adsorbed molecular water. At still higher temperatures surface dehydroxylation reduces the OH^- density on the MgO surface and hence, reduces the luminescence intensity.

The low frequency of the luminescence bands was explained by the low coordination of OH^- ions on the surface [363]. The shoulder at low wavelengths in Figure 33 suggests the existence of more than one type of surface OH groups and/or possible vibronic interactions.

This luminescence of surface hydroxide ions is most probably the origin of the "fluorescence" background in laser excited Raman spectra of oxide surfaces (see section I.A.1.c.) which, for instance, showed a pronounced structure for alumina surfaces due to different types of hydroxyl configurations [298].

On thoria, emission bands at 540 and 630 nm (excited by $\lambda_{ex} > 300$ nm) were analogously assigned as electronic transitions of surface OH groups [362].

In conclusion, luminescence spectroscopy may well be applied for the study of electronic properties of surface OH groups and their coordination, although only few experimental data are available today.

3. Magnetic Resonance Spectroscopy

a) Electron spin resonance (ESR)

The ESR technique is a classical technique and has been used extensively for the study of paramagnetic species that can exist on solid surfaces. These may comprise surface metal complexes, surface defects and adsorbed molecules or ions. Electron spin resonance spectra permit conclusions on the nature of paramagnetic surface groups, their surface bonding characteristics and their geometrical configuration. However, since the separation of the responsible energy levels is only of the order of magnitude of 1 cm^{-1}, ESR spectra are extremely sensitive to environmental changes near the paramagnetic center. Of particular interest with respect to surface studies is the extremely high sensitivity of the ESR technique, since it permits the detection of very low concentrations of surface groups.

Various reviews have appeared in the past which describe the application of the ESR technique to surface studies [16, 29–31, 364]. Only few typical examples of ESR spectra of surface functional groups will therefore be discussed here.

Electron donor and acceptor properties of oxide surfaces and their possible relation to acid and base properties have been studied extensively by means of chemisorption of suitable probe molecules. This subject has been reviewed by Flockhart [364].

An interesting application of ESR spectroscopy is the characterization of surface transition metal complexes and their interaction with coordinated molecules [185]. For instance, Beck and Lunsford [258] have investigated the nature of the active sites for ethylene polymerization over Phillips catalysts (see also section II.A.1.a.iv, and II.A.2.a.ii), using NO as a probe molecule.

Figure 34 shows typical ESR spectra of the Cr species on the SiO_2 surface which were observed after various treatment conditions. After heating the samples in vacuum at 673 K for 4 h the weak axial signal *a* with $g_\perp = 1.970$ and $g_\parallel = 1.90$ was observed which is to be assigned to Cr^{5+} ions (γ-phase chromium ions). When the catalyst was reduced in CO at 673 K, the ESR spectrum displayed a rather broad, symmetric absorption with a line width of 300–400 G and $g = 1.960$ (spectrum *b*, Fig. 34). This spectrum is characteristic of clustered Cr^{3+} ions (or the β-phase). The reduced samples only gave rise to weak Cr^{5+} signals, the intensity of which decreased strongly with reduction time. Exposure of the partially reduced catalyst to NO led to an ESR signal with $g_\perp = 1.967$ and $g_\parallel = 1.902$ (spectrum *c*, Fig. 34) which Beck and Lunsford [258] assigned to a $Cr^{3+}(NO)_2$ complex, which was confirmed by its ir spectrum (see section II.A.1.a.iv). The spin density of this species went through a maximum with reduction time, indicating that the Cr^{3+} state can be further reduced to Cr^{2+}, which was also detected by spectroscopy as a dinitrosyl complex [258, 264]. It was concluded, that the $Cr^{3+}(NO)_2$ complexes must be isolated and, in contrast to others [255–257], Beck and Lunsford [258] provided evidence for their function as active sites for ethylene polymerization.

The free radical species present on carbon black surfaces have been studied successfully by the ESR technique [365] although the interpretation of the spectra is not always unequivocal.

The *optical detection of magnetic resonance* (ODMR) spectroscopy may find application for the study of surface groups. Only one example for this technique as applied to surface studies is known to the authors, in which the adsorption of SO_2 on zeolite 4A was investigated [366].

The *spin labelling technique* may be helpful in studies of surface groups which do not have unpaired electrons.

b) Nuclear magnetic resonance (NMR)

Various aspects of the uses of NMR spectroscopy for the study of surfaces, adsorbed species at interfaces and surface groups have been reviewed by

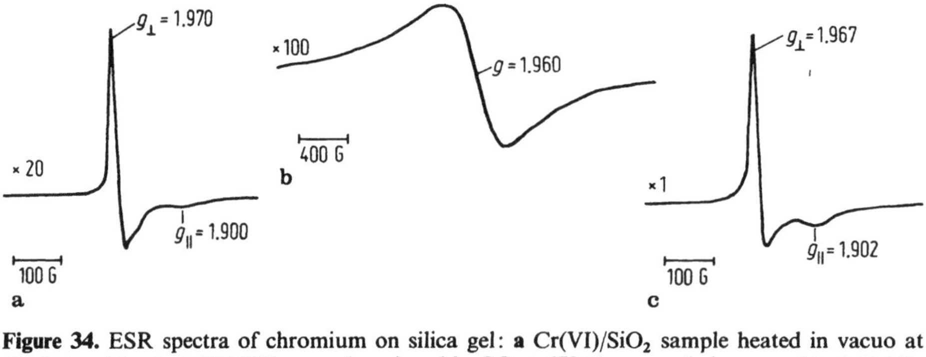

Figure 34. ESR spectra of chromium on silica gel: **a** Cr(VI)/SiO_2 sample heated in vacuo at 673 K for 4 hr; **b** Cr(VI)/SiO_2 sample reduced in CO at 673 K; **c** sample b exposed to 1.33 kPa NO at 298 K for 1 hr, and gas phase NO removed. (Reproduced with permission from ref. [258])

Resing [25], Fripiat [26], Pfeifer [27], Duncan and Dybowski [366a], and in a chapter of reference [16]. In addition, an extensive literature review has been given by Derouane *et al.* [28].

Assume two identical protons at a distance r which interact with an external magnetic field (Zeeman effect) and with each other (dipolar interaction). The latter interaction leads to an additional splitting of the proton energy levels and, as a consequence, the NMR spectrum of this system will consist of a pair of absorption lines separated by

$$\Delta H = \frac{3}{2} g_n \beta_n (1 - 3\cos^2\theta) \frac{1}{r^3}, \tag{8}$$

which is a measure of the strength of the dipolar interaction. In this equation, g_n is the nuclear g factor ($g_n = 5.585$ for a proton), β_n the nuclear magneton and θ is the angle between the direction of the external magnetic field vector and r, the radius vector between the two protons.

In polycrystalline catalysts or adsorbents with high surface areas, the angle θ can take all possible orientations and the experimental spectrum is the envelope of spectra corresponding to the multitude of orientations. Moreover, the distance r between interacting surface protons or protons in surface functional groups has normally a broad distribution, so that the individual lines can usually not be resolved, particularly for groups which are chemically linked to the surface and hence are strongly hindered with respect to molecular reorientation. In this rigid lattice case, motional narrowing of the NMR linewidth does not occur. Usually, chemical shifts and hyperfine splitting cannot be resolved by conventional techniques in proton NMR spectra of such solid samples. Only wide-line NMR spectra can therefore be obtained.

However, structural information can be obtained even in such unfavorable cases for polycrystalline materials by van Vleck's method of moments [367]. For a polycrystalline sample where the angular term is averaged over all directions, the second moment M_2 is given by

$$M_2 = \frac{3}{5} g_n^2 \beta_n \frac{1}{N} I(I+1) \sum_{j,k} \frac{1}{r_{jk}^6} \tag{9}$$

where I is the nuclear spin, N the number of identical nuclei per unit cell in the sample, and r_{jk} denotes the distance between two nuclei j and k. Since r_{jk} appears in the denominator to the sixth power, only those nuclei which are in close proximity (<1 nm) need be taken into consideration. Model calculations can then provide information regarding the average distance between nuclei (*e.g.* protons) on the catalyst surface provided that paramagnetic impurities do not influence the line shape.

High resolution NMR in solids has become possible by means of the recently developed multiple pulse techniques [368, 369]. By applying intense multiple-pulse cycles of electromagnetic radiation, static magnetic field inhomogeneities as well homonuclear and heteronuclear dipole interactions can be essentially averaged. The line width of absorption bands in solid samples can thus be reduced by several orders of magnitude, so that chemical shifts of nonequivalent nuclei can be resolved, and typical powder patterns can be detected, which then permit the chemical shift tensor to be determined.

Besides proton NMR spectroscopy, ^{13}C NMR has most frequently been applied to the study of surfaces. The main advantage of *^{13}C NMR spectroscopy* as compared to proton NMR is the much larger chemical shift (order of magnitude 300 ppm) which results in better resolution. Moreover, ^{13}C has a much smaller nuclear magnetic moment than the proton, so that dipolar interactions are significantly lower. Using FT-^{13}C NMR, medium resolution spectra of weakly adsorbed molecules (motional narrowing) at natural ^{13}C abundance (1.108%) can be obtained relatively easily [370–375]. However, even with ^{13}C NMR, the molecule or group under consideration must undergo isotropic rotation on a timescale of less than 10^{-5} s, in order to average ^{13}C—^{1}H dipolar interactions to small values, while dipolar interactions between ^{13}C nuclei are negligible at natural abundance. The conventional ^{13}C NMR technique can therefore not be applied to the study of strongly chemisorbed molecules or surface functional groups since their molecular reorientational motion is too slow.

In order to eliminate the ^{1}H—^{13}C dipolar interaction, so as to measure only chemical shifts the *proton decoupling* technique can be applied [369]. Proton decoupling is achieved by applying a rf field at the proton resonance frequency which must be stronger than the local magnetic field. The decoupling may readily be combined with *cross polarization* [369, 377], which produces a valuable increase in signal to noise ratios. The enhancement factor in this experiment is given by the ratio of the ^{1}H and ^{13}C gyromagnetic ratios $\gamma_H/\gamma_C \approx 4$.

Still another technique to achieve line narrowing in solid samples is based on the familiar motional averaging and is known as *spinning at the magic angle* [369, 378]. Such spinning the sample effectively averages out the dipolar interactions and/or the anisotropy in the chemical shift tensor. The angular dependence of the dipolar interactions is given by $(1 - 3\cos^2\theta)$ as shown by equation (8), where θ is the angle between the external magnetic field and the radius vector joining the two interacting nuclei. If the nuclear spin system is rotated about an axis inclined at an angle α to the applied field at an angular velocity ω, the $\cos\theta$ term can be written as [379]

$$\cos\theta = \cos\alpha\cdot\cos\psi + \sin\alpha\cdot\sin\psi\cdot\cos\omega t\,, \tag{10}$$

with ψ representing the angle between the radius vector between the two nuclei and the rotation axis. The time independent part of the $(1 - 3\cos^2\theta)$ function then becomes

$$\frac{1}{2}(3\cos^2\alpha - 1)\cdot(1 - 3\cos^2\psi)\,.$$

Hence, when $3\cos^2\theta = 1$ (*i.e.* $\alpha = 54.7°$, the magic angle), the dipolar interaction term is equal to zero and only the isotropic interactions will be detected in the spectrum. In other words, only the isotropic chemical shift

$$\langle\sigma\rangle = \frac{1}{3}(\sigma_{\parallel} + \sigma_{\perp}) \tag{11}$$

will be measured, where $\sigma_{\parallel}$ and $\sigma_{\perp}$ are the principal values of the chemical shift tensor parallel and perpendicular to the applied magnetic field. Sidebands in the spectra at integral multiples of ω on either side of the central spectrum are produced by the time dependent term in equation (10). For rotational frequencies greater than the static linewidth, the satellites do not interfere with the main spectrum, and their intensity becomes weaker with increasing rotational frequency. For ^{13}C with a shift range of approximately 300 ppm, a spinning frequency of about 2.0–2.2 kHz suffices at a carbon resonance frequency of 15 MHz. Figure 35 gives an example for adsorbed $^{13}CO_2$ demonstrating the described features of the technique.

i) Surface hydroxyl groups. NMR studies of surface hydroxyl groups have frequently been carried out with the goal to obtain information on their acidic and catalytic properties. O'Reilly *et al.* [381] and Hall and coworkers [382] studied silicas and silica-aluminas which had been heat-treated *in vacuo* at 773 K for extended periods of time. The wide-line NMR signal of the surface protons was unexpectedly narrow and had a Lorentzian line shape. The half-band width was independent of temperature in the range 63 to 553 K. It was therefore suggested that the protons were rigidly bonded to hydroxyl groups which were randomly distributed over a small fraction of the possible sites on the surface. The chemical shift of -3 ± 2 ppm relative to water indicated that the hydrogen-atoms are similar in character to those in typical alcoholic hydroxyl groups.

The hydroxyl group densities could be determined within $\pm 20\%$ experi-

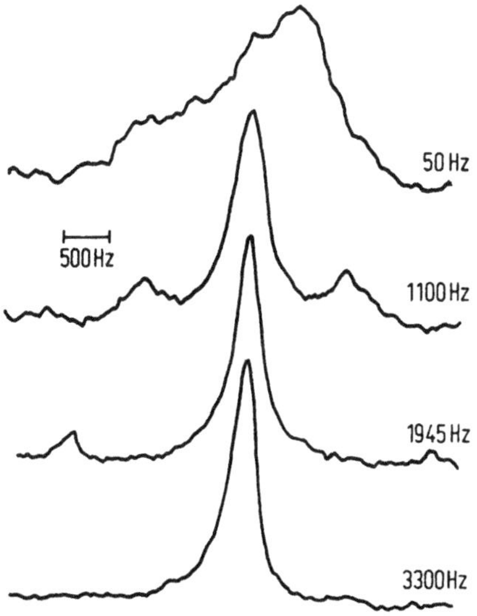

Figure 35. ^{13}C NMR spectra of $^{13}CO_2$ adsorbed on a commerical Na^+-mordenite contained in a rotor spun at the magic angle. (Reproduced with permission from ref. [380])

mental error and the obtained values compared very well with those measured by deuterium exchange techniques (see section II.B.1.a.). Since the proton NMR signals from the nonacidic silica gels and the acidic silica-alumina catalysts of alumina contents below 20% were qualitatively indistinguishable, it was suggested that the majority of the protons on silica-alumina was non-acidic, with an upper-limit for Brönsted acidity of $<3 \times 10^{13}$ acidic protons per cm^2.

Bermudez [383] proposed a method for evaluating the number of surface hydroxyl groups from NMR line widths. It was assumed that rapid proton exchange occurs between adsorbed molecular water and surface hydroxyl groups. The observation of a single NMR line at room temperature is good evidence for the validity of this assumption. Bermudez deduced a linear relationship between the line-width of the NMR signal for silicas containing 0.5 to 3 monolayers of molecularly adsorbed water and the concentration of the latter. The slope of this straight line equals $2T_s/[OH]$, where T_s is the transverse relaxation time of silanol protons and [OH] is the surface concentration of silanol groups.

The method of second moments was applied to characterize surface hydroxyl groups of various oxide catalysts. Kvilividze *et al.* [384–386] and Egorov *et al.* [387] determined second moments for silanol groups on silica surfaces and concluded that isolated groups and pairs (vicinal and geminal) of hydroxyl groups could be detected and their relative proportions were estimated.

Webster *et al.* [388] suggested the presence of clusters of OH groups on the surface of partially hydroxylated MgO.

Line shape and second moment calculations for various surface models of silica and decationated Y-zeolites were carried out by Freude *et al.* [389]. Their results as compared to experimental data ruled out a statistical distribution of spins in a volume or on a plane and rather suggested the presence mainly of pairs of hydroxyl groups in decationated Y-zeolite with a proton-proton distance of 0.37 nm.

Enriquez *et al.* [390] analyzed the line-shapes of proton NMR signals of hydroxyl groups in "rigid lattice" conditions, on rutile. They concluded that the most probable hydrated planes were in order of decreasing importance, the (110) and (101) planes. On these planes the OH groups were assumed to occupy two different sites corresponding to overall charges of opposite sign (see section II.A.1.a.i) and the preferred hydrogen-bond donor sites were supposed to be those groups bearing an overall positive charge.

Recently, Cirillo *et al.* [391] investigated the behavior of surface protons on an 8% molybdena-alumina catalyst. Second moments for surface hydroxyl groups were calculated based on the model of the Al_2O_3 surface proposed by Knözinger and Ratnasamy [73] (see section II.A.1.a.i). The decrease in the percentage of hydroxyls observed by proton NMR with increasing temperature of dehydroxylation suggested a strong interaction between the hydroxyl protons (which escaped NMR detection but which were shown to be present by D_2-exchange measurements) and the ^{27}Al ions exposed at the surface by the removal of water. Restoration of these hydroxyls to detection by

NMR uppon adsorption of D_2O supported this hypothesis. Although new hydroxyl groups were created during reduction of the catalysts, the fraction detectable by NMR decreased. This was attributed to the introduction of paramagnetic centers by reduction, *e.g.* Mo^{5+}. Model calculations, however, showed that the number of hydroxyls contributing to the NMR signal larger than possible were the paramagnetic centers and hydroxyl groups uniformly distributed over the catalyst surface. A clustering of surface hydroxyls was therefore proposed which would consequently suggest the formation of islands (perhaps having three-dimensional structures) of molybdate species on the surface of the alumina support.

In a very detailed study of surface hydroxyl groups on silica, alumina and silica-aluminas, multiple pulse techniques have been applied by Schreiber and Vaughan [392]. They determined the density of hydroxyls from the initial magnitude of the free induction decay observed after a 90° pulse, information on the local environment and arrangement of hydroxyl groups was obtained from an analysis of the free induction decay lineshape, and information on their chemical bonding was deduced from the components of the chemical shift tensor as measured by multiple pulse techniques. For pure silica, the free induction decay lineshape was Lorentzian with a full width at half height of 1600 Hz. An average proton-proton distance after calcination at 773 K of greater than 0.4 nm was estimated which does not exclude the presence of geminal hydroxyl groups. The observed lineshape was almost completely determined by dipolar interactions. This could be proved by the application of an eight pulse cycle which reduced the linewidth to only 250 Hz. The resulting spectrum is shown in Figure 36, which is a typical chemical shift powder pattern broadened by bulk susceptibility effects. The solid line through the data points in Figure 36 is a computer fit for an axially symmetric chemical shift tensor with a Lorentzian broadening function. After correcting for the bulk susceptibility effects, the principal values of the chemical shift tensor were $\sigma_\perp = -6.5$ ppm (-366 Hz) and $\sigma_{||} = +0.4$ ppm ($+23$ Hz) relative to tetramethylsilane. The anisotropy ($\sigma_{||} - \sigma_\perp$) of this chemical shift tensor is quite small as compared to hydroxyl protons in other media. This was interpreted as being due to an unusual chemical bonding

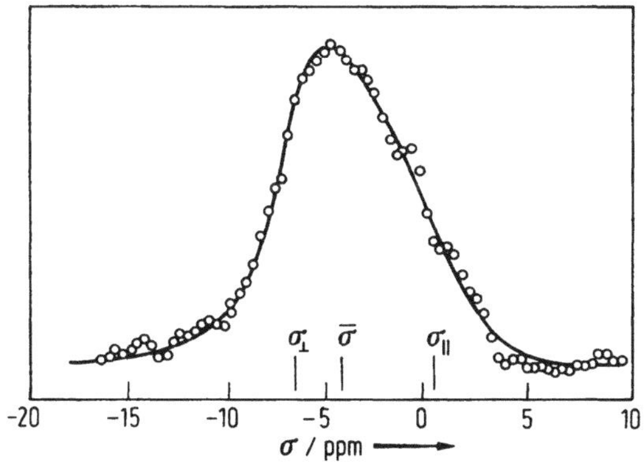

Figure 36. Powder pattern using eight-pulse cycle of 100% SiO_2 calcined sample. The signal amplitude is plotted *vs* chemical shift σ/ppm relative to TMS. (Reproduced with permission from ref. [392])

Table 14. Surface hydroxyl group densities on calcined silica-aluminas of varying compositions as determined by NMR techniques. (Reproduced from ref. [392])

SiO_2/wt %	$S_{BET}/m^2\,g^{-1}$	OH group densities/OH nm^{-2}	
		SiOH	AlOH
100	576	1.6	—
90	693	1.5	—
75	615	1.2	—
50	727	1.0	2.7
25	839	0.2	2.4
10	768	—	3.4
0	571	—	3.1

on the silica surface or to a partial averaging by an angular motion of the O—H vector (*i.e.* wagging or rotation of the hydroxyl bond around the oxygen).

In contrast to silica, a Gaussian lineshape was observed for the proton resonance signal of hydroxyl groups on alumina and the second moment of 2.8 G^2 was almost completely accounted for by heteronuclear $^1H—^{27}Al$ dipolar interactions. The very small homonuclear $^1H—^1H$ dipolar broadening implies that the protons are widely separated on alumina after calcination at 773 K. Similar conclusions had been drawn later by Stübner *et al.* [393].

For silica-aluminas (0–100 % SiO_2), Schreiber and Vaughan [392] were able to determine the surface concentrations of AlOH and SiOH groups. Their results for catalysts outgassed at 773 K are summarized in Table 14. It appears to be of substantial significance that AlOH groups were not detected on the 75 % silica sample, while nearly 75 % of the protons on the 50 % silica sample appear to be AlOH groups. Earlier work by O'Reilly *et al.* [381] and Hall *et al.* [382] on a 89 % SiO_2 sample had limited the percentage of surface protons in AlOH groups to less than 20 %. Hence, a clear qualitative change appears to occur in the nature of the hydroxyl groups between the 75 and 50 % silica samples, with the 100, 90 and 75 % being similar and exhibiting no detectable AlOH groups while the 50, 25, 10 and 0 % SiO_2 samples have the bulk of their protons attached in close proximity to an aluminum ion. The absence of AlOH groups in silica-aluminas containing less than about 20 % Al_2O_3 was also noted by infrared spectroscopy [138] (see section II.A.1.a.i).

ii) Surface acidity. Various attempts have been undertaken to distinguish between Brönsted and Lewis acidity by means of NMR techniques using basic probe molecules, and to estimate the site densities of the two types of acid sites on catalyst surfaces.

Kiviat and Petrakis [394] measured wide-line proton NMR spectra of pyridine adsorbed on molybdena on alumina catalysts. The physically adsorbed pyridine had considerable mobility on both calcined and reduced samples, while chemisorbed pyridine showed much broader spectra. Pear-

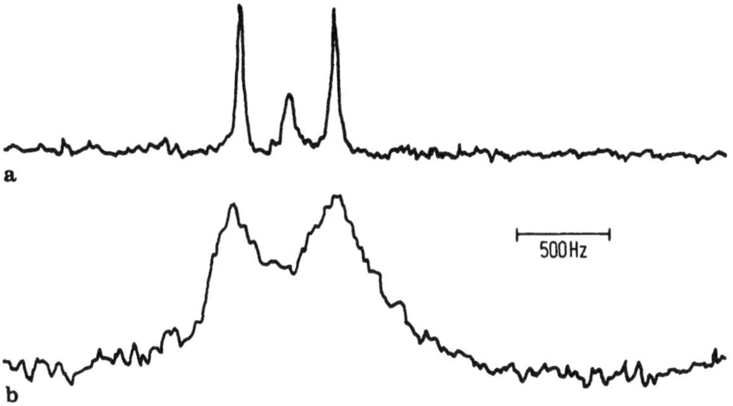

Figure 37. ^{13}C NMR spectra of pyridine. **a** On SiO_2 degassed at 523 K. Fractional coverage 0.77. Spectrum recorded at 301 K, 10^4 scans. **b** On Al_2O_3 degassed at 523 K. Fractional coverage 0.86. Spectrum recorded at 353 K, 2.6×10^5 scans. (Reproduced with permission from ref. [397])

son [395] used perdeuterated pyridine as a probe molecule and measured the wide-line NMR spectra of surface protons on alumina. A considerable narrowing of the proton signal was observed in the presence of perdeuteropyridine, which was taken as evidence for the existence of protonic acidity on the alumina surface. This conclusion was in contrast to general experience and an alternative interpretation of the line narrowing by pyridine adsorption was given [396]. It was suggested that an increased proton mobility could also be induced by hydrogen bonded pyridine and pyridinium ions must not necessarily be formed to explain the effect (see also section II.A.1.a.ii). Gay and Liang [397] could not detect any measurable Brönsted acid sites on alumina by ^{13}C NMR of adsorbed pyridine. A ^{13}C Cross-polarization magic angle spinning NMR study was recently described by Dawson *et al.* [397a] for pyridine chemisorbed on γ-Al_2O_3.

^{13}C NMR spectroscopy of probe molecules may be a useful means of studying surface proton acids, since rather large chemical shifts occur upon protonation of aromatic and heterocyclic amines in aqueous solution. As an example Figure 37 shows the ^{13}C NMR spectra of pyridine adsorbed on silica and alumina. The fairly narrow lines and rather small chemical shifts relative to liquid pyridine, suggested that on SiO_2 no interactions stronger than hydrogen-bonding occured. In contrast, the spectra for pyridine adsorbed on Al_2O_3 are much broader. Computer least-squares fitting allowed the chemical shifts of C-2, C-3 and C-4 to be measured with only poor accuracy. The C-2 line has a shift of $+0.2 \pm 0.3$ ppm relative to the liquid. This did not show the large 7.8 ppm upfield shift observed in acid solution, consistent with the absence of any measurable Brönsted acidity and the coordination of pyridine onto Lewis acid sites. In addition, the steric hindrance for this coordinative interaction could convincingly be demonstrated by the comparative use of 2,6-dimethylpyridine and 3,5-dimethylpyridine as probes [397]. As shown in Figure 38, the two probe molecules give largely different line-widths although their basicity in aqueous solution

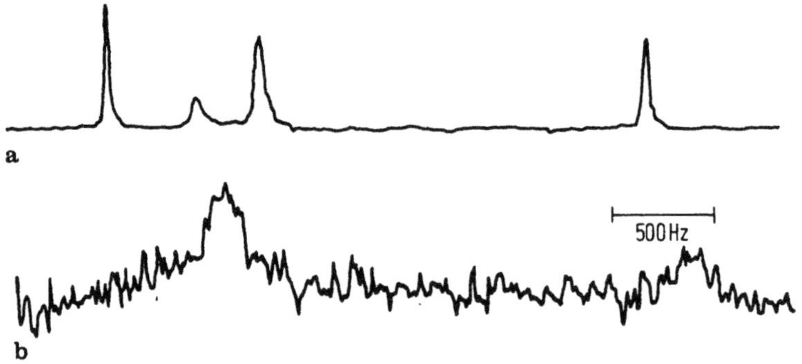

Figure 38. **a** 2,6-Dimethylpyridine on Al_2O_3 degassed at 523 K. Fractional coverage 0.95. Spectrum recorded at 353 K, 1.8×10^4 scans. **b** 3,5-dimethylpyridine, same conditions. Fractional coverage 0.73, 8.5×10^4 scans. (Reproduced with permission from ref. [397])

is almost identical. This is clear evidence for a strong steric effect which prevents the 2,6-disubstituted pyridine from approaching a Lewis site and hence, its high mobility leads to motional narrowing.

For the distinction of simultaneously existing Brönsted and Lewis sites, Liang and Gay [398] recommended the use of 4-ethylpyridine as a probe. For this molecule the chemical shifts at C-2 were found to be similar for both H^+ and BF_3 adducts. The shift at C-4, however, is consistently larger for the protonated species. This makes it principally possible to determine both Lewis and Brönsted acid sites according to

$$\sigma_{obs} = f_H \cdot \sigma_H + f_L \cdot \sigma_L + (1 - f_H - f_L) \cdot \sigma_n \, , \tag{12}$$

where σ_{obs} is the observed chemical shift, σ_H, σ_L and σ_n are the chemical shifts of a protonated probe molecule, of a probe molecule coordinated to a Lewis acid site and adsorbed on a nonacidic site, respectively, and f_H and f_L are the fractions of molecules interacting with Brönsted and Lewis sites, respectively. Knowledge of σ_H, σ_L and σ_n is required to estimate f_H and f_L, and some values have been tabulated by Liang and Gay [398].

Freude *et al.* [399] have recently studied the protonic acidity of HY zeolites. They suggested the use of the ratio n_A/τ as a relative measure for the acidity of surface OH groups, if n_A denotes the number of protons involved in a particular process and τ is the mean residence time of a hydroxyl proton at a lattice oxygen. Measurements of the pyridine induced mobility of hydroxyl protons showed an increasing acidity (about one order of magnitude) for an increasing degree of decationization from 36 to 91 % or for decreasing activation temperature from 773 K to 573 K. The ratio n_A/τ measured by proton exchange in the system zeolite/benzene increased in the same manner with the degree of decationization. Mestdagh *et al.* [400] and Freude *et al.* [401] had previously attempted to correlate turnover numbers or junp frequencies of the proton on a supercage oxygen in the zeolite framework with acidity and catalytic activity.

Based on proton spin relaxation measurements, Rauscher *et al.* [402]

recently developed a dynamic model for the protonation of pyridine in decationated Y-zeolites. It is assumed that after a mean time of 5×10^{-7} s at 313 K the pyridinium ion desorbs as a pyridine molecule hopping to another OH group, where it is converted into a pyridinium ion again. Less frequently, *i.e.* after an average time of 10^{-5} s, a jump as a pyridinium ion may occur which carries the proton of the OH group to another oxygen of the zeolite skeleton.

Correlations between proton acidity and dynamic NMR parameters have been searched for over the years [26]. However, a simple general relationship between proton mobility (jump frequency, namely in the absence of an adsorbate) and surface acidity does not seem to exist.

iii) Functionalized surfaces. Gay [403] studied the nature of methoxide species formed during chemisorption of methanol on MgO by NMR spectroscopy using proton decoupling and cross polarization techniques. His results confirmed earlier infrared studies of the same system.

Formate species are formed on chemisorption of formic acid on Y zeolites. Duncan and Vaughan [404] studied this system in much detail by multiple

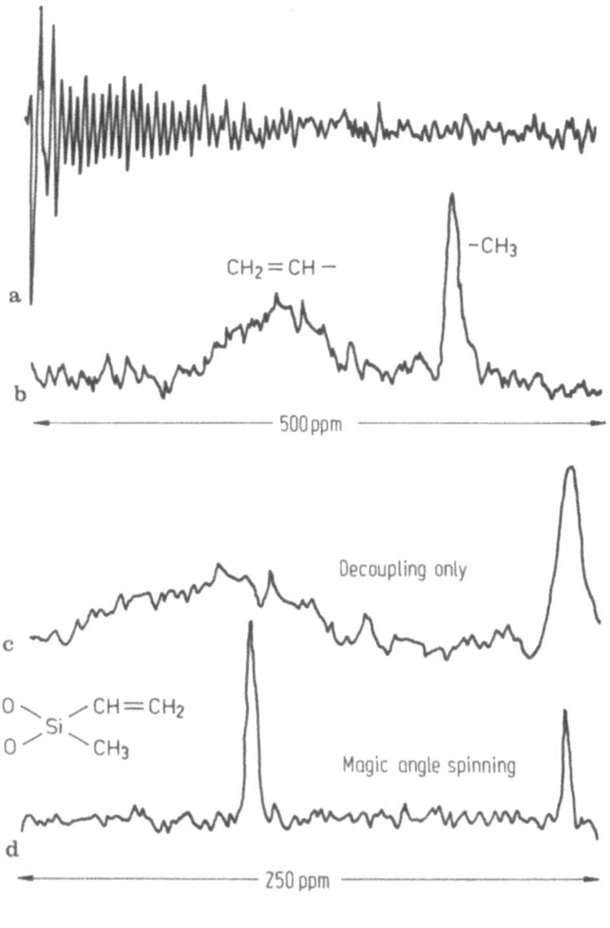

Figure 39. ^{13}C NMR spectra for asbestos reacted with vinylmethyldichlorosilane: **a** free induction decay of **b** and **c**; **b** decoupling but no spinning, 500 ppm sweep width; **c** same, 250 ppm width; **d** magic angle spinning and proton decoupling. (Reproduced with permission from ref. [406])

pulse NMR techniques. It could be determined that two different forms of formate species are produced, namely a unidentate and a bidentate form, which can be distinguished by the symmetry of the ^{13}C chemical shift powder patterns. Cross-polarization NMR techniques were used to exploit differences in the $^{13}C—^{1}H$ dipolar couplings to separate the two groups. The ratio of bidentate to unidentate species is about 52:48 on NH_4—Y zeolite and about 83:17 on the ultrastable H—Y zeolite at a coverage of 0.3 monolayer. The strength of the $^{13}C—^{27}Al$ dipolar interaction suggested that the bidentate formate ions are bonded to the Al atoms of the zeolites, while the unidentate form appeared to be linked to the Si atom. The ^{1}H NMR spectrum of the carbonyl hydrogen, isolated by a dipolar-difference technique, is shifted downfield from that of neat formic acid, indicating that the hydrogen has become more acidic.

The high potential of the combined use of the various newly developed NMR techniques for the characterization of surface functional groups has been demonstrated by recent studies of modified silica [405, 406] and modified chrysotile asbestos [406, 407]. The ^{13}C spectra of asbestos reacted with vinylmethyldichlorosilane as obtained by the proton decoupling and magic angle spinning techniques are shown in Figure 39. The spectrum without spinning clearly does separate the methyl and vinyl ^{13}C resonances, but magic angle spinning clearly shows the 2:1 ratio for the vinyl/methyl group intensities.

Figure 40 shows a series of proton decoupled ^{13}C NMR spectra of modified silicas [405] bearing $\geq$Si—O—Si$(OH)_n(C_6H_5)_{3-n}$ ($n = 0, 1, 2$), $\geq$Si—O—Si$(OH)_2CH_2C_6H_5$, and $\geq$Si—O—Si$(OH)_2CH_2C_6H_4SO_3H$ (spectra b, c, d, e and f). The signal for aromatic carbons is a broad resonance from about 0 to 250 ppm (relative to tetramethylsilane) for all compounds. The signal for the methylene groups is more narrow and occurs near 30 ppm.

The main characteristics of these spectra are that the shape is approximately the same for all compounds and that this shape changes from low temperature to room temperature. This change in shape must be due to some rotational motion which is slow at low temperature and fast enough at room temperature to average out part of the chemical shift anisotropy. Model calculations of chemical shift powder patterns suggested that a rotation around the O—Si bond (spectra b, c, and d in Fig. 40) or around the Si—C bond (spectra e and f in Fig. 40) with the phenyl rings at a fixed position with $\alpha \approx 25°$:

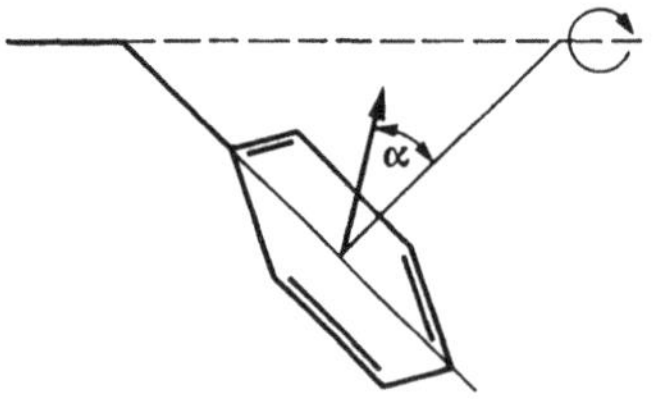

is the most likely rotational motion. From these considerations regarding the dynamic behavior of the surface groups, additional information about the

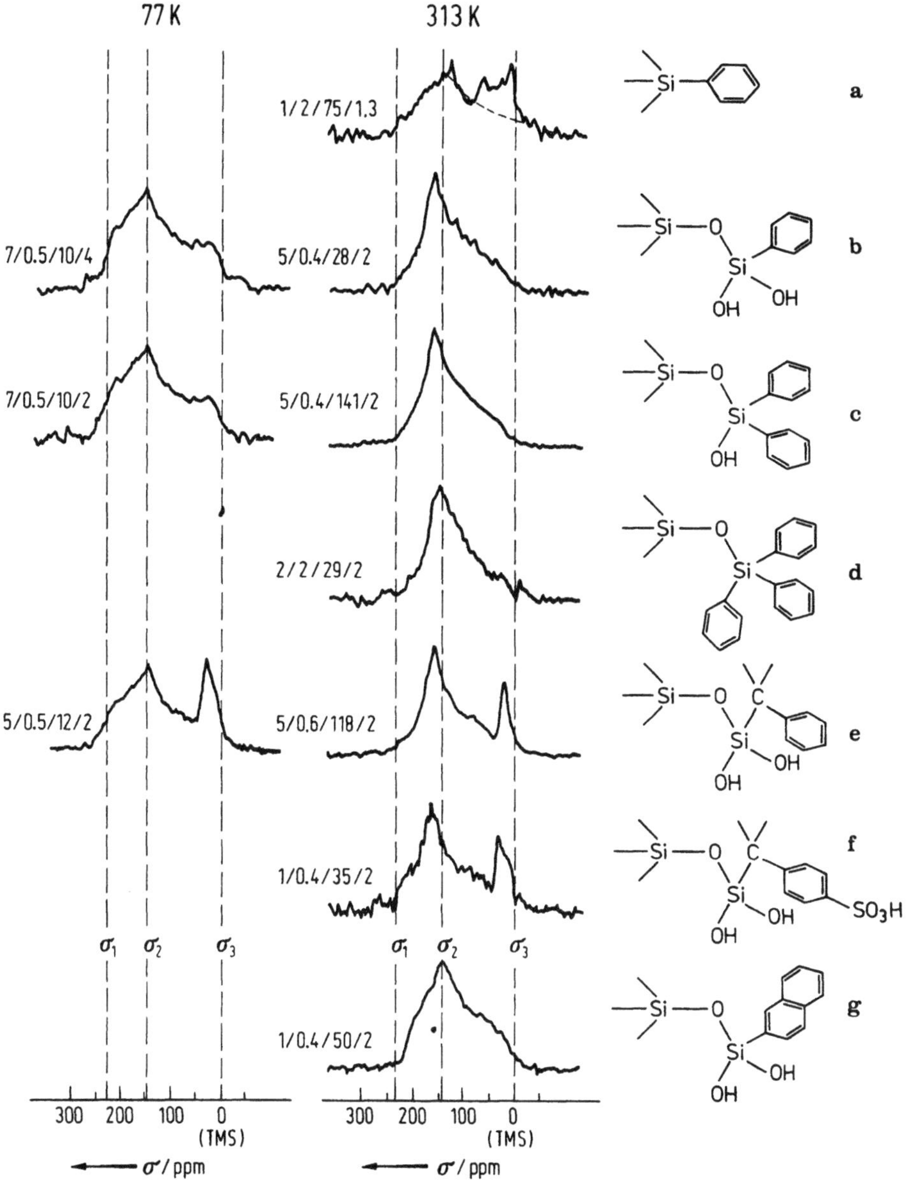

Figure 40. Proton decoupled ^{13}C NMR spectra of surface modified silica gels (the ≡SiO to the left on the drawings represent the silica surface). Resonance frequency 15 MHz. Numbers indicate cross-polarization time (ms)/repetition time(s)/thousands of accumulations/width of Lorentzian smoothing (ppm). (Reproduced with permission from ref. [405])

surface-adsorbate structure can be obtained [405]. The modification of the silica is carried out by reaction with chlorophenylsilanes. In principle, this can result in one or two bonds to the surface, *e.g.*

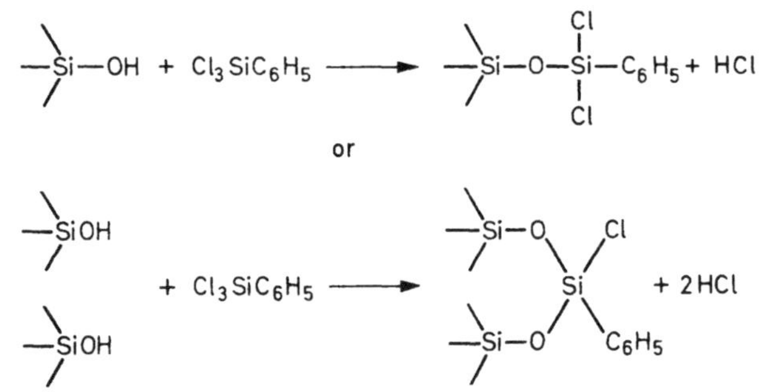

After the reaction, OH is substituted for Cl under hydrous conditions. The doubly bonded group has only one mode of large scale motion; *i.e.* reorientation around the Si—C (phenyl) bond. This motion would not conform with the line shape and its temperature dependence within the limits of resolution. Hence, all or the majority of functional groups must be joined by a single bond to the surface. No information on the number of bonds to the surface could be obtained in the case of phenylmethylsilanol groups.

Surface modification is frequently carried out with reagents having long hydrocarbon (methylene) chains connecting the surface and the functional group. For those materials one would expect appreciable motional freedom

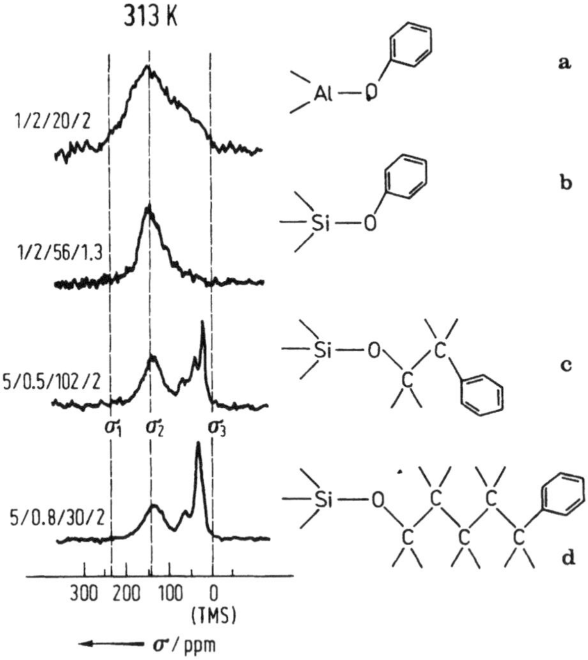

Figure 41. Proton decoupled ^{13}C NMR spectra of surface modified silica gels (the ⩾SiO to the left on the drawings represent the silica surface). Resonance frequency 15 MHz. Numbers indicate cross-polarization time (ms)/repetition time (s)/ thousands of accumulations/ width of Lorentzian smoothing (ppm). (Reproduced with permission from ref. [405])

and hence, much narrower NMR signals than observed for the more rigidly anchored groups discussed above. Figure 41 compares the proton decoupled ^{13}C NMR spectra of surface modified silicas containing methylene chains of increasing length [405], namely $\geqslant Si-O-C_6H_5$, $\geqslant Si-O-(CH_2)_2C_6H_5$ and $\geqslant Si-O-(CH_2)_4C_6H_5$. It is interesting to compare spectrum b in Figure 41 ($\geqslant Si-O-C_6H_5$) with the spectra of $\geqslant Si-O-Si(OH)_2C_6H_5$ or $\geqslant Si-O-Si(OH)_2CH_2C_6H_5$ (spectra b and e in Fig. 40). Evidently the reorientation which takes place for the phenolic ligand averages the chemical shift anisotropy much more effectively than the reorientations in phenylsilyl ligands. Comparison of spectrum b in Figure 41 with theoretical chemical shift powder patterns suggested rotation in the $\equiv Si-OC_6H_5$ group about the Si—O bond with a fixed angle, $\alpha = 45°$, between the Si—O—C plane and the phenyl plane. Figure 41 also shows that the spectrum of phenoxide species linked to Si is very different from that of an $Al-OC_6H_5$ group (spectrum a), for which the broad chemical shift dispersion has the form characteristic if nonreorienting phenyl groups.

Inserting a methylene chain between the surface anchoring and the phenyl group results in the expected increased degree of reorientational freedom [405] (spectra c and d in Fig. 41). A semi-high resolution spectrum is obtained, in which the oxymethylene, methylene and aromatic carbon signals are clearly separated. This confirms a model which assumes the chainlike functional groups to represent freely moving, outstretched cords anchored in one end. No evidence for entanglements between neighboring groups can be extracted from the spectra.

Recently, Maciel and Sindorf [408] reported ^{29}Si chemical shifts which were measured on silica gel surfaces. The same group studied the products obtained by reaction of methylchlorosilanes with silica gel surfaces by means of ^{29}Si and ^{13}C cross-polarization and magic-angle spinning NMR spectroscopy [409].

To conclude this section, it may be relevant to recall a recent prognosis by Resing *et al.* [406]: "Proton decoupled ^{13}C NMR, with magic angle spinning where appropriate, has shown its utility in surface analysis of chemisorbed molecules and of the substrate itself. Molecular orientations and motions can be determined. The drawback at the moment is sensitivity. But a ^{13}C probe for operation at 10 K has been reported [410] which promises a gain of twenty in signal strength. T_1 has not been a problem in surface studies thus far. *The future for surface analysis looks very good.*"

4. Photoelectron Spectroscopy

In x-ray photoelectron spectroscopy (XPS), the kinetic energy E_K of photoelectrons which are emitted from core levels of surface atoms on x-ray irradiation (photon energy $h\nu$) is measured [16, 32, 34, 411]. The experimental determination of E_K permits the electron binding energy E_B to be determined through the relation

$$E_K = h\nu - E_B \,. \tag{13}$$

The electron escape depth is typically 2–4 nm for oxides [412]. The topmost atomic layers contribute therefore preferentially to the photoelectron signal. Assuming a homogeneous distribution of an element within the topmost 10 nm layer of a solid on an area of 0.5 cm^2, approximately 1 wt% will be the lower limit of detectability. If the element of interest is distributed on the surface only within a chemisorption layer, typically less than 10% of a monolayer would be detectable. For quantitative measurements [411, 413] of surface concentrations or concentration ratios, the ionization cross sections of the relevant atomic levels must be known and the energy dependent escape depth of electrons has to be considered. Quantitative analysis may become difficult or even impossible for porous materials (*e.g.* zeolites), when part of the elements or groups of interest is located within the pores or cavities where they might not be "seen" by the XPS technique. Charging of insulating materials can pose severe problems in the correct determination of binding energies.

The XPS technique permits an elemental analysis of solid surfaces and chemisorption layers to be carried out on the basis of the element specific E_B-values. In addition, "chemical shifts", line splittings and shake-up satellites can be used to determine valence states of the various elements and to characterize their chemical environment.

For catalyst studies and investigations of surface functional groups, a reaction chamber should be attached to the spectrometer via a transfer system. Also, in situ spectrometers which enable XPS studies while surface reactions or chemisorption processes proceed at elevated pressures (typically 100 Pa), would be extremely helpful.

The XPS technique was widely used for the characterization of promoted and unpromoted molybdate [414–421] and tungstate [422] catalysts and the valence states of the group VIb metal and of the promoter elements were determined after various treatments. Iwasawa *et al.* [423] synthesized fixed Mo olefin metathesis catalysts by the facile reaction of $Mo(\pi\text{-}C_3H_5)_4$ with acidic surface hydroxyl groups [250] (see also section II.A.1.a.iv). An XPS study of the Mo surface species after various treatments enabled to correlate the catalytic activity with the presence of Mo^{4+}. Yermakov *et al.* [424] characterized tungsten catalysts which were analogously synthesized by

Table 15. XPS and chemical analysis data for various W/SiO_2 catalysts after treatment with H_2S at 673 K. (Reproduced with permission from ref. [424]. After treatment with H_2S, samples were heated 473 K under vacuum)

No.	Catalyst before H_2S treatment	E_B/eV		Atomic ratio S:W	
		W4$f_{7/2}$	S2p	XPS	Chemical
1	$W(\pi\text{-}C_4H_7)_4/SiO_2$	32.8	162.2	2.3	2.9
2	Catalyst 1 reduced at 873 K	33.2	162.4	1.4	1.5
3	Catalyst 2 oxidized by O_2 at 673 K	32.2	162.0	2.0	2.9
4	Catalyst oxidized by O_2 at 73 K	32.3	161.9	2.5	2.7

anchoring $W(\pi\text{-}C_4H_7)_4$ on silica surfaces, in their oxidized, reduced and sulfided states. The S:W surface ratios of sulfided catalysts were determined from a quantitative analysis of S2*p* and $W4f_{7/2}$ peaks. A comparison of these XPS data with a chemical analysis is shown in Table 15. The particular preparation method guarantees a high dispersion of the W species in the oxidized precursor state, so that an agglomeration to bulk WS_2 sulfidation is largely prevented. From their XPS data in conjunction with electron microscopic studies, Yermakov *et al.* [424] suggested the following reaction scheme for the conversions of tungsten species on silica

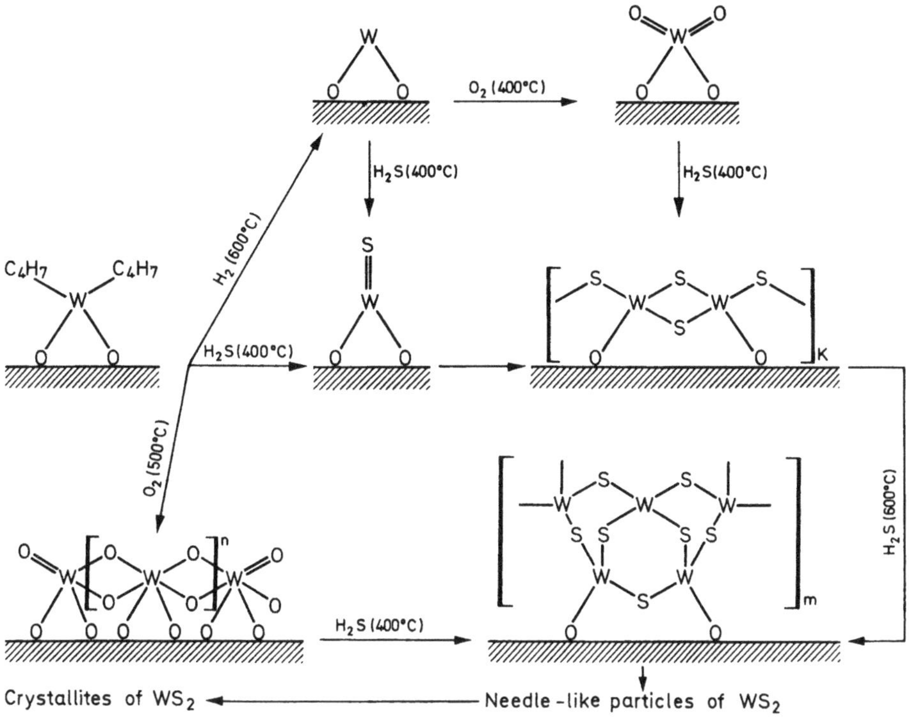

Tereczki *et al.* [425] have studied the type of bonding of nitrogen on active carbons which were treated with NH_3 at elevated temperatures. As shown in Figure 42 two N1*s* peaks can be distinguished at binding energies of 400 to 400.5 eV and 399 eV. These can be assigned to amine or amine-like and pyridine-like surface groups, respectively. The pyridine-like species can be described as *e.g.*

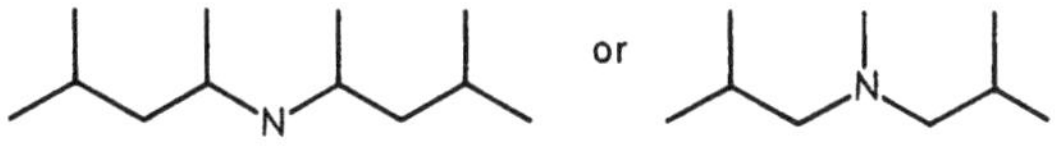

An additional N1*s* peak was observed at 398 eV after fusion with KCN and nitro groups (E_B = 406–407 eV) were formed on reaction with HCO_2. It was reported by Tereczki *et al.* [425] that the catalytic oxidation activity was significantly enhanced by the incorporation of N, especially of the

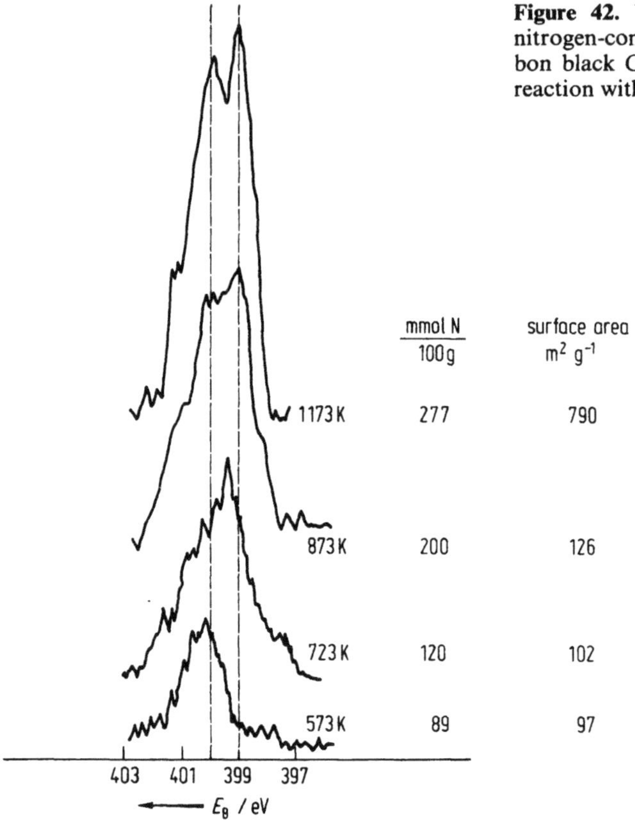

Figure 42. N1*s* photoelectron peaks of nitrogen-containing surface groups on carbon black CK3 which were produced by reaction with NH_3

pyridine-like groups. This example nicely demonstrates the ability of the XPS technique to detect different chemical environments of a given element in various surface functional groups.

Two distinct nitrogen species were also detected on Spherisorb-5-Amino and Spherisorb-5-Nitrile by Carley *et al.* [426]. The two materials are stationary phases used in HPLC columns, which are usually prepared by reacting the appropriate alkoxysilanes with the hydroxyl groups of silica surfaces (see section II.A.1.a.iv). The N1*s* binding energies for Spherisorb-5-Amine at 398.5 and 400 eV were considered to be due to NH_2 and NH species. Quantitative analysis of XPS intensities showed that the estimated nitrogen concentration very closely resembled the concentration of originally present surface hydroxyl groups, which suggests a high degree of OH replacement. However, the accuracy of quantitative XPS analysis for such surfaces of unknown topography might be doubtful. A possible explanation for the formation of NH species on the amine-modified silica surface was given by Knox [426a] when he suggested reaction of the silanol group with the amine group of the alkoxy-silane:

$$\equiv SiOH + NH_2(CH_2)_x Si(OEt)_3 \longrightarrow \equiv SiNH(CH_2)_x Si(OEt)_3 + H_2O$$

Thus, the XPS technique should be extremely helpful for the characterization of functionalized oxides which are used for the heterogenization of complexes, clusters and enzymes.

5. Thermal Desorption Spectroscopy (TDS)

Thermoanalytical methods have been applied in the past to study the thermal stability of surface groups, particularly of hydroxyl groups, and to measure their surface density. Kondo and Muroya [427] observed three endothermic peaks at approximately 470, 670 and 770 K, when they followed the dehydroxylation of a silica gel by *differential thermal analysis* (DTA). These peaks corresponded to small weight losses of the samples which were detected by *thermogravimetric analysis* (TGA). By comparison with the effects on the infrared spectra in the hydroxyl stretching region produced by thermal dehydroxylation, the DTA peaks were attributed to the OH species having wavenumbers at 3230, 3460 and 3520 cm^{-1}, respectively. In a similar study, Krauss and Naumann [428] followed the dehydroxylation of various silica samples by means of the combined use of TGA and infrared spectroscopy. They concluded that the desorption of molecular water was represented by a TGA peak at approximately 490 K, while condensation of two types of mutually interacting OH groups led to TGA peaks at approximately 650 and 950 K ($\tilde{\nu}_{OH} = 3550$ and 3640–3680 cm^{-1}, respectively). Isolated OH groups ($\tilde{\nu}_{OH} = 3750\ cm^{-1}$) were assumed to desorb at 1470 K.

Thermal desorption spectroscopy TDS — originally called temperature programmed desorption (TPD) [38] — is the most frequently used and perhaps most useful and informative thermoanalytical technique for the characterization of surface functional groups and chemisorbed species. This technique involves measurement of the desorption rate of surface species into an inert carrier gas (or into vacuum) as a function of the linearly programmed temperature [429]. The evolution of the desorbed gases into the carrier gas can easily be followed by conventional gas chromatographic detectors (thermal conductivity or flame ionization detectors) which may be combined with quadrupole mass spectrometers for analysis of the chemical composition of the desorbed material. Alternatively, TD spectra can also be obtained from the first derivative of TGA curves [429]. The practical and theoretical aspects of the method have been discussed in various articles [38–40, 430]. The desorption rate is given by

$$-\frac{dn}{dt} = k_d n^x = n^x \nu_x \exp(-E_d/RT)\,, \tag{14}$$

where n is the concentration per unit surface area of the desorbing species, k_d is the desorption rate coefficient, x the order of the desorption process, ν_x the frequency factor and E_d the activation energy of desorption. TDS analysis provides the following information:

(i) the initial surface population by the desorbing species can be estimated from the integral desorption peak area;

(ii) the order of the desorption process can be determined;

(iii) the position T_m of the maximum of the desorption peak on a temperature scale is a measure of the activation energy E_d of the desorption process and hence, of the surface bond energy;
(iv) the appearance of more than one desorption peak is an indication for the presence of energetically or chemically distinct states (typically an energy resolution of approximately 10–20 kJ mol^{-1} can be achieved);
(v) the desorption band shape is determined by the order of desorption x for energetically homogeneous surfaces and it may provide information on the energy distribution for energetically heterogeneous surfaces [431, 432].

Difficulties may arise with porous materials due to intra- and interparticle diffusion [38].

TDS has most frequently been applied for studies of dehydration and dehydroxylation processes of oxide surfaces. Thus, Munuera and Stone [433] have investigated the thermal desorption of H_2O from rutile surfaces. Two TDS maxima were observed, one at 523 K and the other at 643 K. The results of parallel infrared measurements supported the conclusion that the former maximum was due to comparatively weakly adsorbed molecular water whereas the letter arose from desorption of chemisorbed water present on the surface as two types of distinguishable hydroxyl groups. The high-temperature peak was asymmetric towards higher temperatures and the maximum shifted to higher temperatures when the initial degree of hydroxylation decreased. This behaviour is typical for a broad distribution of bond energies. Similarly, the five types of different hydroxyl groups present on the surface of transitional aluminas could not be resolved in TDS, but resulted in only one asymmetric peak the maximum of which shifted to higher temperatures T_m with decreasing degree of surface hydroxylation [432, 434]. Stakebake [435] studied the surface hydroxylation of PuO_2 and observed desorption maxima for the desorption of hydroxyl groups near 600 K. A time-of-flight mass spectrometer was used in these studies to monitor the water partial pressure. The activation energy of desorption was found to be 285 kJ mol^{-1}.

The thermogravimetric method to obtain thermal desorption spectra from the first derative of TG curves has been applied by several groups. Thus, the thermal dehydroxylation of γ-manganese dioxide [436], tin(IV) oxide [437] and chromic oxide [438] was studied. Particularly interesting is the investigation of the surface hydrolysis on four different Cr_2O_3 samples by Morishige *et al.* [438], since it is one of the few examples in which several desorption peaks for surface hydroxyls could be resolved. Four distinct peaks appeared with maxima at about 473(I), 533(II), 578(III) and 683 K(IV) in the TD spectra of the original samples. The corresponding surface hydroxyls of peaks I and II could be regenerated reversibly upon rehydration at room temperature, while the reproduction of the corresponding surface hydroxyls of peak IV required a thermal activation in water vapor. On the contrary, the species corresponding to peak III could never be regenerated upon rehydration.

Modifications of the TDS technique are the *temperature programmed reduction* (TPR) [439] and the *temperature programmed decomposition* (TPDE) [440]. The latter technique has been used successfully for a structural characterization of the surface complexes formed during the thermal decomposition of metal carbonyls (*e.g.* $Mo(CO)_6$) supported on oxide surfaces.

B. Chemical Methods

1. Active Hydrogen

Hydrogen is called active if its acidity surpasses a limiting value, which corresponds to a pK_a of approximately 20. The hydrogen will dissociate to a sufficient extent to allow rapid exchange with hydrogen atoms of water or alcohols. It can also combine with stronger bases than the surface atom to which it is bound. In general, hydrogen bonded to electronegative atoms such as oxygen, sulfur or nitrogen is active, but acetylenic hydrogen might also react.

The methods of determining active hydrogen can be differentiated into isotope exchange methods and chemical reaction methods.

a) Isotope exchange

Hydrogen has two isotopes, deuterium and tritium, which occur in low concentrations in natural hydrogen. Tritium offers the advantage of its radioactivity, which makes detection and quantitative determination fairly easy. However, it is not practicable to use pure tritiated compounds for the isotope exchange. Mostly, water has been used so far for the ion exchange, but labelled methanol can be used as well. Isotope exchange with molecular hydrogen has also been used for the estimation of surface hydrogen.

In exchange reactions of the type

$$\text{surf-H} + D_2O \rightleftharpoons \text{surf-D} + \text{HDO}$$

the so-called isotope effect must be considered. Due to slightly different zero point energies of the X—H, X—D, and X—T bonds (X = electronegative element), there is often an unequal distribution of D (or T) between X—H and water. An analogous situation exists for the distribution of the heavier hydrogen isotopes between surface hydrogen and water. The equilibrium constant

$$K = \frac{[X-D]\,[HDO]}{[X-H]\,[D_2O]} \tag{15}$$

differs from unity (X may also be a surface site). The deviation from unity decreases with increasing temperature; its temperature dependence is used for the production of heavy water by the H_2S/H_2O cycle. The distribution coefficient is unknown in most cases: it differs from 1 by probably not more than 10 to 20% in the case of surface hydroxyl groups and water at room temperature. Its value is 2.26 at 298 K for the H_2O/HDS equilibrium, however [441].

Davydov *et al.* [162] think that the deviation from unity is negligible in the case of deuteration of surface hydroxyl groups in view of the experimental inaccuracies in determinations of active hydrogen.

The problem of the isotope effect can be overcome, however, by exchange with undiluted D_2O which can be obtained in quite high purity at a moderate price. One can force practically complete exchange by using an appropriate excess. After re-exchange with a known excess of ordinary light water, the heavier isotope will be almost completely in the aqueous phase. Thus, the isotope effect will not influence the result in any significant way.

Deuterium exchange: Small concentrations of deuterium in water can be determined by various means [442] (see also section II.A.1.a.i). Exchange with D_2O was first used by Haldeman and Emmett [443] in the determination of the hydroxyl groups on alumina, silica, and silica-alumina. The hydrogen content in a known quantity of deuterium oxide after exchange was determined from the difference in vapor pressure at 273 K, measuring the heat-conductivity of the vapor. Rather large samples were needed. Extinction at 2500 cm^{-1} has been used for the determination of relatively small quantities of HDO in exchange water [161]. The proportion of deuterated hydroxyl groups on a silica wafer which was in exchange equilibrium with the sample was determined from the intensity of the 2760 cm^{-1} absorption peak of OD groups by Davydov *et al.* [162]; the total amount of OH groups was then estimated from the quantity of D_2O introduced.

Since only small quantities of exchange water will usually be available, one could determine deuterium by mass-spectroscopy after conversion to molecular H_2/HD. This reduction can be established with zinc at about 670 K. Still, the problem remains that the mass 2 corresponds to H_2 as well as D, mass 3 is caused by HD, but also by H_3^+. The relative abundance of these species depends on the instrument parameters as well as pressure. Evaluation of the mass spectra may therefore be complicated depending on the desired precision.

Small quantities of deuterium in hydrogen can also be determined using a catharometer. The heat conductivity of HD is significantly lower than that of H_2. By using a standard gas chromatograph with hydrogen as a carrier gas, there is no need for separation of HD from H_2 [444]. Traces of air *etc.* are separated from the HD pulse by an active charcoal column at ambient temperature.

Determination of active hydrogen on the surface of TiO_2 [445] or of silica [446] by this method has been described. The deuterated water, approximately 100 μl, was converted to H_2/HD by injecting it with a syringe onto CaH_2, as described by Arnett and Duggleby [444]. In a more convenient method [447] the CaH_2 is replaced by $LiAlH_4$ or a solution of diborane in tetrahydrofurane (containing $BH_3 \cdot THF$), see Figure 43. After freezing the H_2O/HDO sample with liquid nitrogen, the apparatus is evacuated. The water is then allowed to warm up and to vacuum distill onto the $LiAlH_4$ or $BH_3 \cdot THF$ solution which is cooled to 77 K. Hydrogen evolution is started by letting the reagents warm up slowly. Thus the reaction is not too violent. The gas is brought to atmospheric pressure

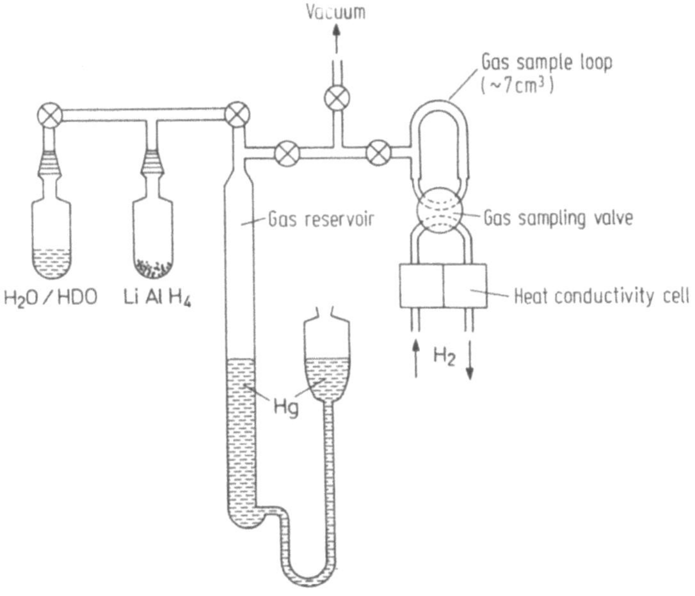

Figure 43. Set-up for the determination of small quantities of deuterium in water. The H_2O/HDO mixture is condensed onto the $LiAlH_4$. The hydrogen formed on warming of the $LiAlH_4$ is collected in the gas reservoir, and the sample loop is filled at atmospheric pressure. The gas sample is injected into the heat-conductivity cell by turning the gas sampling valve. It is advisable to pass the gas through an active-charcoal column on its way from the gas sampling value to the heat conductivity cell (not shown)

using the mercury levelling vessel, and the evacuated sample loop is filled with gas. A pulse of the gas is injected into the gas chromatograph by turning the sampling valve. The gas mixture can be measured repeatedly. The standard deviation observed with a 0.064 M solution of D_2O in water was 4.2%, it was 7.8% with a 0.275 M solution.

A sample of the solid containing approximately 200 μmol active hydrogen is outgassed prior to isotope exchange at a temperature sufficient to remove physically adsorbed water. Temperatures of 370–390 K suffice usually, but this can be ascertained only using infrared spectroscopy (see section II.A.1.a.i). Of course, it is possible to follow the temperature dependence of active hydrogen by raising the outgassing temperature. The sample should not be too small, and the wall area of the exchange apparatus should not be too large, since glass surfaces contain active hydrogen, too.

Hydrophobization of the glass surface, *e.g.* with $(CH_3)_2SiCl_2$, reduces this interference to some extent. Strong leaching of the glass by excessive washing should be avoided. Approximately 0.75 g of water were used for the re-exchange of the deuterated sample.

Another way of determining active hydrogen on solid surfaces is by measuring the weight change on isotope exchange [446]. Obviously the mass of the sample increases by 1 μg for each μmol of hydrogen exchanged by deuterium. Modern vacuum microbalances have a sensitivity of 1 μg or better combined with a maximum load of 1–2 g. Active H contents of

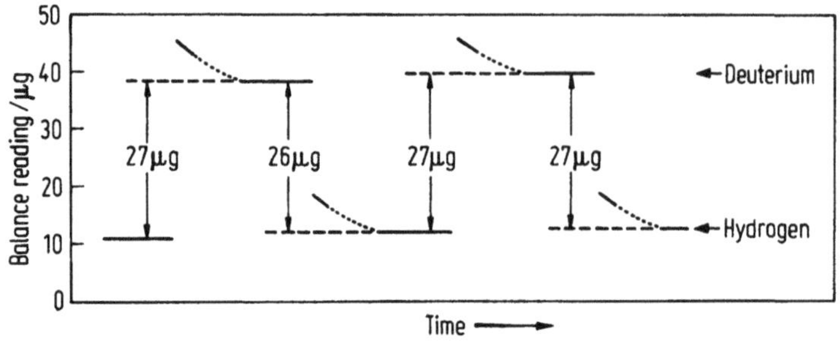

Figure 44. Mass changes on isotope exchange of surface OH groups with D_2O and re-exchange with H_2O. The sample was oxidized silicon carbide with a surface area of 9.5 m^2 g^{-1}. Measurement with an electronic vacuum microbalance Sartorius model 4101

ca. 100 µmol g^{-1} or more can be determined with sufficient precision. The sample is outgassed at a temperature sufficient to remove physically adsorbed water until there is no further weight change. Then D_2O is adsorbed somewhat above room temperature at a pressure a little below saturation pressure at room temperature. After exchange the heavy water is pumped off at the desired temperature, *e.g.* 390 K, to constant weight. Re-exchange with ordinary water should reduce the weight to its original value. The isotope exchange can easily be repeated several times (see Fig. 44).

The problems encountered using this method are caused by static electricity and by adsorption of contaminants, *e.g.* vapors of vacuum grease. The latter can be reduced by having the sample during the measurement at somewhat higher temperatures than the other parts of the apparatus (The tubular heating device should have bifilar winding in order to avoid magnetic effects. One can also turn off the heating current during the actual weighing). Static electric charges can be particularly annoying. The glass tube around the sample must have a conducting and grounded layer, *e.g.* wire mesh, conductive paint or silver or gold mirror. It is better, but experimentally more complicated, to have this conductor on the inside of the tube than on the outside.

It was observed that the hydrogen in the surface of silica, alumina, and silica-alumina exchanges with deuterium gas at elevated temperatures [448]. This method has been elaborated by Hall and associates [382, 449]. The samples were heated in a circulating D_2 stream, and the hydrogen content of the gas was monitored by means of a heat-conductivity cell. The substances also catalyzed equilibration between H_2, HD and D_2. The activation energies for this exchange are different for different OH groups. Isotope exchange started with alumina at 413 K and reached maximum reaction rate at 523 K, with another small peak at 623 K [382]. With silica, exchange set in at 713 K with maximum rate at 883 to 989 K. Silica-alumina was intermediate. The accuracy of the method was stated to be better than $\pm 5\%$, often $\pm 2\%$. By controlled, linear heating, differently bound hydrogen gives distinct peaks. The big draw-back is that usually temperatures well above 370 K have to be used; many substances have already lost much of their hydroxyl population at these temperatures. The temperatures necessary for exchange are not known for other substances than those mentioned. The distribution factor for H and D has been assumed to be unity in these studies. There is also exchange between surface active hydrogen and CD_4 [450].

Tritium exchange: Exchange with tritium-labelled water has been used also in the determination of active hydrogen. The method has been elaborated and described in detail by Unger [227]. The conditioned sample is equilibrated with labelled water, and the decrease in the counting rate of the labelled water or the tritium content of the solid is measured after separation of the water by distillation into a cold trap. In the case of tritiated silica, the sample was dissolved with sodium hydroxide solution, but other methods of collecting the tritium content of the surface can be envisaged. As Unger points out, the problem of the isotope effect has to be considered. He determined the active hydrogen of a standard silica sample using a chemical method (methyl lithium method), and compared this value with the result of tritium exchange. With silica, the tritium exchange values have to be multiplied by a correction factor of 1.27 in order to obtain results comparable with those of chemical determinations of active hydrogen. Obviously, this factor has to be determined for each kind of surface group. It must be mentioned, however, that Unger found much better agreement with butylamine titration values of heat-treated silicas if the uncorrected results of tritium exchange were used [451].

b) Chemical reaction methods

Determinations of active hydrogen by chemical methods consist usually in the combination of the positively polarized active hydrogen with a negatively charged species with formation of gas. Hydrides, mainly $LiAlH_4$, or metal methyl compounds have been used. Gas evolution can be measured with high sensitivity and sufficient accuracy.

The use of $LiAlH_4$ for the determination of active hydrogen on carbon black surfaces has been described by Rivin [452]. It was found that it is of advantage to use a high-boiling solvent for the $LiAlH_4$, *e.g.* diethyl carbitol. The reagent is applied to the pre-dried carbon black by a syringe through a rubber septum. After initial H_2 evolution, the sample is heated to 343 K. The reaction vessel is connected to a gas buret, and the volume of the evolved gas is corrected to standard conditions. One milliequivalent of active H corresponds to 22.4 ml (STP) of H_2. However, it is not advisable to use $LiAlH_4$ with silica, since even quartz powder is attacked with slow evolution of SiH_4 [453]. It has been reported [186] that dry pyrex glass reacts with $LiAlH_4$ solutions, whereas neutral glass seemed to be inert.

A well established method for the determination of active hydrogen in organic compounds is the reaction with the Grignard reagent methyl magnesium iodide (or bromide) according to Zerewitinov. Use of this reagent for the determination of surface hydroxyl groups was first suggested by Deuel and Huber [454], using clays. Later, this method was also employed in the determination of hydroxyl groups on silica [455]. In this paper. CH_3MgI in anisol was used; the evolved methane was purified by low-temperature

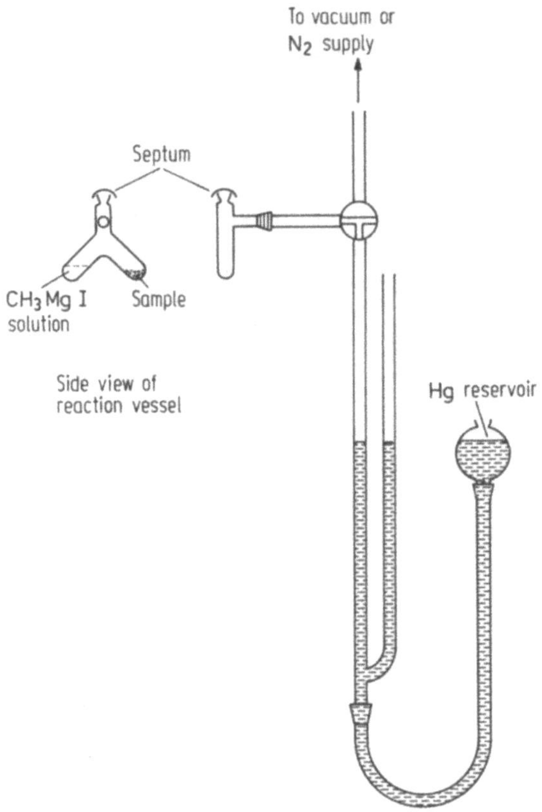

Figure 45. Determination of active hydrogen by reaction with Zerewitinov reagent. The sample is placed in one limb of the reaction vessel which is placed in a sufficiently wide glass tube for outgassing. The openings of the vessel are then closed with a septum and a ground-glass stopper until it is attached to the gas burette. After evacuation and filling with dry nitrogen, reagent solution is injected into the free limb, and — after establishment of thermal equilibrium and of ambient pressure — the solution is poured onto the sample by rotating the vessel. The gas burette, made of 3 mm bore capillary tubing, has a capacity of about 75 µmol gas (at room temperature)

distillation and it was collected for volumetric measurement by use of a Töpler pump. The authors observed a slow continuous methane evolution which was explained by an autolysis of the reagent. Instead of methyl magnesium iodide, solutions of methyl lithium could be used with similar results.

The Zerewitinov method is used in one of the authors' laboratory with good results [456]. The sample is weighed into one leg of a two-legged vessel (Fig. 45), and a series of such vessels is outgassed in a wide tube in high vacuum at the desired temperature. After filling with dry nitrogen, the vessels are closed with a rubber septum. The vessels could as well be charged with the samples in a dry glove box. The vessels are attached to the gas burette, and the reagent solution (CH_3MgI in di-*iso*amylether) is injected into the second leg. The apparatus is filled with dry nitrogen of ambient pressure and, after closing of the stopcock, the reagent is poured onto the sample by rotating the vessel on the ST joint. The evolved gas is measured in the buret after temperature equilibration of the reaction mixture by use of a water-filled beaker. The gas buret is made of capillary tubing of *ca.* 3 mm, bore. The diameter of the capillary and its constancy over its length had been calibrated with mercury prior to assembling the apparatus. It is essential that the system is filled with nitrogen or an other inert gas. The volume of the injected reagent should be as small as possible since the solvent will keep some dissolved methane.

The reproducibility of the method depends on the concentration of active hydrogen; usually it was around 3 to 5%. The method cannot be used with reducible substances such as TiO_2, Fe_2O_3 *etc.* Additional gas is formed and ethane was detected in the reaction gas. Problems arise also with porous substances. Diffusion of the reagent into the pores is slow, and methane may stay adsorbed. With narrow-pore active charcoal no methane evolution was observed in one case although OH and COOH groups were known to exist on its surface. Exclusion of lithium methyl reagent from narrow pores of silicagel has also been observed [457].

The use of ethylmagnesium bromide has also been described [458], as well as that of ethyl lithium, and of triethyl aluminium dissolved in decalin.

Pure trimethyl aluminium was reacted with silica surfaces by Peglar *et al.* [459], who found that all OH groups reacted with formation of methane. It had been reported earlier that hydrogen-bonded OH groups on the silica surface did not react quantitatively [460]. Excess $Al(CH_3)_3$ can be condensed at 193 K. Hanke [461] used dimethyl zinc in the form of its THF adduct $Zn(CH_3)_2 \cdot 2\,THF$. The adduct (b.p. 355–356 K) has the advantage of being less volatile than pure $Zn(CH_3)_2$ (b.p. 317 K). Excess reagent was trapped by cooling with liquid nitrogen, and the methane was collected for measurement by use of a Töpler pump. Nondek [462] developed a pulse method in which the methane was transported by a helium stream to a heat-conductivity cell. A determination can be performed in approximately two hours. The question of reduction of oxides like TiO_2, Fe_2O_3 or MoO_3 has not been considered by the author, although results with Fe_2O_3 or Cr_2O_3 appear somewhat high. The use of pure $Zn(CH_3)_2$ for determination of OH groups on the TiO_2 surface had been described by Boonstra and Mutsaers [463]. The excess of reagent was frozen out at 118 K in this study.

Use of gaseous reactant, such as $Al(CH_3)_3$ or $Zn(CH_3)_2$ seems to be advantageous in the case of porous substances. The main draw-back of these substances is their pyrophoricity which asks for more elaborate apparatus. Experiments in one of the authors laboratory have shown that in the reaction of silica with CH_3MgI one Mg atom is bound for each active hydrogen [463]; preliminary experiments showed also that after the reaction of $Zn(CH_3)_2$ with CeO_2 surfaces one Zn atom was bound for each active hydrogen atom. The metals were brought into solution with dilute hydrochloric acid and then titrated complexometrically. The reproducibility of the titration values was not satisfying, however.

Determination of B after the reaction with BCl_3, or of Al after reaction with $AlCl_3$ or $Al(CH_3)_3$ is not an indicator of the content of active hydrogen, since these molecules may give surface concensation reactions with one or two of their substituents. This will be discussed in section III.B.2.c.

Active hydrogen and adsorbed water on oxide surfaces have also been titrated with sodium or potassium naphtenide solutions [464]. The reaction is fast and quantitative, the authors claim that a determination can be done in 40 minutes. The end point in the titration is recognized by the appearance of the green color of reagent or — still more sensitive — by the appearance of a permanent ESR signal in the reaction mixture (an ESR tube was fused to the reaction vessel). Only one hydrogen atom of water reacts with the reagent. The organic anion radical will reduce any reducible sample, however.

2. *Acidic and Basic Groups*

It is well known that the surface of a solid may have acidic or basic properties — or both — which are used for a great variety of heterogenously catalyzed reactions. Because of its importance the subject of surface acids or bases has been repeatedly dealt with in reviews [186, 187, 465, 466]. Therefore, it will suffice to describe this topic relatively shortly in this chapter. Moreover, in addition, surface acidity or basicity is not necessarily due to surface functional groups, it may be an intrinsic property of the surface. It is customary to distinguish between Brönsted acids capable of donating protons, and Lewis acids which are electron-pair acceptors. In the normal case, Lewis acid centers are lattice cations exposed on the surface because of their incomplete coordination. Analogously, anions in the surface, *e.g.* oxide ions, are Lewis bases. We do not consider as functional groups these lattice ions which are exposed in the surface and which are coordinatively unsaturated. Functional groups with atoms which are foreign to the structure, however, can possess acidic or basic properties as well. Carboxylic groups on carbon surfaces or amines grafted to silica surfaces are good examples.

The oldest method of determining acidic or basic groups consisted in equilibration with appropriate bases or acids, respectively. Samples are agitated with an excess of aqueous (or alcoholic) base or acid, and the decrease in concentration is determined in an aliquot of the supernatant. By changing the acidity of the conjugate acid to the base used, groups of different acidities can be distinguished in favorable cases, *e.g.* with carbon-containing surface oxides [1]. Sodium bicarbonate, sodium carbonate, sodium hydroxide and sodium ethoxide were used in this study. It is necessary, to use a concentration of the reagent which is sufficient to reach the plateau of the ion exchange adsorption isotherm. Concentrations of 0.05 to 0.1 N are usually sufficient. Also, sufficient time for equilibration should be provided. Direct potentiometric titration is frequently quite tedious because of the very slow equilibration, especially with porous substances. In many cases, the substance may be attacked by aqueous alkali; this is the case with silica or with alumina. Instead of adsorption from more or less alkaline solutions, ion exchange adsorption from ammonium acetate [467, 468] or calcium acetate [469], has also been used with titration of the liberated acetic acid against bromphenol blue. It has been held against this method that Lewis acids may form Brönsted acids in aqueous media by polarization of adsorbed water molecules. Of course, acids which are stronger acids than the hydronium ion, H_3O^+, cannot be differentiated by titration in aqueous media.

The differentiation of sites of different acidities has been attempted using the adsorption of amines and other bases from the gas phase. Irreversible adsorption of ammonia [470], pyridine or quinoline [471] has been taken as a measure of strong acid sites. The heat of desorption has also been used as indicator of acid strength [472]. A convenient way of determining the activation energy for desorption is provided by temperature-programmed desorption (TPD) [38] (see section II.A.5).

A method for the rapid determination of adsorbed ammonia uses oxidation of the NH_3-treated and outgassed sample by a 2% O_2-in-He stream; the oxygen content of the effluent gas stream is monitored by an oxygen analyzer [473]. It was found that the ammonia oxidation occured at different temperatures with different samples, *e.g.* at 775 K with fluorided alumina and at 875 K with silica-alumina. The authors suggested tentatively that NH_3 bound to Lewis or Brönsted sites can be distinguished. However, unambiguous distinction between adsorption on Brönsted or Lewis acid sites is only possible by infrared spectroscopy (see section II.A.1.a.ii and iii).

A very widely practized determination method for acidic sites uses the adsorption of indicator dyes which change color upon protonation. The color changes observed on adsorption of indicator dyes on acidic surfaces were first described by Weil-Malherbe and Weiss [474]. Walling [475] used several indicators with different strengths of their conjugate acids and could thus estimate the acid strength of the most acidic sites on a variety of solids. A quantitative determination of acid sites was attempted by Tamele [476] who used titration with a stronger base, *e.g.* *n*-butylamine, to displace the indicator from the acidic sites. The amine was applied dissolved in a nonpolar solvent. Benesi [477] improved this method equilibrating the solid with the base first and adding the indicator afterwards. The underlying assumption is that the indicator will not be adsorbed in the protonated form if the acid sites with an acidity stronger than that of the indicator acid are blocked by the stronger base *n*-butylamine. Using a series of indicators, an acidity distribution can be established.

The surface reaction is

$$\text{Surf.-H} + \text{B} \rightleftharpoons \text{Surf.}^- + \text{BH}^+$$

(B is the indicator base). Since the activities of H^+, B and BH^+ are not known, only the Hammett acidity function H_0 [478] can be used instead of pK_a values of the surface acids:

$$H_0 \equiv -\log\left(a_{H^+} \times f_B/f_{BH^+}\right), \tag{16}$$

f_B and f_{BH^+} being the activity coefficients of adsorbed indicator base and indicator acid, respectively. With the assumption that the dissociation constant K_a of the indicator acid is not changed in the adsorbed state, one gets

$$H_0 = pK_a + \log \frac{[B]}{[BH^+]} \quad (17)$$

([B] and $[BH^+]$ are the concentrations of the indicator in its basic and its acidic form).

The pK_a values of the most frequently used Hammett indicators are listed in many articles [186, 187, 465, 466]. It is tacitly assumed that Hammett indicators react only with Brönsted acids. In principle, the free electron pair of the indicator base could as well react with a surface Lewis acid, although this reaction might be barred in many cases for reasons of steric hindrance. Mizuno *et al.* [479] found indication in the UV spectra of *p*-nitroaniline adsorbed on SiO_2/H_2SO_4, on alumina or on silica-alumina that the absorption maxima have different positions for *p*-nitroaniline bound to Brönsted or to Lewis acid sites.

The titration is performed in dry solvents (benzene may be replaced by toluene, *iso*-octane, decalin or other non-polar solvents) with carefully dried samples. The detrimental effect of water has been described in detail in ref. [187]. Since direct, drop-wise titration is not possible because of slow equilibration, Benesi [477] added consecutively rising quantities of *n*-butylamine solution to a series of samples, and determined the end point by successive approximation. It was suggested that equilibration could be expedited by an ultrasonic treatment of the suspension [480]. It is difficult with many indicators to determine the color change visually. Application of the usual Hammett indicators has often yielded acid strength distributions which are flat over a wide range of pK_a values. Furthermore, these results were in contradiction to other physical or catalytic measurements. Therefore, spectrophotometry has been applied to ascertain that the adsorbed indicator had an absorption spectrum similar to that of the authentic protonated dye [355, 481] (see section II.A.2.a.i). A few new fluorescence indicators were introduced in the place of Hammett indicators [481]. The absorption bands of organic dyes can also be shifted merely by physical adsorption. This was shown for the H_0 indicators benzalacetophenone and anthraquinone [482]. There is a tendency to overtitrate using these Hammett indicators; substituted arylmethanols seemed to be better suited as indicators [482]. The use of aryl methanols (aryl carbinols) instead of Hammett indicators had been proposed by Hirschler [483]. Triaryl methanols react with strong proton acids according to

$$R_3C{-}OH + H^+ \rightleftharpoons R_3C^+ + H_2O\,.$$

The color change from the colorless triaryl methanol to the colored carbenium ion is much easier to detect visually. For these indicators, an H_R acidity function was defined by Deno *et al.* [484]:

$$H_R \equiv -\log a_{H^+} + \log a_{H_2O} + \log (f_{R_3C^+}/f_{R_3COH}) \quad (18)$$

This equation contains also the activity of water, and obviously results will be influenced by traces of moisture still more strongly than with Hammett indicators. However, Hirschler [483] claimed better correlation of the titration values with catalytic activity of silica-alumina cracking catalysts when H_R indicators were used instead of H_0 indicators.

This method must be approached with caution, however, due to observations by Take *et al.* [485, 486] and by Deeba and Hall [487] who found that equilibrium between the solutions of the bases and the solid surface can be established extremely slowly. *n*-Butylamine once adsorbed on the surface of weakly acidic silica is not desorbed even if strongly acidic silica-alumina is introduced into the system; apparently the heat of desorption is too large for rapid establishment of equilibrium [485]. In column chromatography experiments a pulse of *n*-butylamine was adsorbed completely at the top of a silica-alumina column [487]; if the indicator was introduced later on, it was adsorbed as a colored band below the zone saturated with butylamine. On the other hand, if the indicator was injected first, it stayed at the top of the column and did not move even after prolonged washing with benzene. In separate experiments, the quantity of *n*-butylamine was determined which was necessary to saturate the acid sites completely. This value agreed quite well with the result of the standard titration using 4,4′,4″-trimethoxytriphenyl methanol ($pK_a = 0.82$) as indicator, much better than with the H_0 indicator butter yellow ($pK_a = 3.3$). Deeba and Hall [487] concluded that it is not possible to obtain a correct acidity distribution by the standard Benesi titration, whereas it can be obtained by their method if not only *n*-butylamine but other, weaker bases are employed as well. This method has also the advantage that it can be applied to colored substances. The acidity distributions determined in this way resembled those obtained by temperature-dependent irreversible adsorption of gaseous bases such as pyridine. Titration by the Benesi method can give the total number of acid sites, however.

Taniguchi *et al.* [488] found that adsorption of ammonia from the gas phase prior to the adsorption of the Hammett indicators works as well, in principle, as titration with *n*-butylamine. However, the quantity of acid sites corresponding to a given Hammett indicator was larger by a factor of 2.2 when ammonia was used. The authors concluded that this difference is a consequence of different steric requirements of the bases, each adsorbed *n*-butylamine molecule will cover also neighboring acid sites.

Determination of *basic sites* on a surface analogous to the Benesi titration has also been described [187, 489]. The method was improved upon by including a greater number of indicators [490]. Benzoic acid has been employed as a titrant, but use of trichloroacetic acid permits to cover also very weakly basic sites [491]. Determinations of basic sites by this method have been described much less frequently in the literature than titration of acidic sites. In view of the findings that equilibria are not always rapidly established in the determination of acidity distribution, some scepticism is warranted also for the titration of basic sites.

Acidic sites can also be determined by characteristic reactions. *Carboxylic groups* on oxidized carbon surfaces can be determined by esterification with methanol in the presence of hydrogen chloride at reflux temperatures [492]. Methylation can also be accomplished with dimethyl sulfate [1] or with diazomethane [493]. These agents methylate *phenolic hydroxyl groups*, too; it was possible, however, to distinguish between carboxyl and phenolic groups, because carboxylic esters are easily hydrolyzed, *e.g.* with hot dilute hydrochloric acid, whereas phenol ethers are stable under these conditions [492, 493]. The methoxyl content of the methylated samples can be determined with quite good accuracy by the standard Zeisel method [494]. Methoxyl determinations of samples treated with etheric diazomethane solutions before and after hydrolysis treatment yielded carboxyl and phenol contents [1, 469, 492, 493]. Methylation with diazomethane has also been successfully employed with acidic surface OH groups on titania [495].

3. Specific Adsorption

Specific adsorption of a reagent on certain sites or functional groups can be used for the determination of the concentration of these sites. The meaning of the term "specific adsorption" is that the free energy of interaction between the reagent molecules and certain adsorption sites is much more favourable than for the other parts of the surface of the solid. In the most favorable cases, the reagent will be held so strongly on the sites in question that an excess which is adsorbed by much weaker forces can be pumped or washed off. It is then easy to determine the bound reagent analytically and to calculate the number of "specific" adsorption sites if the stoichiometry of the reaction is known.

The number of specific interactions is unlimited, of course. Two characteristic examples will be discussed.

a) Adsorption of water on hydrophilic adsorption centers

Physisorbed water molecules interact with the surface either by van der Waals forces or by hydrogen bonds, if there are donor or acceptor sites for hydrogen bonds present. The interaction energy is higher for hydrogen bonds than for van der Waals attraction, of course. However, the hydrogen bonds are not as strong as to allow separation of hydrogen-bonded water molecules by simply pumping off the more weakly bound water. A determination of hydrophilic adsorption centers on otherwise hydrophobic surfaces is possible by an evaluation of the adsorption isotherms, however, provided that the concentration of the hydrophilic centers, *e.g.* hydroxyl groups on the surface is not too high and that they are isolated.

Water vapour adsorption isotherms on hydrophilic substances, *e.g.* alumina, are sigmoid, *i.e.* of type II of the Brunauer, Deming, Deming and Teller classification [496]. Adsorption of water on entirely hydrophobic surfaces, *e.g.* the surface of graphitized carbon black, results in a type III adsorption isotherm (Fig. 46). This type of adsorption isotherm is observed when the energy of mutual interaction of the adsorbed molecules is significantly higher than the energy of interaction of the adsorbed molecules with the

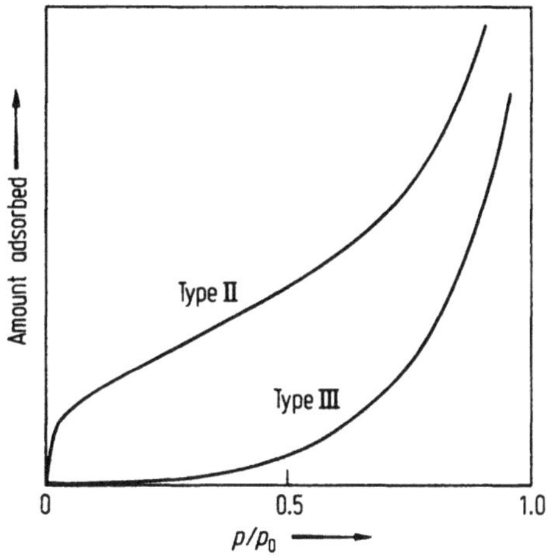

Figure 46. Type II and type III adsorption isotherms

adsorbent surface. This adsorption isotherm is characterized by a very small slope at relative pressures below 0.5.

If the hydrophobic surface carries a few hydrophilic centers, water adsorption on these sites is superimposed on the type III isotherm. It is possible to evaluate the composite isotherm, using the BET adsorption equation. In the deduction of BET equation [497] for the determination of the monolayer capacity in nonspecific physisorption, it is assumed that the heat of adsorption is constant in each layer, and that it is higher in the first layer than in the following layers. It is assumed, furthermore, that the heat of adsorption in the following layers does not differ significantly and that it is equal to the heat of liquefaction of the adsorptive. In the case of hydrophilic centers on a hydrophobic surface, the heat of adsorption of water molecules on these centers is definitely higher than on the other parts of the surface. Consequently, application of the BET equation to the adsorption isotherm yields the quantity of water adsorbed on the hydrophilic centers as "mono-

Table 16. Active hydrogen and hydrophilic adsorption centers

Sample	Active hydrogen/		Hydrophilic Ads. centers/	Ref.
	Zerewitinov	Deuterium exchange		
	μmol g^{-1}	μmol g^{-1}	μmol g^{-1}	
Diamond (20 m^2 g^{-1}), outgassed at 1173 K	18	—	19	[498]
—, oxidized with O_2 at 693 K	62	63 ± 3	64	[446], [498]
—, treated with H_2 at 1173 K	7	—	7	[498]
SiC (9.5 m^2 g^{-1}), O_2 at 293 K	70	54	53	[446], [500]
—, treated with HF[a]	37	—	30	[500]
Aerosil (145 m^2 g^{-1})[b]	1050	980	1120	[446], [501]
Aerosil (330 m^2 g^{-1})[b]	2300	2210	2160	[446], [501]

[a] Outgassed at 423 K
[b] Outgassed at 413 K

layer capacity". It was shown that this quantity agrees very well with the quantity of active hydrogen determined by deuterium exchange or by the Zerewitinoff method of diamond powder [446, 498, 499], for silicon carbide [446, 500] or for Aerosil silica.

The results are compared in Table 16. The samples had been outgassed at 383 K prior to water vapor adsorption. The isotherms were determined gravimetrically using an electronic microbalance (Sartorius, Type 4101). "Monolayer capacities" determined at 293 K and at 274.5 K agreed very well, in general. One molecule of water is adsorbed on each hydrophilic center at "monolayer coverage".

It is astonishing that this method worked also for the surface of Aerosil, a silica prepared by flame hydrolysis of $SiCl_4$. The silica surface consists of silanol groups, $\geq$Si—OH, and siloxane bonds, $\geq$Si—O—Si$\leq$. Siloxane bonds are hydrophobic [502] which is ascribed to *dp* π-bonding between oxygen and silicon; everybody knows that silicones are hydrophobic whereas polyethers are hydrophilic. This approach cannot be applied to the surfaces of metal oxides such as Al_2O_3, SnO_2 *etc.*

Of course, more basic adsorptives, such as ammonia, could be used as well. Using stronger bases, it should be possible, in principle, to distinguish between groups of different acidities. This subject has been discussed already in section II.B.2. The energy differences involved are usually too small to distinguish stronger acids by the approach just described for the determination of hydrophilic centers on hydrophobic surfaces. As mentioned earlier, some differentiation is possible using temperature-programmed desorption.

b) Adsorption on coordinatively unsaturated metal ions

Philips catalysts (see also sections II.A.1.a.iv and II.A.2.a.ii) are prepared by impregnation of silica with chromic acid, activation and reduction. The best reductive agent is carbon monoxide. It has been shown that the catalytically active sites in olefin polymerization are Cr(II) ions which are bonded to two silanolate groups [256, 260].

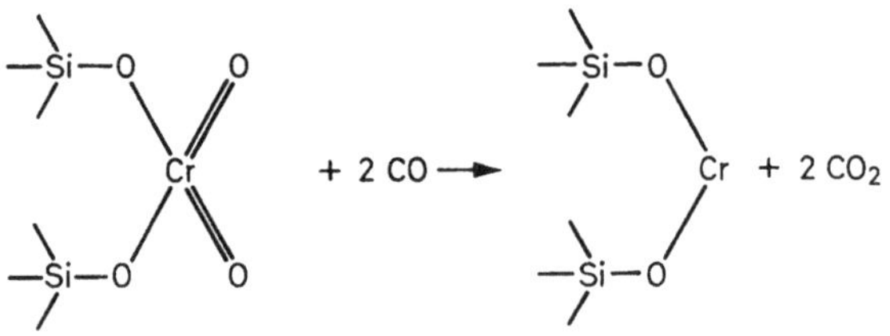

Only about 10% of the Cr(II) ions produced in the reaction step are catalytically active [503]. The chromium(II) ions on the silica surface are extremely sensitive to oxygen Silica-gels covered with Cr(II) are used to remove efficiently traces of oxygen from inert gases. The residual O_2 partial has been determined as low as 10^{-10} Pa (10^{-9} ppm at normal pressure) [504]. Chromium(II) is oxidized to chromate(VI) in this reaction with emission of organgered light (chemoluminescence) [505, 506].

Lewis bases are strongly adsorbed on the coordinatively unsaturated metal ions. The number of active sites has been estimated by titration with carbon monoxide [503]. The stoichiometry of strong CO chemisorption is apparently 1 CO molecule per Cr(II) ion on the surface [263, 507]. Zecchina *et al.* [263] observed also bridging carbonyl groups indicating close neighbourhood of chromium atoms. However, highly active catalysts are obtained only at low Cr concentrations. Adsorption of other Lewis bases has also been studied. Ammonia [508], water [509], alcohol [509], ethers [509] and trimethylsilanol [509] form quite stable, purle-colored complexes with a ratio of Cr: ligand equal to 1:2. When the adsorptive are offered in excess, additional molecules are adsorbed much more weakly. The blue surface complexes have probably a Cr:L ratio of 1:4. Nitric oxide forms also a dark brown 1:2 complex [258, 264, 510], which has a very low electron spin density [510]. The conclusion is that NO^+ ions are formed on low-spin chromium(0). Nitrogen dioxide forms a 1:2 complex at low temperatures ($T \leqq 218$ K) [510]. Nitrous oxide, however, is adsorbed in a 1:1 complex [510], as is hexamethyldisiloxane, an extremely weak donor [509]. These surface complexes are able to bind an additional CO molecule. Finally, it was shown that olefins form blue 1:1 and violet 1:2 complexes [503]. 1:1 and 1:2 complexes can be distinguished by their reflection spectra [503, 509]. Cyclopentadiene forms a brown adsorbate with one pentahapto cyclopentadienyl ligand per chromium atom. One hydrogen atom is also bound to the chromium atom [503]. On heating to 423 K, this hydrogen is given off, and chromium is oxidized to a formal oxidation state of +3.

Similar coordinatively unsaturated metal compounds on silica gel have also been described with manganese [511] and with vanadium [512, 513]. It is assumed that vanadium(III) is linked to the surface as a tripod. Oxygen is chemisorbed by this purple species at room temperature forming a dark blue peroxide [512]:

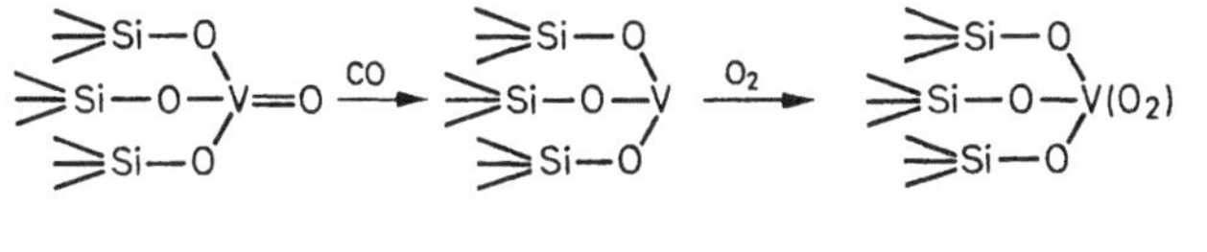

Otherwise this species is very similar in its chemical reactivity to Cr(II) on silica.

It follows from these examples that strong and very weak Lewis bases are specially chemisorbed on coordinatively unsaturated metal ions in most cases with a definite stoichiometry. Quantitative measurements of the chemisorption reaction allow easy determinations of metal centers on the surface. These determinations can be done volumetrically, gravimetrically, by chromatographic frontal analysis, or by elemental analysis of the sample, depending on the system to be studied.

Adsorption of CO has been studied repeatedly on surfaces which had been reacted with transition metal carbonyls and decarbonylated by heating in vacuo (see sections III.B.2.f and III.B.3.e).

3. Generation of Surface Groups

A. Primary Generation of Simple Groups

The atoms in the surface of a solid are coordinatively unsaturated. It is obvious that they have a tendency to bind other atoms or atom groups available in the neighbouring phase. Under normal conditions, reactive species are oxygen and water, eventually also carbon dioxide. Other reactive elements or compounds, *e.g.* the halogens or ammonia, can also be chemisorbed on the surfaces.

The binding of the foreign species will usually be similar in its nature to the binding forces within the solid, that is covalent or ionic. (Metals are not considered here).

It follows, that oxygen will be bound under normal conditions by such substances as carbon, silicon or silicon carbide, whereas oxides will bind water on their surfaces, forming hydroxyl groups. Oxides of sufficiently basic nature will also chemisorb carbon dioxide. It seems sensible to discuss a few examples separately.

1. Surface Oxides on Carbon

Solid, elemental carbon occurs in several allotropic forms, graphite, diamond and a few other forms of lesser importance which were discovered in the recent years. Due to the layer structure of graphite, foreign elements can be bound only at the edges of the hexagon layers, and at defects in the outermost layers. Basal planes of graphite remain clean of adsorbed oxygen for several days in a vacuum of 10^{-7} Pa when other surfaces would be contamined [514]. The specific surface area of well-crystallized graphite is usually quite small, and in consequence of the anisotropic morphology the surface properties are dominated by the basal planes. Electronic interaction of adsorbed molecules with the π-electron states in the basal carbon layers plays certainly an important role in the adsorption process, but it does not lead to irreversible bonding.

Much larger surface areas are frequently observed with less well-ordered carbons derived from graphite. These paracrystalline (also called microcrystalline) carbons show turbostratic disorder, that is the layers are stacked parallel without any further relative ordering. Usually, the layers have relatively small diameters, although they are larger in many cases than the "crystallite size", L_a, determined from the broadening of the (*hk*) reflexes in X-ray diffraction. The L_a parameters correspond only to the average diameters of coherently scattering, planar and relatively defect-free parts of the layers. High-resolution lattice imaging electron microscopy showed that the layers are often bent or warped. High surface area carbons of technical importance are carbon blacks and active charcoals. Cokes have usually relatively small surface areas.

The colloid-chemical behavior of high surface area carbons is dominated by the surface groups. It is well-known that the carbon surface can have either acidic or basic properties [515]. These properties are due to acidic and basic surface oxides, respectively. Acidic surface oxides are always formed when a carbon is heated in oxygen to temperatures of approximately 573 K or higher [1, 3]. Acidic surface oxides are also produced by the action of strongly oxidizing solutions at room temperature (or higher). Acidic surface oxides have thus been produced with concentrated HNO_3, aqueous solutions of NaOCl, $KMnO_4$, $(NH_4)_2S_2O_8$, or with O_3 in CCl_4 and with other oxidants. Electrochemical oxidation is also possible [516]. The oxygen-containing surface groups decompose on heating the carbon to high temperatures, evolving CO_2, CO and H_2O. Heating to about 1270 K is necessary to remove all chemisorbed oxygen. Basic surface oxides form when carbon is heated to 1270 K in vacuo or in inert gas and is exposed to oxygen only after having been cooled to room temperature under inert conditions. It was found that some oxygen is chemisorbed immediately on exposure of the carbon to oxygen, but an additional quantity of oxygen is chemisorbed when this carbon is immersed in prototropic liquids [517]. Water has sufficient acidity [517]. The second oxygen uptake was found to be equal to the first, "dry" chemisorption.

Positively charged surface sites are formed in the second reaction which bind anions from the solution (OH^- from pure water) [518]. The carbons have anion exchange properties. It was concluded on the basis of chemical evidence that the basic surface oxides have pyrone-like structures [519]. This is shown in Figure 47. The carbonyl group and the hydroxyl group need not necessarily be situated on the same carbon hexagon or in close neighbourhood. It is sufficient that they can be connected with resonance structures as shown in Figure 47. It follows from the formation conditions that is not possible to have carbons under ambient conditions which do not carry basic surface oxides. Covering the surface with acidic surface oxides diminishes the concentration of basic surface oxides to some extent, but they can never be avoided.

The constitution of acidic surface oxides is more diverse. Carboxylic groups and phenol-like hydroxyl groups have been detected and verified by several chemical reactions [1, 492]. More carboxylic groups are formed by "wet" oxidation at low temperatures than by oxidation with oxygen at 570–670 K [1]. Carboxylic anhydrides form under anhydrous conditions [1]. A part of the carboxylic groups show reduced acidity. This has been explained

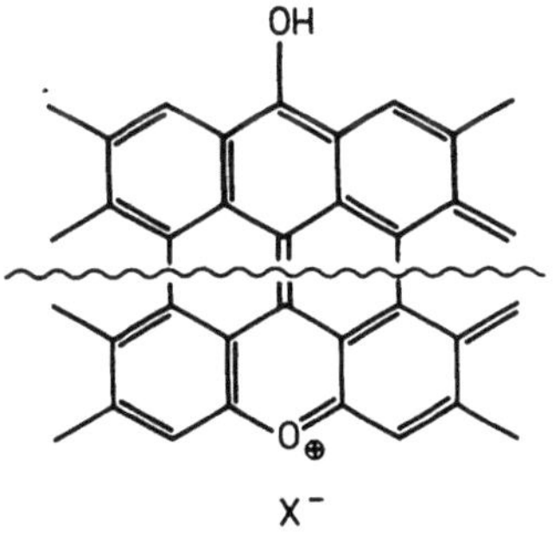

Figure 47. Model of basic surface oxides on carbon. The wave line indicates that the two oxygen functions can be separated by some distance; it is only necessary that a suitable resonance structure can be constructed. X^- symbolizes an exchangeable anion

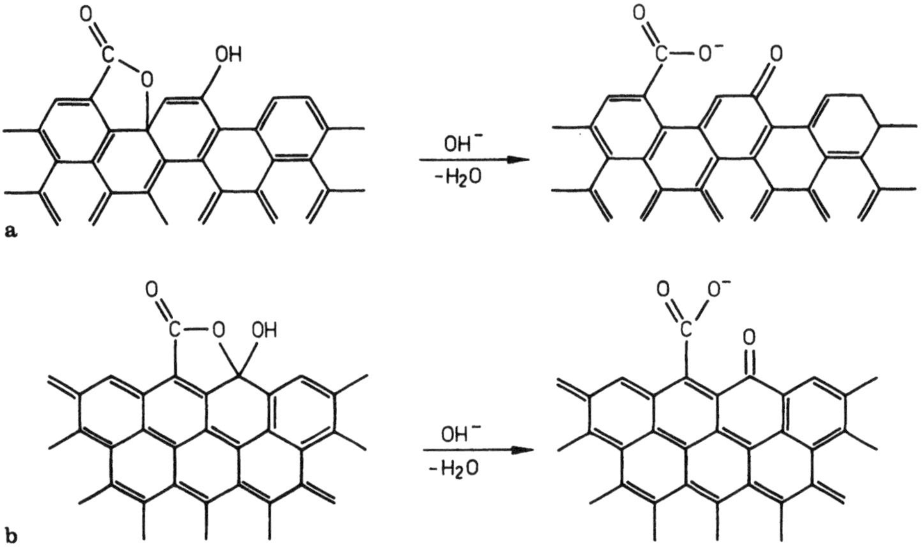

Figure 48. Model of fluorescein-like lactone **a** and lactol **b** on a carbon surface

by lactone structures similar to fluorescein (or phenolphthalein) [520]. However, it seems more plausible that the structures are lactol-like as in *o*-carboxylic acids of aromatic ketones [492], as shown in Figure 48.

In addition to carboxylic groups and phenolic groups, a very weakly acidic group has been observed in oxidized carbons, which reacts with sodium ethoxide forming the salt of a hemi-acetal [492]:

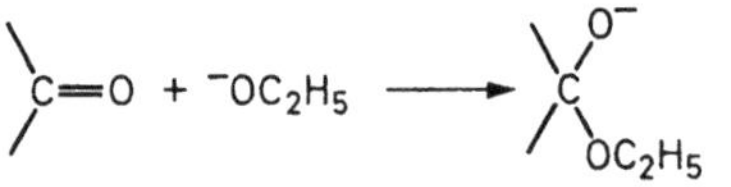

Invariably, more oxygen is chemisorbed on oxidized carbons than can be accounted for by the quantity of these acidic groups (and the basic groups always present, too). Very little is known of the chemical nature of this oxygen. Very likely, some is bound in ether-like bridges,

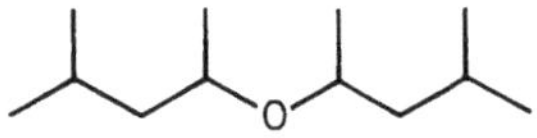

and in the form of carbonyl groups which do not react with sodium ethoxide.

Of course, carbonyl groups on the edge of the carbon layers may be joined pairwise by a resonance system, similar to that described for the basic surface oxides, giving rise to quinone-like behaviour. Analogously, two hydroxyl groups may react similar to hydroquinones. Such structures produce redox properties similar to the quinone/hydroquinone system. The presence of such structures has been inferred from the electrochemical reduction/oxidation behavior of carbon-containing surface oxygen [521–525]. There is also chemical evidence for the existence of quinone structures [452, 526, 527].

Surface oxides have also been prepared by etching of the carbon in a microwave plasma discharge in an oxygen atmosphere [525]. It is assumed that the same surface groups form as are present in acidic surface oxides. Such surface groups are also formed when oxygen is chemisorbed on a graphite which had been extensively milled under argon [528].

2. Surface Halogen on Carbon

Halogens react with carbon surfaces either by direct chemisorption or by substitution of hydrogen. The surface of many carbons which have been prepared from organic precursors is covered with hydrogen. When such carbons are treated with chlorine at elevated temperatures, *e.g.* 723 K, chlorine is bound with evolution of hydrogen chloride [3]. Chlorine uptakes of 7.5 mmol Cl g^{-1}, corresponding to a Cl content of ~20 %, have been observed with some carbon blacks. Even with graphitized carbon black, an uptake of 300 μmol Cl g^{-1} has been observed [3]. Bromine is chemisorbed to a lesser extent.

Nikitina *et al.* [529] treated graphite, which had been milled under argon at 77 K, with chlorine. The chlorine uptake corresponded to 1.1×10^{15} Cl atoms per cm^2 of prism face. Chlorine was also found on graphite which had been milled under carbon tetrachloride [530].

Treatment of diamond powder of 20 m^2 g^{-1} surface area with chlorine at 370 to 670 K resulted in a Cl uptake of 205 μmol Cl g^{-1} [531].

Chemisorption of fluorine on carbon is more difficult to control due to the high reactivity of fluorine. Well-crystallized graphite tends to form graphite fluoride at higher temperatures, *e.g.* near 875 K. In this compound fluorine is bonded to — more or less — all carbon atoms of the layers, giving an ideal composition $(CF)_n$. The interlayer distance of the carbon sheets increases considerably [532]. A compound $(C_2F)_n$ can be obtained under specific reaction conditions [532a].

Controlled fluorination of graphite surfaces has been described by Cadman *et al.* [533], who obtained small partial pressures of F_2 or F by microwave discharges in SF_6 or C_2F_6. CF, CF_2 and CF_3 groups were identified using XPS. Essentially analogous results were found when diamond was treated in the same way [534]. Treatment of diamond powder with fluorine gas at atmospheric pressure at room temperature resulted in a chemisorption of fluorine which corresponded closely to the chlorine uptake at 673 K [531]. IR spectroscopy indicated the presence of CF_2 groups.

Bonding of halogen to a carbon surface should make possible the grafting of organic groups by treatment with the corresponding lithium compounds.

3. Sulfur on the Carbon Surface

Treatment of carbon with sulfur at high temperatures, *e.g.* 873 K [535], results frequently in a substantial sulfur uptake, especially with porous carbons. It was shown, however, that only a relatively small quantity of sulfur is chemisorbed; polymer sulfur is formed in the pore system which cannot revert to soluble α-sulfur due to steric hindrance [536]. Sulfur chemisorption can also be achieved by treatment with CS_2 at 873 K [537]. It cannot be excluded that polymer sulfur is also formed by this treatment.

Carbons with chemisorbed sulfur show enhanced adsorption capacity for Hg^{2+} ions [538]. This is probably due to the presence of surface —SH groups.

4. Surface Groups on Silica

The surface of silica consists of ≥Si—OH groups, usually but incorrectly called silanol groups, and siloxane groups, ≥Si—O—Si≤. The existence of silanol groups in various surroundings, *e.g.* isolated, vicinal, geminal (=$Si(OH)_2$) has been proven by infrared spectroscopy (see section II.A.1.a.i). The quantity of silanol groups on the surface of silica remains usually well below the value expected for complete hydroxylation of the surface [495]. The conclusion is that a part of the silica surface consists of exposed siloxane bonds as well. Different forms of silica differ in the degree of hydroxylation.

On heating of a silica sample, neighbouring silanol groups condense with evolution of water vapour. The newly formed "siloxane bonds" are very reactive unless the silica is heated to high temperatures. The reason is that the direction of the *sp*^{3} hybrid orbitals used in bonding of hydroxyl groups to the surface silicon atoms is fixed if the silicon atoms are bound to the bulk by two or three siloxane bonds. In consequence, a strongly polar surface group is formed, and a normal siloxane bond can form only when the SiO_2 network softens sufficiently at high temperatures to allow a re-orientation of the SiO_4 tetrahedra:

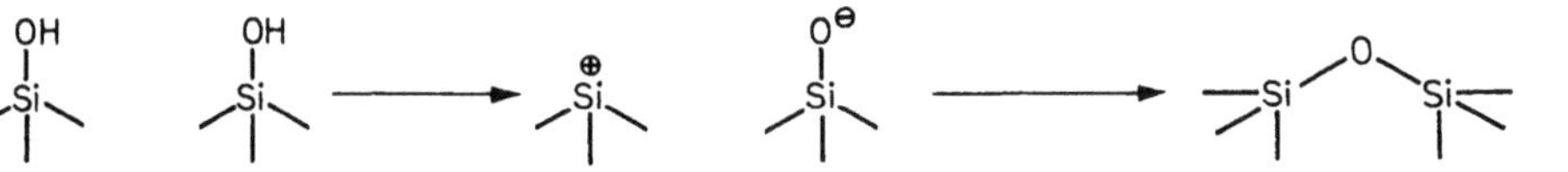

The re-orientation starts at approximately 720 K, and is complete at approximately 1120 K [539]. The polar surface groups are frequently called "strained siloxane bonds". They are very reactive towards many molecules with polarized bonds, giving rise to dissociative chemisorption. Reaction with water results in rehydroxylation, reaction with ammonia gives rise to the formation of ≥SiOH and ≥$SiNH_2$ groups, (see section II.A.1.a.iv). Siloxane bonds which have been annealed at high temperatures cannot react with water or analogous molecules, since the fixed valence direction of the silicon atoms allows bonding of only one oxygen atom in this position.

5. Surface Groups on Metal Oxides

The freshly cleaved surface of oxides is polar. It consists of coordinatively unsaturated metal ions and coordinatively unsaturated oxide ions. Consequently, these surfaces will react with water under dissociation with formation of two hydroxide ions (or hydroxyl groups) [445, 540]

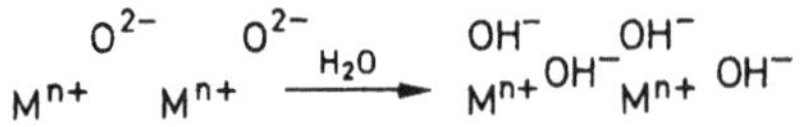

Other molecules with strongly polarized bonds will react analogously. It has been pointed out that this reaction will reduce the electrostatic potential in the surface in accordance with Pauling's electrostatic valence rule [445].

This reaction produces OH groups which will have different coordination to cations — as shown in the reaction scheme above — depending on the crystal face. In the scheme shown above, OH^- groups coordinated to one and to two cations are produced. This difference has been proven by IR spectroscopy (see section II.A.1.a.i). An OH group will be the more acidic. the more metal ions are coordinated to it. One can distinguish acidic and basic OH groups by chemical means on many oxides [540]. The polarizing effect on the OH groups depends on the charge and the size of the cations. In consequence, the basic or the acidic character may be more pronounced, depending on the metal constituent of the oxide. This gives rise to the range observed in the zero-points of charge [541].

The surface groups, mainly the OH functions, are used for the preparation of derivatives with many reagents, the so-called surface modification, to be described below.

B. Reactions of Surface Groups, Secondary Generation of Functional Groups

The surface groups formed spontaneously on the surfaces of solids or produced by simple treatment can be used to produce derivatives or more complex surface groups by chemical reactions. Frequently the surfaces are functionalized with a view on certain applications. Such surfaces are often called *modified surfaces*, the reactions used in their preparation are known as *surface modification.*

1. Modified Carbon Surfaces

The functional groups in the surface oxides of carbon have been identified largely by the preparation of derivates. Analogous reactions have also been used to attach larger and more elaborate molecular entities to the surface. It is obvious, that methods which are standard practice in organic chemistry have been applied to the surface chemistry of carbon.

a) Derivatives of surface carboxylic groups

It is trivial that salts with a large number of cations can be formed because of the acidic character of the —COOH group. The most obvious organic derivatives of carboxylic groups are carboxylic esters. They are formed by refluxing the oxidized carbon with alcohols using an acid catalyst, *e.g.* HCl gas [1, 3, 492, 542] or BF_3 [543]

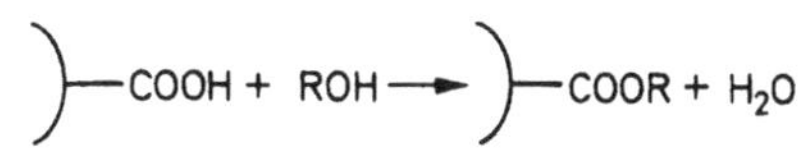

Mostly, methanol has been used. The same degree of methylation as after refluxing with methanol was found after blowing a stream of methanol vapour in dry nitrogen over the carbon at 395 K [3]. Gas phase esterification has been described to be the more efficient, the longer the alkyl chain of the alcohol is [544]. However, the authors used reaction temperatures which were, in each case, a few degrees above the boiling point of the alcohol which increases with chain length, of course.

Methyl esters are also formed in the reaction with a solution of diazomethane in dry ether [1, 492, 493] or with dimethylsulfate [1, 545]. As has been pointed out in section II.B.2, these reagents will also methylate acidic hydroxyl groups, *e.g.* phenolic hydroxyl groups.

A very thorough methylation using diazomethane in dimethylformamide followed by treatment with methyl iodide in the presence of the methylsulfinyl anion has been described for humic acids [546]; also non-phenolic hydroxyl groups were methylated.

Surface esters are susceptible to hydrolysis reactions, of course, but they are quite stable if no acidic or basic catalysts are present. Carbons esterified with *n*-hexanol or with *n*-octanol were hydrophobic and resistant to hydrolysis [544].

Acyl chlorides of caroxylic acids, *i.e.* —COCl groups, are produced by the reaction with thionyl chloride [1, 547], $SOCl_2$, phosphorus pentachloride, PCl_5, or similar reagents

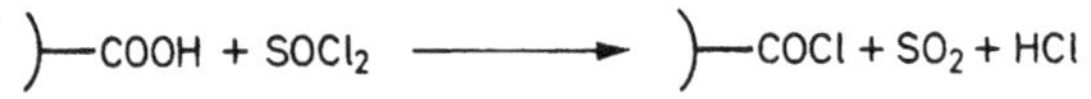

The advantage of using thionyl chloride is that the reaction products SO_2 and HCl are volatile.

Circumstantial evidence indicates that neighbouring carboxyl groups will form carboxylic anhydrides in the reaction with $SOCl_2$ [547]. The samples were usually refluxed with pure $SOCl_2$. However, this reagent tends to decompose at reflux temperature, giving SCl_2, S and other reaction products which might contaminate the surface. Therefore, reaction with a boiling solution of $SOCl_2$ in benzene or toluene might be preferable [548–550]. The acyl chloride groups are highly reactive and can be used for further syntheses.

Acid amides can be obtained by reaction of —COCl groups with ammonia or amines. Ammonium salts of carboxylic groups on carbon form $—CONH_2$ groups above 383 K [551]. At higher temperatures nitriles, —CN, may form.

b) Derivatives of surface hydroxylic groups

Hydroxyl groups in the carbon surface have mostly phenolic character, that means they are weakly acidic. The reaction with diazomethane or dimethyl sulfate has been mentioned above. OH groups react with acyl chlorides with formation of esters

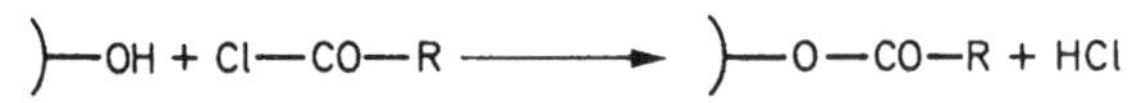

The reaction has been described with acetyl chloride [1] and with *p*-nitrobenzoyl chloride [1, 492]. The reaction with trimethylchlorosilane, Cl—$SiMe_3$, or trichlorosilanes, Cl_3Si—R with R = $—(CH_2)_3Cl$, $—(CH_2)_2—Ph$, is analogous [552].

Phenolic as well as carboxylic OH groups will react in these reactions [547]. The trichlorosilanes have been applied in boiling benzene; it is necessary to work under absolutely anhydrous conditions since otherwise polymers may form by condensation of silanols [552].

Acetic anhydride has also been used successfully for acetylation of surface OH groups [469].

Phenols react also with 2,4-dinitrofluorobenzene [1, 492].

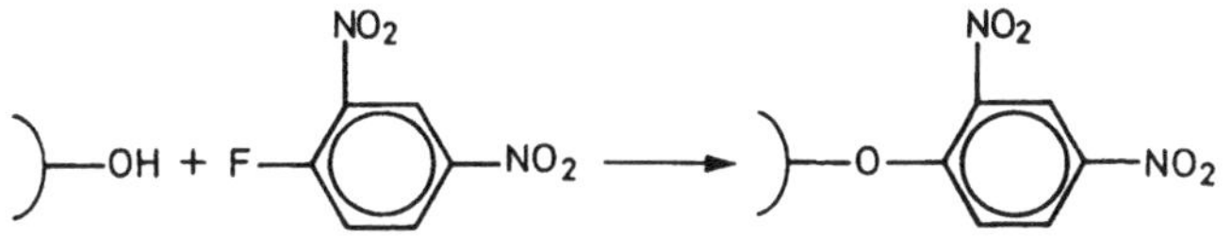

c) Derivatives of surface carbonyl groups

Carbonyl groups, also such of quinones, can be reacted with 2,4-dinitrophenylhydrazine [492, 552] or similar reagents.

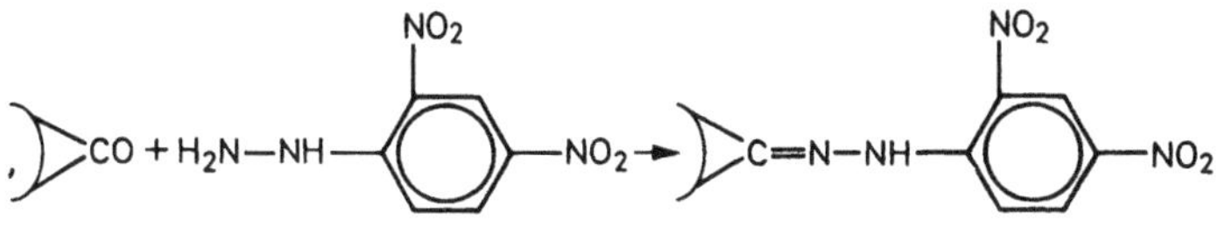

The reaction with hydroxylamine has also been described [469].

d) Other methods of functionalizing carbon surfaces

The surface of carbon blacks is especially rich in hydrogen, but also in other carbons hydrogen is bound to peripheral carbon atoms of the layers. As mentioned in section III.A.2, this hydrogen may be substituted by halogens, in particular chlorine. Substitution by sulfochloride in the reaction with chlorosulfuric acid has been reported [553]. Some of the hydrogen may be replaced by lithium in a transmetallation reaction with butyl lithium in the presence of N,N,N',N'-tetramethylethylenediamine [553]. Lithium organic compounds will also react with carbonyl and carboxyl functions, of course. Nitration of carbon black surfaces with a mixture of concentrated nitric and sulfuric acid or with nitryl tetrafluoroborate, $NO_2^+ \cdot BF_4^-$, has also been reported [554].

Surface groups on the carbon surface are removed by heating to approximately 1250 K. The resulting surface is reactive at lower temperatures, and will chemisorb oxygen or chlorine (cf. section III.A.1. and 2.). Mazur *et al.* [555] concluded that half-filled or empty orbitals should stick out of the edges of the carbon layers, and that olefins should add onto such sites. Carbon fibers ($240\ m^2\ g^{-1}$) were outgassed at 1295 K and 10^{-3} Pa; they adsorbed — after cooling — olefins irreversibly, *e.g.* *iso*-octene, allene, cyclopentadiene, acrylyl chloride, acrylic methyl ester. Subsequent oxygen uptake decreased stoichiometrically. Ethylene and propylene were chemisorbed to a surprisingly small extent, however. The surface treated with acrylyl chloride was able to bind β-amino anthraquinone as shown in the reaction scheme of Figure 49 [555]. This idea was taken up by Oyama

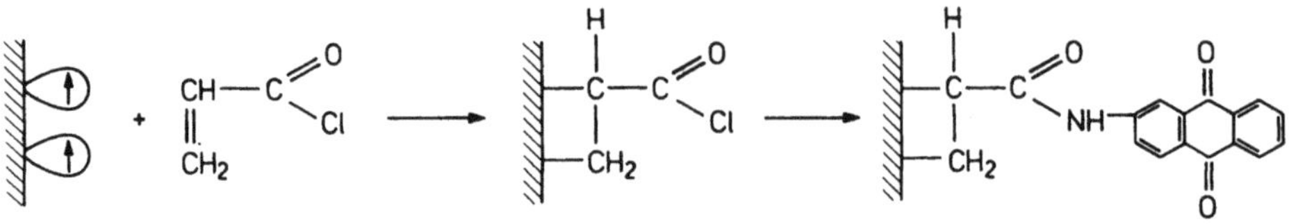

Figure 49. Reaction scheme for addition of acrylyl chloride to surface of outgassed carbon and subsequent reaction with β-aminoanthraquinone

et al. [556] who etched disks of pyrographite in an argon plasma and exposed the reactive surface to ammonia or amines, which were chemisorbed.

e) Anchoring of complex functions or molecules onto carbon surfaces

The most convenient way for connecting complex molecules to the carbon surface seems to be the use of the carboxylic and hydroxylic functions of an oxidized surface as anchoring points. Miller and associates [549, 557] used acyl chlorides on the carbon surface to bind optically active (S)-(—)-phenylalanine methylester via its amino group. This carbon was used as a chiral electrode in electrochemical reduction or oxidation reactions; it produced a few percent excess of an optically active stereoisomer from an inactive starting compound.

Condensation of amine functions with surface —COCl groups has also been used to anchor 4-aminomethyl pyridine to the surface [550]. The nitrogen atom of the pyridine ring was then used to fix ruthenium(II) to the surface, by replacing water in the complex $[Ru(NH_3)_5H_2O]^{2+}$

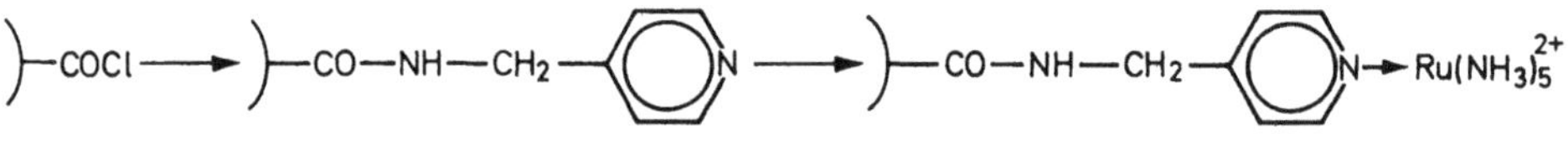

Tetra(aminophenyl)porphyrine has also been grafted onto the carbon surface via —COCl groups [558]. The remaining free amino groups were then reacted with 3,5-dinitrobenzoyl chloride. The N1*s* peak areas in XPS for nitrogen of amino and of nitro groups indicated that two of the amino groups of the porphyrine compound had reacted, in average, with surface COCl-groups. The grafted porphyrine ring was successfully metallized with $CoCl_2$ in acetonitrile as shown by cyclovoltammetric measurements [548, 558].

Zwier and Burke [559] reported recently that they succeeded in grafting *n*-octyl groups onto a carbon used for HPLC by treating first with $SOCl_2$ and then with the *n*-octyl Grignard reagent. The authors concluded from the C1s photoelectron spectra that also phenolic hydroxyls had been substituted by chlorine.

Another way of grafting larger molecules to the carbon surface consists in the formation of amide bonds using either carboxylic functions on the carbon, or amine that has been irreversibly adsorbed on plasma etched carbon [556]. The condensation reaction between —COOH and $—NH_2$ function is facilitated by addition of carbodiimides, usually dicyclohexylcarbodiimide (DCC), which binds the water, released in the reaction, forming a disubstituted urea [560]

$$R{-}N{=}C{=}N{-}R + H_2O \longrightarrow RHN{-}CO{-}NHR$$

Benzidine was thus covalently bonded to the surface of a carbon which had been plasma-etched in oxygen [525]

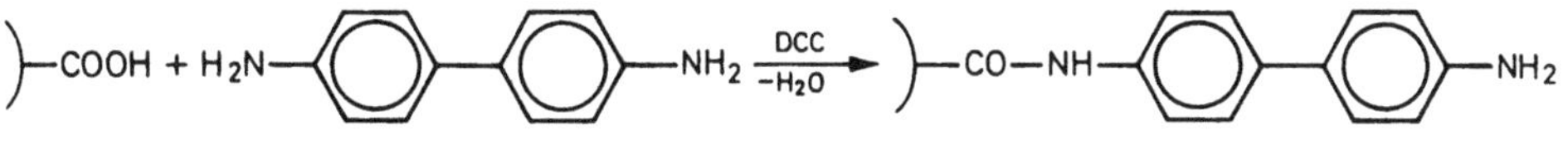

Fixation of *n*-alkylamino groups to oxidized carbon resulted in a hydrophobic surface [561, 562]. Grafting of 4-aminomethylpyridine provided a ligand, again, for binding the pentammine ruthenium(II) complex (see above).

Carbon which had been plasma etched in argon was treated with ammonia or with ethylene diamine. Nicotinic acid was condensed to the NH_2-functions creating again a complex ligand for the fixation of the pentammine ruthenium(II) complex [556]

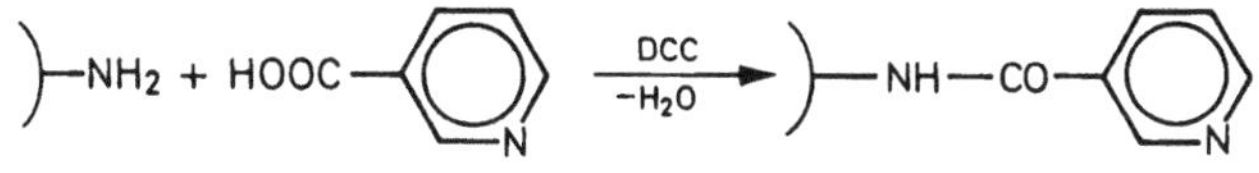

The free carboxyl group of the aqua-[pentahapto-hydrogenethylenediamine-tetraacetato(3-)]-ruthenium(III) complex has been reacted with NH_2 groups on a carbon surface in an analogous fashion [563].

Stable Cu(II) complexes are also formed by carbons with grafted ethylene diamine entities; the surface hydroxyl groups were condensed with $(MeO)_3Si{-}(CH_2)_3{-}NH{-}CH_2{-}CH_2{-}NH_2$ [552].

Ferrocene has been covalently bound to the surface of glass-like carbon which had been etched in an oxygen plasma, using the following reaction sequence [564]

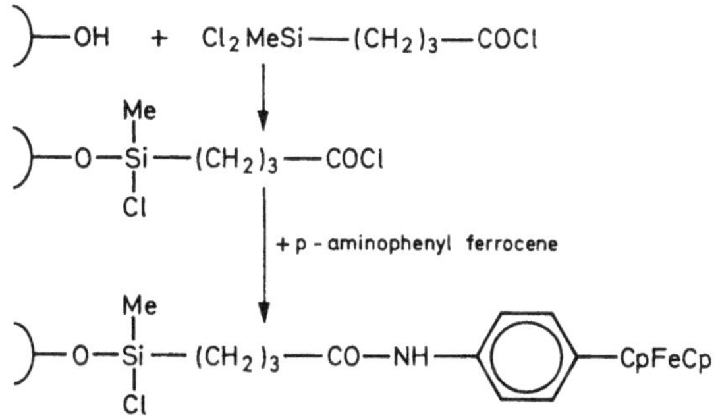

Of course, the 4-(methyldichlorosilyl)butyryl chloride can also react with two neighbouring surface hydroxyl groups; it is also possible that it forms a surface ester with its acyl chloride group; the remaining Si—Cl functions will also react with amines. Reviews of the use of chloro- or alkoxysilanes for functionalizing carbon electrodes have been published recently [565, 566].

These articles describe also other reactions used in grafting functional groups onto electrodes.

Another possibility is to use cyanuric trichloride (trichloroazine) as a multifunctional reactive chloride. It will give a condensation reaction with surface hydroxyl groups. The remaining reactive chloride functions can be used for the binding of alcohols, alkyl or dialkyl amines or hydrazine derivatives [567]

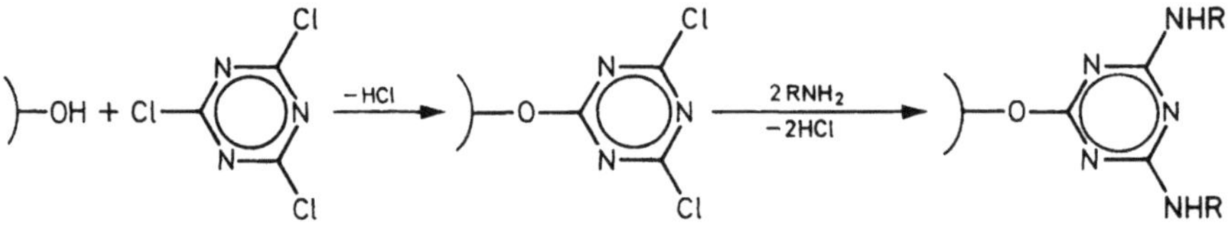

Hydroxymethylferrocene has also been grafted onto the cyanuric chloride-treated carbon [568]. It was shown by measurements of the contact angle of water drops sitting on the treated surface, that it is better first to condense two amine molecules to the cyanuric chloride and to have this molecule react with the oxidized carbon surface [561, 562]. The surface was more hydrophobic when the reaction was done in this sequence. The alkylamine-grafted surface is the more stable towards hydrolysis the longer the alkyl chains are [561, 562]. Reactive dyes on the basis of trichloroazines have been reacted with oxidized carbon fibers (95 m^2 g^{-1}; the excess of unreacted dye was determined spectrophotometrically, and a surface coverage of 43 % was estimated from the quantity of fixed dye [561].

Oyama *et al.* [569] found that *β*-ferrocenyl ethylamine was spontaneously and irreversibly adsorbed on graphite electrodes. The mechanism of this reaction is not understood, possibly the amine functions react with the carboxyl groups on the surface.

Donnet *et al.* [570, 571] found that free radicals, formed by thermolysis of azoisobutyronitrile or lauroyl peroxide, *e.g.*

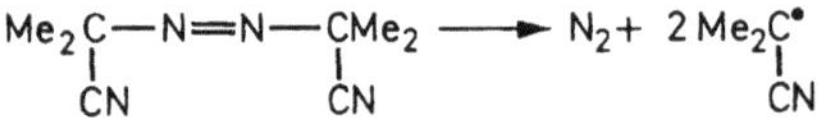

are bound on the surface of carbon black. It is assumed that surface quinones play a role in this reaction.

f) Anchoring of macromolecules onto carbon surfaces

Grafting of macromolecules onto carbon black surfaces has been described in a review article by Donnet *et al.* [572]. The reactions were mostly done with styrene monomer or polystyrene, but isoprene has also been used. Polystyrene is bound on the surface after an induction period by thermal polymerization of the monomer at 368 K in the presence of carbon black [573]. It is assumed that quinones are reduced to semiquinone-like radicals during the induction period. Radicals of the growing polymer may combine with the surface radicals. Grafting of the polymer is also initiated by thermolysis of lauroyl peroxide groups, which form in the reaction of quinones with lauroyl peroxide radicals [574]

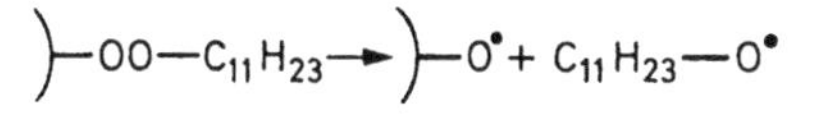

Anionic polymerization of styrene can be induced by lithium organic compounds. The "living polymer" may graft onto a carbon black carrying carbonyl groups and carboxyl methyl esters [575]

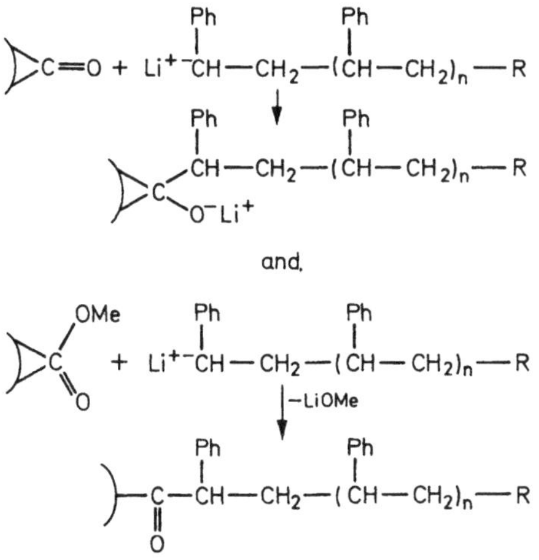

The grafting of the living polymer is the more efficient, the shorter the polymer chain is.

Cationic grafting of poly(tetrahydrofuran) onto the carbon black surface has been described by Tsubokawa *et al.* [576]. Carboxyl chlorides on the surface were treated with anhydrous $FeCl_3$ and tetrahydrofuran at 303 K:

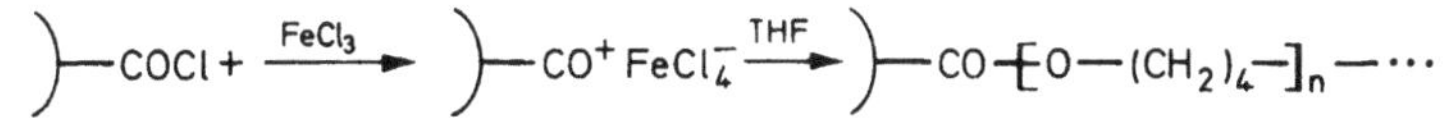

2. *Modification of Silica Surfaces*

The hydroxyl groups on the surface of silica are weakly acidic. Their pK_a value has been estimated to be near 7 [134, 179]. They are susceptible, in principle, to the same reactions as carboxyl groups on a carbon surface. Most of the reactions employed for the modification of silica surfaces make use of the hydroxyl (silanol) groups, *e.g.* chlorination (replacement of OH by Cl) and esterification or other condensation reactions. The siloxane bonds, being much less reactive, have been used comparitively rarely for the binding of surface groups. Surface esters, Si)—O—R, are susceptible to hydrolysis, in principle; therefore, bonding of organic groups to the surface via silicon-carbon bonds is preferable in many applications. Much of the literature on surface functional groups on silica has been reviewed in the books by Iler [577] and by Unger [227].

a) Surface halides and hydrides

Surface chlorides: These are formed in the reaction of hydroxylated silica with chlorides such as $SOCl_2$, SCl_2, S_2Cl_2, $COCl_2$, CCl_4. A comparative study [578] has shown, that $SOCl_2$ is the reagent of choice which reacts most easily and with an acceptable rate. The advantage of this reagent is that gaseous reaction products are formed:

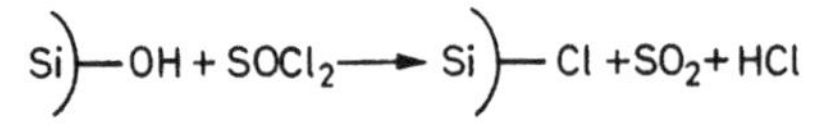

Originally, the reagent has been applied under reflux conditions [579, 580]. However, the silica is often darkened because $SOCl_2$ decomposes partly under normal reflux forming various decomposition products, *e.g.* S, S_2Cl_2, SO_2 [581]. Furthermore, it was found that the reaction remains incomplete under these conditions; less than half of the SiOH groups were converted to Si—Cl; the remaining active hydrogen added up to the original silanol content (see Table 17). Complete chlorination can be accomplished by using a solution of $SOCl_2$ in benzene [582], as suggested originally by Deuel [583]. The ratio of benzene to thionylchloride has to be 4:1 [578]. Almost all hydroxyl groups were replaced when gaseous $SOCl_2$ was used at 673 K, 3,5 Cl atoms were introduced per nm^2 [584].

Chlorination with refluxing S_2Cl_2 resulted in similar chlorine contents as obtained with pure $SOCl_2$ [578]. Peri [121] reported that CCl_4 vapor reacts with surface silanols at 623 to 873 K, forming $COCl_2$, HCl and Si—Cl. Likewise, Hair and Hertl [585] described complete substitution of the OH groups by chlorine in the reaction with CCl_4 at 673 K using Cab-o-Sil which had been outgassed at 1073 K. However, a substantial part of the silanol groups originally present on a silica surface are lost by condensation at the high reaction temperatures. Experiments in which Aerosil silica was heated with CCl_4 to 423–623 K either in gas phase or in an autoclave, resulted in a very incomplete reaction [578]. The results of these and a few other experiments are presented in Table 17. Maximal chlorination to 3.4 Cl nm^{-2} was observed at 873 K [584]. However, setting in just near 673 K, siloxane bonds are attacked by CCl_4

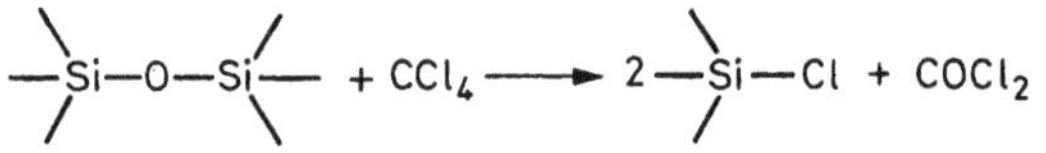

This reaction becomes important at 873 K [584]. The silanol groups remaining on the surface at 970–1170 K are quantitatively converted to Si—Cl by Cl_2 gas [121]; this reaction is slow and incomplete at 673 K [585]. Reaction of hydroxylated silica with $SiCl_4$ leeds also to fixation of reactive chlorine in the form of $=SiCl_2$ and $-SiCl_3$ groups [121, 582, 584, 585].

The Si—Cl groups are very reactive. They hydrolyze rapidly in moist air. Reaction with NH_3 leeds to formation of Si—NH_2 [121]. Alcohols form alkoxy esters [583]. Lithium organic and Grignard compounds can be reacted with Si—Cl, grafting the organic radical to the silica surface

Table 17. Reaction of Aerosil (222 m^2 g^{-1}) with various chlorinating and fluoriding reagents (data from ref. [518])

Reagent and reaction conditions	Reaction		Bound Cl or F/	Active H/	sum of Cl(F) + active H/
	temperature/K	time/hr	μmol g^{-1}	μmol g^{-1}	μmol g^{-1}
—, outgassed	373	24	—	1057	1057
$SOCl_2$, reflux	348	72	402	621	1023
$SOCl_2$, gas phase	433	14	332	662	994
$SOCl_2$, 1:4 in benzene, reflux	350	78	1100[a]	0	1100[a]
S_2Cl_2, reflux	411	71	370	n.d.[b]	n.d.
CCl_4, autoclave	623	43	184	804	988
CCl_4, gas phase	623	43	217	746	963
$COCl_2$, gas phase	453	17	83	n.d.	n.d.
HF in pyridine	295	18	3500	148	3648
SF_4, gas phase	373	9.5	2275	119	2394

[a] Other Aerosil sample, 138 m^2 g^{-1}, 1275 μmol g^{-1} active H.
[b] n.d. = not determined.

[580, 585, 587]. The experience of one of the authors (H.P.B.) is that Grignard reagents do not readily react with surfaces chlorinated with $SOCl_2$. It is probably advantageous to use $SiCl_4$ as a linking agent as described by Locke *et al.* [585]. These authors have also used chlorinated silica in a direct Wurtz synthesis using butyl bromide

$$\mathrm{Si)\!-\!Cl + Br\!-\!C_4H_9 \xrightarrow{2\,Na} Si)\!-\!C_4H_9 + NaCl + NaBr}$$

Surface fluorides: Replacement of Si—OH groups by Si—F is more difficult, because SiO_2 and HF react to give SiF_4. When Aerosil was treated with a solution of HF in pyridine, a weight loss of 2.7% was observed [578]. The fluorine content of the surface corresponded to 9.5 F nm^{-2}, that is a surface completely covered by fluorine. Similar results were obtained with gaseous sulfur tetrafluoride (see Table 17). Remarkably, these samples were hydrophobic at first; they were wetted by liquid water only in the course of about 30 minutes.

Surface bromides and iodides: These have been described by McDaniel [588].

Surface SiH-groups: These are formed in the thermolysis of Si—OCH_3 groups at or above 875 K [589]. There was also IR evidence for SiH_2 groups. Partial conversion to Si—H has been observed on treatment of chlorinated silica with hydrogen at 975–1075 K [586]. Si—H groups are fixed to the surface in the condensation reaction with $HSiCl_3$ (*vide infra*).

b) Surface esters and surface amides

Surface esters form in the condensation reaction of alcohols with silanol groups on the silica surface:

$$\mathrm{Si)\!-\!OH + HOR \rightleftharpoons Si)\!-\!OR + H_2O}$$

The reaction is easily accomplished by refluxing the silica with primary or secondary alcohols boiling above 375 K, preferably above 400 K. Using a distillation column, the water formed in the reaction is removed from the equilibrium by drawing off the azeotrope [590, 591]. The surface esters are hydrophobic and kinetically quite resistent to hydrolysis if long-chain alkanols are used. For esterification with small-chain alcohols, an autoclave should be used. As Broge [592] reported, repeated autoclaving with anhydrous methanol leeds to a very high degree of methylation. Quantitative methylation of silica hydrogels has also been achieved by prolonged treatment with methanol in a Soxhlet extractor [593]. Quantitative determinations of the degree of esterification under various reaction conditions have been described by Stöber *et al.* [594]. Secondary alcohols can be grafted to the silica surface analogously, but tertiary butanol will dehydrate to butene under the reaction conditions [595]. This alcohol can be bound by reaction with the chlorinated surface [583].

Esterification with ethylene glycol or with 1,3-propanediol in an autoclave has been described [596] as well as the reaction with aminoalcohols, *e.g.* ethanolamine (reflux or autoclave) [597]. Phenols have been bound to the silica surface by refluxing in pure form or in solution in decalin [598]. Mostly, 1.5 to 2 organic hydroxy compounds were bound per nm^2. Silica esterified with 3-hydroxypropionitrile has been prepared for use as a stationary phase in chromatography [599]. Possible reaction mechanisms for esterification of surface silanols are discussed by Low *et al.* [600]. Utsugi *et al.* [601] concluded from the retention of optical activity on esterification with chiral secondary alcohols that the OH groups of the silica, not of the alcohol, leave in the form of water.

Methylation of silanol groups can also be accomplished using diazomethane in diethyl ether, but the reaction remains incomplete [580]. Surface esters form also on reaction of chlorinated silica surfaces with alcohols [583] or with sodium alcoholates [602, 603].

Similar to surface esters are the *surface amides* formed in the reaction of chlorinated silica with amines:

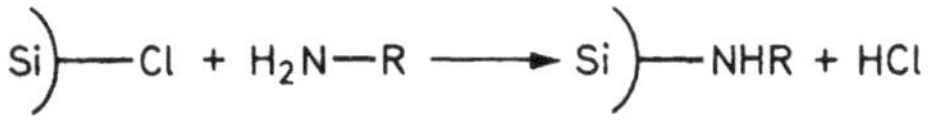

A large variety of such compounds with R= alkyl chain or alkyl chains terminating in a functional group have been prepared by Brust *et al.* [604] for chromatographic purposes. Such surface groups are quite sensitive to hydrolysis; rapid loss of functional groups was observed at *pH* values smaller than 3 or greater than 8.

c) Condensation with chloro- and alkoxy silanes

The most widely used reactions for the modification of silica surfaces involve condensation with clorosilanes or alkoxysilanes. Reaction with trimethylchlorosilane and/or with dimethyldichlorosilanes is widely used for making silica hydrophobic

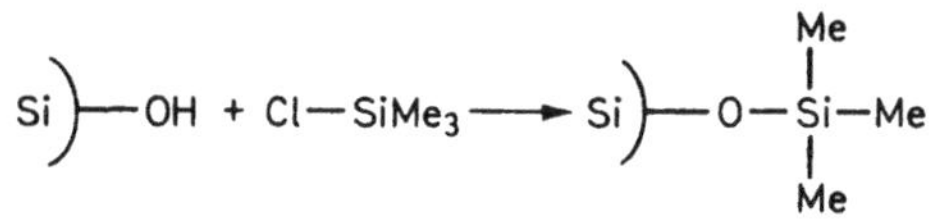

and

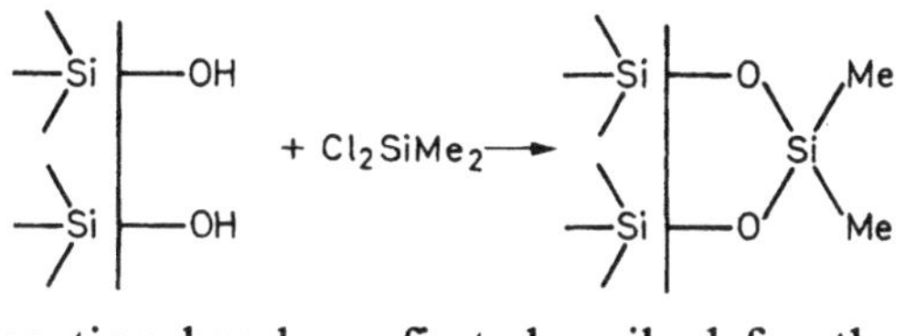

The reaction has been first described for the surface of silica by Kohlschütter *et al.* [605] and by Stöber [606]. Since then, a large number of studies, mainly of spectroscopic nature, have been published. Pyrogenic silicas such as Aerosil are treated in this way in a large scale to produce hydrophobic fillers and anti-caking agents.

For grafting of organic groups, usually compounds of the type X_3SiR are used (X=Cl, OMe or OEt, R=organic group). Of course, also inorganic halides such as PCl_3 [130, 607], BCl_3 [130, 607–609], $AlCl_3$ [139, 607, 609], $GeCl_4$ [130, 205, 307, 607], $SnCl_4$ [130, 607], or $TiCl_4$ [607, 610, 611] react in a similar way with HCl evolution.

One important question to be discussed first is: how many of the reactive halogen atoms or alkoxy groups of one reagent molecule react with the silica surface? This question can be answered by determining the quantities of bound reagent and of the remaining chloride functions on the reagent. In addition, infrared spectroscopic observations of the changes in the surface silanol population are helpful. Koltsov and Aleskowskii [610] concluded that $TiCl_4$ reacted with three OH groups on a fully hydrated silica gel, whilst only two OH groups reacted per fixed $TiCl_4$ molecule when the silica was partially dehydrated. However, nearly all other authors agree that tripodal bonding to a silica surface is very unlikely because of the steric requirements (cf. Peri [612]).

Berg and Unger [613] reacted silica gel with the phenylchlorosilanes $PhSiCl_3$, Ph_2SiCl_2 and Ph_3SiCl, and determined the ratio of phenyl groups (from carbon analysis) and chlorine on the surface. No chlorine was found after reaction with the monochloro compound. Ph_2SiCl_2 had reacted monofunctionally, and $PhSiCl_3$ formed two siloxane bonds to the surface with one chlorine atom remaining on the newly introduced silicon atoms. However, when the silica gel had been heat-treated at or above 975 K $PhSiCl_3$ condensed only monopodically; obviously, the distance between the remaining silanol groups is too large to allow bipodal fixation. Peri treated a partially dehydroxylated silica aerogel with $SiCl_4$ [121, 612] or with $AlCl_3$ [612]. Two molecules of HCl were released for each molecule of $SiCl_4$ or $AlCl_3$ taken up; also from the ratio of bound aluminum and chlorine; it was concluded that these compounds react bifunctionally by 90%, provided the silica gel has not been outgassed at too high temperatures, *i.e.* above 875 K. A monopodal binding was prevalent when the silanol population had been reduced to <1.2 OH nm^{-2}. The results of these experiments led to a model of the silica surface containing mainly geminal hydroxyl groups (see section II.A.1.).

Somewhat in contrast to these results, Hertl [614] found approximately 60% of bifunctional and 40% monofunctional binding in the reaction of methyltrimethoxysilane, $MeSi(OMe)_3$, with Cab-o-Sil. Studying the stoichiometry of the reactions of BCl_3, Me_2SiCl_2, or $TiCl_4$ with Aerosil, Armistead *et al.* [58] determined a ratio of bipodal to monopodal binding of approximately 70:30 for a fully hydroxylated surface with 4.6 OH nm^{-2}. After dehydroxylation at 775 ± 50 K, one molecule of reagent reacted per silanol group. Koltsov *et al.* [615] concluded that trichlorosilane, $SiHCl_3$, reacts with two silanol groups. The reaction with $SiHCl_3$ offers the advantage that both, Si—H and Si—Cl, can be determined with good precision. Hydrogen can be measured volumetrically after treatment with alkaline solutions, using the Zerewitinov apparatus (see section II.B.1.b.):

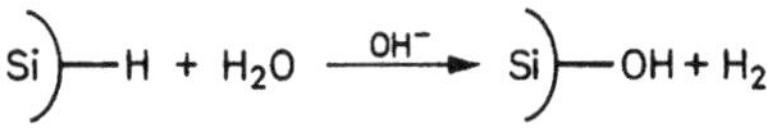

Janssen [578] used this reaction to determine the ratio of monofunctional and bifunctional reaction on Aerosil which had been outgassed at 453 K. It follows from the data in Table 18 that 935—580 = 355 µmol g^{-1} of $-SiHCl_2$ groups and 580—355 = 225 µmol g^{-1} of $=SiHCl$ groups had been bound (39% of reacted silane), and that $1030 - 355 - (2 \times 225) =$ = 225 µmol g^{-1} of unreacted OH groups remained on the surface. Obviously, the unreactive OH groups were not accessible to the reagent because of steric hindrance by the groups already bound. Such groups could possibly be geminal OH groups. A reaction of one trichlorosilane with a geminal OH pair

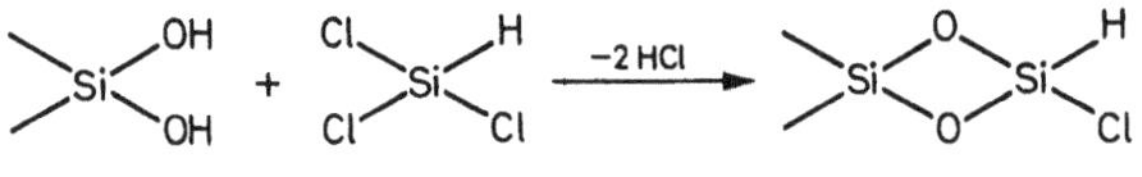

is extremely unlikely, because the four-membered ring is highly strained. Fibrous silica with the SiS_2 structure is only stable under anhydrous conditions; upon access of water, the Si_2O_2 four-membered rings are opened with formation of silanol groups [616].

Table 18. Reaction of Aerosil (222 m^2 g^{-1}) with trichlorosilane[b,c]

Active H after outgassing at 373 K:	1060 µmol g^{-1}
Active H after outgassing at 453 K:	1030 µmol g^{-1}
Si—H after reaction with $SiHCl_3$	580 µmol g^{-1}
Si—Cl after reaction with $SiHCl_3$	935 µmol g^{-1}

[a] Corrected for the 4.7% weight increase by the binding of $SiHCl_3$.
[b] The silica had been outgassed at 453 K prior to reaction.
[c] Data from ref. [578].

Hertl claimed that hydrogen-bonded silanol groups were not affected by reaction with $Me_2Si(OMe)_2$ [617] or with $(Me_3Si)_2NH$ [618]. Lower reactivity of hydrogen-bonded OH groups has also been described by Yates *et al.* [619] for the reaction with trimethyl aluminum, whereas Kunawicz *et al.* [611] found just the opposite. With another sample of Aerosil. 45% of the fixed silane was bound bipodally. These results agree roughly with earlier measurements with BCl_3 [620] where a Cl:B ratio of 1.35–1.44 was observed for Aerosil outgassed at 373 to 473 K; one can calculate 56–65% bifunctional reaction from this ratio.

The conclusion is that tripod-like bonding of silanes with three reactive groups is quite unlikely except, perhaps, in the case of very narrow-pore silica gels. On other surfaces, there is bipodal as well as monopodal fixing of the reagent in a ratio which depends on the nature of the silica surface. The higher the outgassing temperature, the more is the binding by only one siloxane bond prevalent. The unreacted Si—Cl or SiOR bonds will hydrolyse to Si—OH when the sample is exposed to moisture and, depending on the distance of the grafted groups, condensation between neighbouring $-Si(OH)_2R$ or $=Si(OH)R$ groups might be possible. It was shown that commercial silanized silicas contain more active hydrogen than the parent silicas [621]. Various authors suggested an after-treatment with trimethylchlorosilane or hexamethyldisilazane in order to minimize the concentration of silanol groups.

Trimethylsilyl groups can also be grafted to the silica surface by reaction with hexamethyldisilazane, $Me_3Si—NH—SiMe_3$ [622, 623]. The reaction has an even faster rate than that with trimethylchlorosilane [624].

Similar reactions with trimethylaluminum have been repeatedly described [609, 611, 619], methane is evolved in the reaction. All authors observed also formation of $\geq$Si—Me groups when the silica had been partially dehydroxylated. This is caused by reaction with "strained" siloxane bridges:

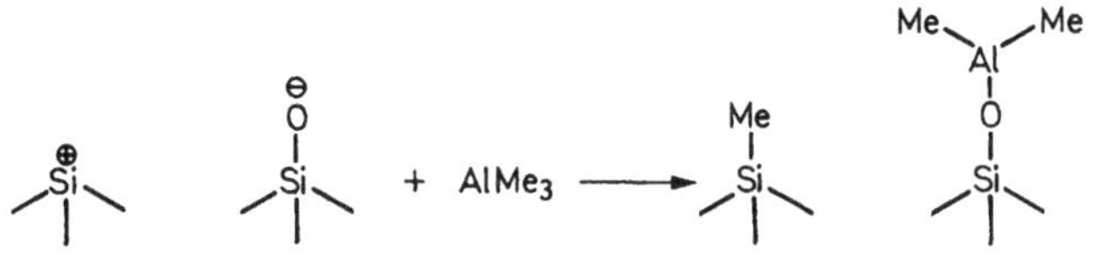

Of course, $R—SiX_3$ molecules with polar Si—X bonds will react in an analogous way [202, 203] (see also section II.A.1.a.iv). A large number of papers have appeared describing the modification of silica surfaces by reaction with trichloro- or trialkoxysilanes carrying various organic groups. The reaction has been used for the preparation of stationary phases in high performance liquid chromatography (HPLC), for special effects on silicon electrodes (which are covered by a thin layer of silicon oxide) and for the immobilization of catalytically active moieties on silica carriers.

Conversion in the reaction with chlorosilanes is more complete at elevated temperatures, *e.g.* 420 to 460 K [227, 624, 625]. In general, the use of alkoxy compounds might seem preferable over that of chlorosilanes because of the formation of less aggressive alcohol instead of hydrogen chloride.

However, it was reported that under identical reaction conditions (boiling xylene solution) more than twice the quantity of $X_3Si{-}(CH_2)_2{-}PPh_2$ was bound on dehydrated silica when X was Cl instead of —OMe; with X = OEt still less phosphine was bound [626]. Also Engelhardt and Mathes [627] found that trialkoxy silanes are less reactive than trichlorosilanes. The modification of silica surfaces for HPLC has been repeatedly reviewed [227, 628, 629].

Chuiko *et al.* [630] treated Aerosil with methyl-chloromethyldichlorosilane, $Cl_2SiMe{-}CH_2Cl$ or with phenylaminomethylethoxysilane, $EtO{-}SiMe_2{-}CH_2NHPh$. Abel *et al.* [631] were probably the first to graft long alkyl chains onto a hydroxylated surface using alkyltriehlorosilanes.

Such surfaces are hydrophobic, they are used in so-called reversed-phase chromatography using more or less polar solvents. On the other hand, by grafting organic moieties containing functional groups, mostly in terminal position, special interactions with solvates can be used to advantage. To-day many modified-silica adsorbents are commercially available (for a list *see* [227]), in particular with the following organic groups to mention only a few:

phenyl	SIL—Ph
3-aminopropyl	$SIL{-}(CH_2)_3{-}NH_2$
3-(2-aminoethylamino)-propyl	$SIL{-}(CH_2)_3{-}NH{-}(CH_2)_2{-}NH_2$
2-cyanoethyl	$SIL{-}(CH_2)_2{-}CN$
1,2-dihydroxy-3-propoxy-propyl	$SIL{-}(CH_2)_3{-}O{-}CH_2{-}$ $-CH(OH){-}CH_2OH$

["SIL" is used as an abbreviation for a silicon atom bonded to the silica surface by an undefined number of siloxane bonds, as has been done repeatedly in the literature].

Burwell and Leal [632] found that 3-aminopropyl-silica was a good adsorbent for CO_2 and for SO_2. The reaction with 2-(4-pyridyl)-1-trichlorosilyl-ethane, $Cl_3Si{-}(CH_2)_2{-}$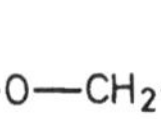N, as well as with octadecyltrichlorosilane, $Cl_3Si{-}C_{18}H_{37}$, has been described by Knox and Pryde [633]. 1,2-Dihydroxy-3-propoxypropyl- and 1-hydroxy-2-amino-3-propoxypropyl-groups are obtained when silica condensed with 1,2-epoxy-3-propoxypropyltrimethoxy silane is treated with dilute acids or concentrated ammonia solution, respectively [634], *e.g.*

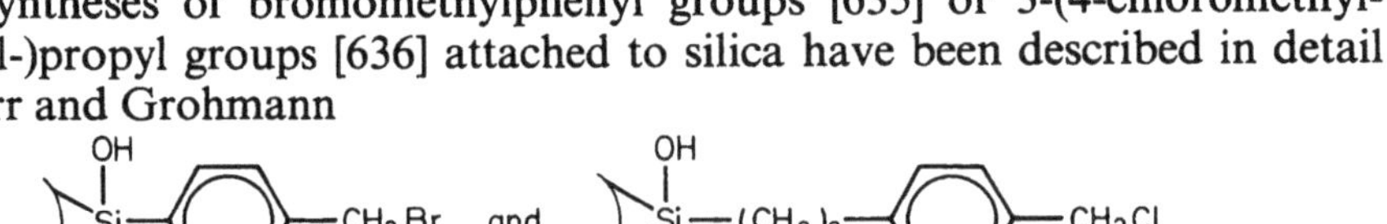

The syntheses of bromomethylphenyl groups [635] or 3-(4-chloromethyl-phenyl-)propyl groups [636] attached to silica have been described in detail by Parr and Grohmann

These compounds were used in Merrifield syntheses of proteins (*vide infra*). Upon hydrolysis these compounds give hydroxymethylphenyl (benzyl alcohol) groups as terminal groups [637].

3-Aminopropyltrimethoxysilane has been substituted in the amino group with chiral substituents, giving *e.g.*

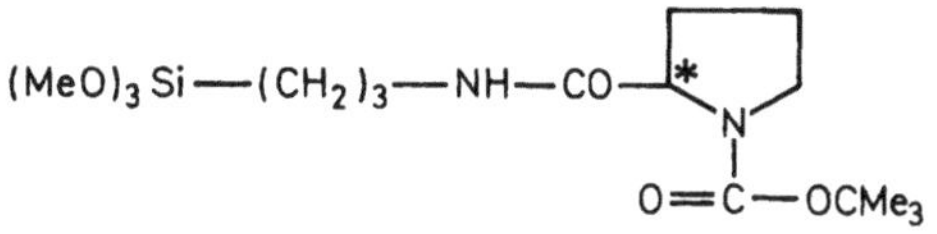

Chromatographic separation of the enantiomers of dansyl amino acids has been achieved using silica treated with such chiral silanes [638].

The amide of L-proline with $(EtO)_3Si—(CH_2)_3—NH_2$ reacted with silica has been used for ligand exchange chromatography; Cu^{2+} was strongly adsorbed from copper tetrammine solutions and it remained on the surface when the mobile phase was acetonitrile/water/ammonia. Various amino acids were successfully resolved into their enantiomers on such column packings [639].

A ferrocene moiety was grafted onto the SiO_2 layer on a silicon electrode using (1,1′-ferrocene-diyl)dichlorosilane [640].

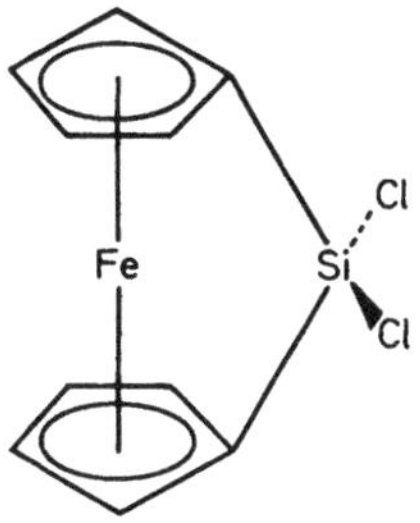

The reagent had been synthesized from 1,1′-dilithioferrocene and silicon tetrachloride.

Preparation of silica with grafted organotin hydride groups has been described by Schumann and Pachaly [641]:

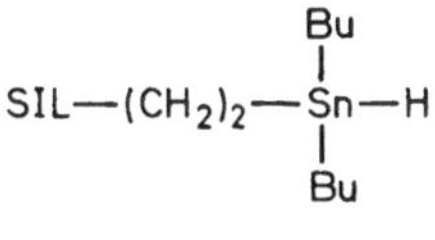

This species can be employed for heterogeneous hydrogenation of organic halides to hydrocarbons. The organotin hydride can be regenerated from the halide formed in the reaction by treatment with di-isobutylaluminium hydride, although a certain loss of activity has been observed.

The grafting of the organosilanes is usually performed by refluxing of the silica with their solutions in dry toluene. It is necessary that the silica as well as the other reagents are absolutely dry, lest polymeric siloxanes form by condensation of silanol groups resulting from the hydrolysis of the trichloro- or trialkoxysilane. Even when solutions of 3-aminopropyltriethoxysilane in anhydrous toluene were used, formation of nodules on planar (oxidized) silicon wafers could not be avoided, as shown by scanning electron microscopy [642]. Very smooth organosilane films were obtained when the wafers were treated with the vapor over the boiling silane. The influence of reaction conditions for maximal attachment of organic functional groups has been studied by Fulcher *et al.* [643]. Treatment with solutions in toluene at reflux temperature resulted in higher loadings than immersion at room temperature in solutions in dry acetone.

The organic moieties require more space on the surface than a silanol group. Only a fraction of the silanol groups present on a fully hydroxylated surface can react, therefore, even when the silane is offered in excess. Stöber [644] noticed already that not more than 2.5 trimethylsilyl groups per nm^2 could be grafted to the surface of Aerosil silica. Similar results corresponding to a binding of 4.0 $\mu mol\ m^{-2}$ were reported by Kiselev *et al.* [645] and by Unger *et al.* [646, 647]. This implies that only approximately half of the silanol groups on the surface can react with $ClSiMe_3$. The size of the $-SiMe_3$ groups is sufficient to cover the surface and its unreacted silanol groups completely. The surface is rendered hydrophobic. In order to reduce the quantity of unreacted silanol groups, the silica (Aerosil) has been heated to 670 K and then refluxed with methanol, prior to treatment with the silane [648, 649]. Only 1.1 silanes per nm^2 reacted with this surface. Without this pretreatment, usually 1.4 to 1.9 organic functional groups are grafted per nm^2 [634]; a list of observed coverages is presented in Unger's book [227]. A loading of 2.0 per nm^2 has been described for $-(CH_2)_3-PPh_2$ groups [228].

The surface area of a fine particle size silica will be significantly changed by the binding of a layer of a thickness which is not insignificant with respect to the particle radii [621, 647]. Obviously, the pore structure of meso- or microporous substances will be influenced, too [647].

Stöber [644] reported that the dissolution of silica in neutral water was strongly inhibited after grafting of trimethylsilyl groups. However, the rate of dissolution increased with contact time. Obviously, there is a slow degradation of the trimethylsilyl cover by hydrolysis. According to Unger [629] the modified silicas with larger groups are fairly stable towards buffered solutions in the *pH* range 1 to 9, although deterioration of packings in chromatographic columns was observed after a few months. Above *pH* 9, the silica is dissolved as silicate. It must be kept in mind that these surface compounds are only kinetically stable in aqueous environment due to the limited accessability of the surface to water molecules. Murrell [228] found, however, that a phosphinated surface to which a rhodium carbonyl complex had been bound was stable in concentrated acetic acid at 423 K.

d) Chemical reactions with anchored functional groups

The organic groups grafted to the silica surface can be further modified. Obviously, amine or phosphine functions can be used for complex formation with transition metal ions in solution. Thus, Pd^{2+} has been bound on $SIL-(CH_2)_3-NH_2$ [650] as well as Cu^{2+} on $SIL-(CH_2)_3-NH-CH_2-$

$CH_2—NH_2$ [231, 651] (see Figure 18). In the latter experiment binding of copper to one as well as to two surface-bound ethylene diamine moieties was observed. Binding to two "en" functions is preferred when the organic moieties are available in sufficiently close neighbourhood. Bonding of numerous other transition metal ions to immobilized "en" has been described [652]. Complexes of $CoCl_2$, $NiCl_2$, $RhCl_3$, and $PdCl_2$ with SIL—$(CH_2)_3$—PPh_2 have also been prepared [653].

Phenyl groups grafted to the silica surface by reaction with $PhSiCl_3$ have been converted to phenylsulfonic acid by treatment with oleum [654, 655]. Only 42% of the phenyl groups present were sulfonated [655]. The product had cation exchange properties. Locke *et al.* [586] sulfonated benzyl groups anchored to the surface employing 20% chlorosulfonic acid in acetone since concentrated sulfuric acid may split Si—C bonds. Chloroform or benzene have also been taken as solvent [587]. The benzyl groups were subjugated to a Friedel-Crafts reaction, using acetic anhydride and aluminum chloride, and to chloromethylation with Cl—CH_2—OMe and zinc chloride [586]. Weigand *et al.* [656] converted anchored *n*-butyl groups to —C_4H_8—SO_2Cl by UV irradiation in the presence of SO_2 and Cl_2.

Amine groups on modified silica have been reacted with carboxylic acids or with acyl chlorides with formation of amide bonds [657]. Amide formation is facilitated by addition of dicyclohexylcarbodiimide (DCC) [560] as has been described in Section III.B.1.e. By forming amides with optically active acylated amino acids stationary phases for the liquid-chromatographic resolution of enantiomers were obtained [658]. The optimal reaction condition for the reaction of SIL—$(CH_2)_3$—NH_2 or SIL—$(CH_2)_3$—NH—CH_2—CH_2—NH_2 with *p*-nitrobenzoyl chloride was found to be refluxing in toluene with a two- to threefold excess of the acyl chloride followed by a "curing" at 343–353 K *in vacuo* [643]. Ferrocene carboxylic acid and tetrathiofulvenyl carboxylic acid have been grafted to a SIL—$(CH_2)_3$—NH_2 layer on a silicon electrode, using the DCC technique [642]. Reaction with succinic anhydride produced a carboxylic function [642]:

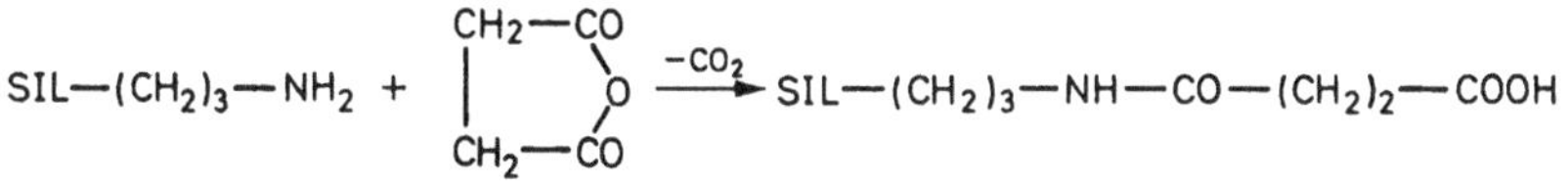

A crown ether has been attached to the silica surface by condensation of the aminopropyl group with chloroformylbenzo-15-crown-5 [659]

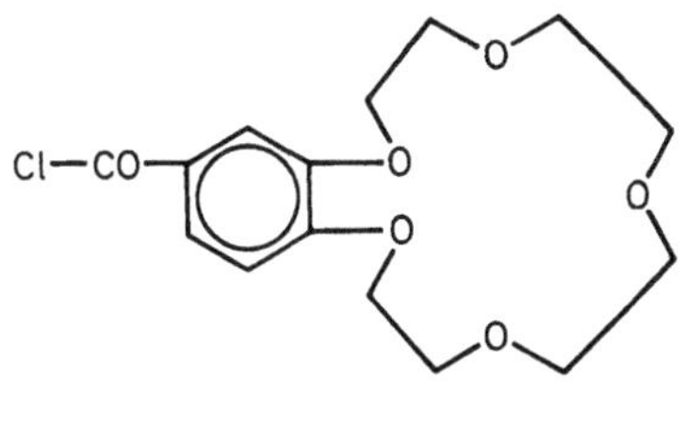

A secondary amine was formed in the reaction with 2,4-dinitrofluorobenzene or 2,4-dinitro-1,5-difluorobenzene [642]. In the reaction with p-isothiocyanatobenzenesulfonic acid a derivative of thiourea was obtained [642]

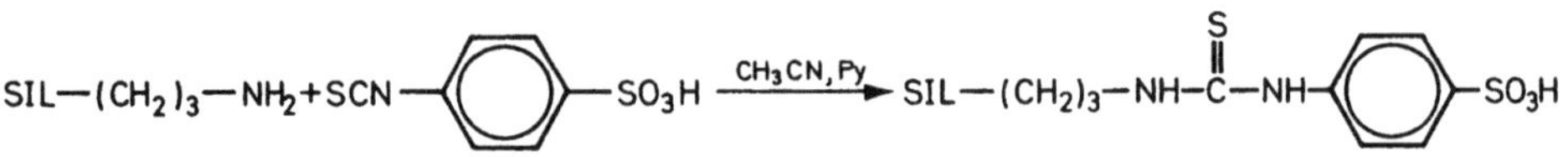

Schiff bases are produced in the reaction of the amine function with aldehydes or ketones, *e.g.* with salicyl aldehyde [660]

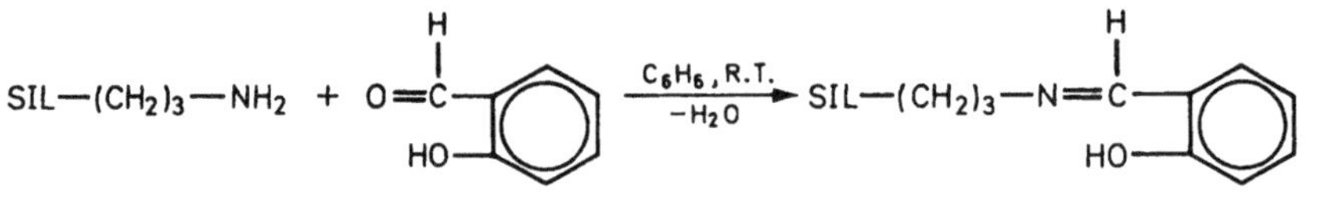

Terminal isothiocyanato groups were synthesized by reaction of the amine groups with thiophosgene, $CSCl_2$, or (better) with carbon disulfide in the presence of DCC and pyridine [642]

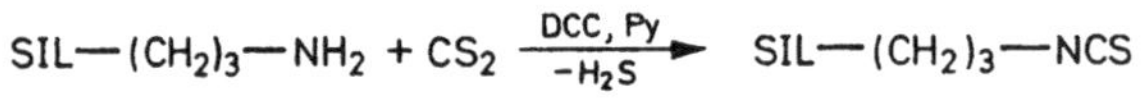

Further reaction with amines in the presence of pyridine leads, again, to derivatives of thiourea.

The terminal chloromethyl group in

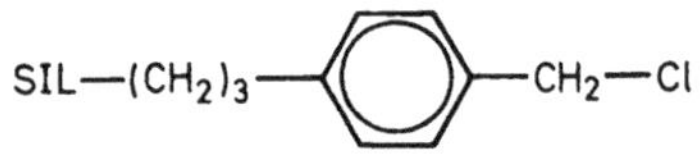

can be hydrolyzed to a benzylalcohol group with sodium carbonate in THF/H_2O, a nitrile group was obtained with potassium cyanide in THF/H_2O [661]. Reaction with dibenzylamine has also been described [662].

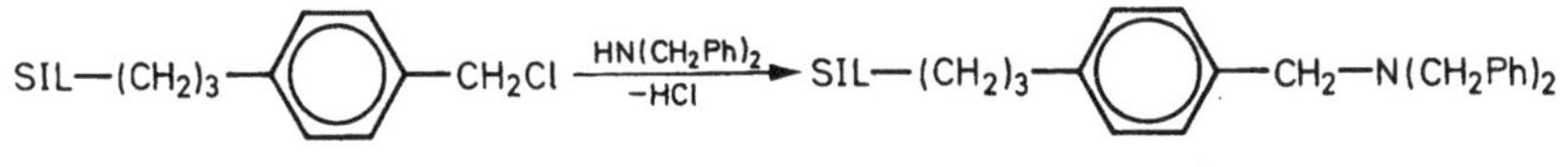

Novotny *et al.* [661] oxidized a grafted olefin to a glycol using 25% hydrogen peroxide in acetic acid:

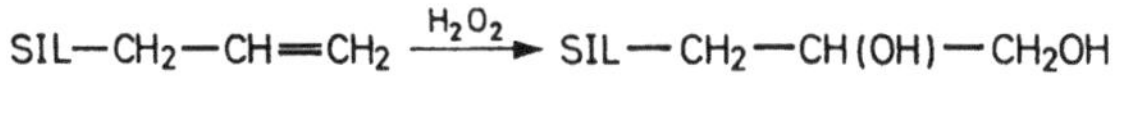

A phosphine ligand on the silica surface has been prepared by reaction of a chloropropyl chain with potassium diphenylphosphide [662]

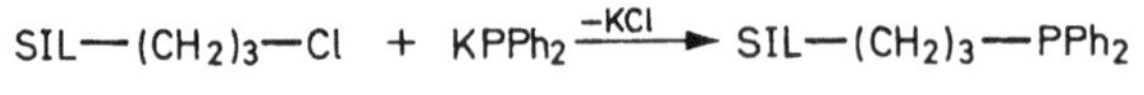

Phenyl groups bonded to silica have been nitrated, the nitro groups were then reduced to amino groups with hydrazine [663]

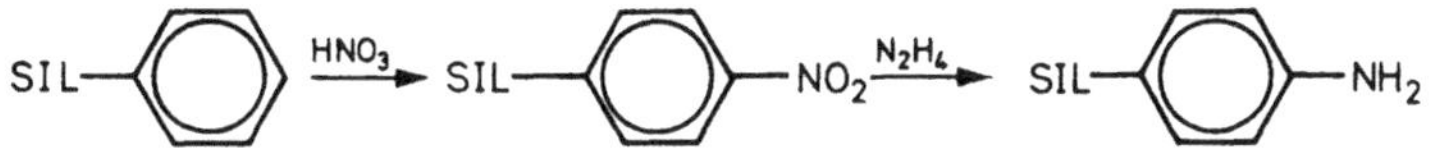

With nitrous acid the diazonium salt was formed which could be coupled with suitable aromatic compounds. Using *e.g.* dimethylaniline, the dye stuff butter yellow grafted to the surface of Aerosil or silica gel was prepared [663]

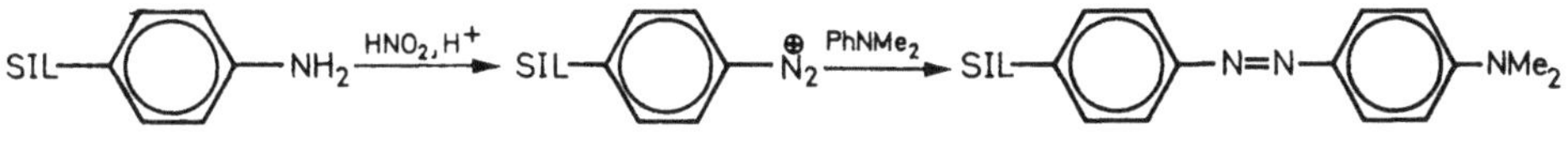

The sample was yellow under neutral solution and turned red on acidification. The dye was dissolved off the silica gel in alkaline solutions. Using a similar approach, Hill [664] and Sugawara *et al.* [665] reduced the nitro groups of *p*-nitrobenzoylchloride-treated SIL—$(CH_2)_3$—NH_2 with sodium dithionite, and coupled with 8-hydroxyquinoline after diazotation:

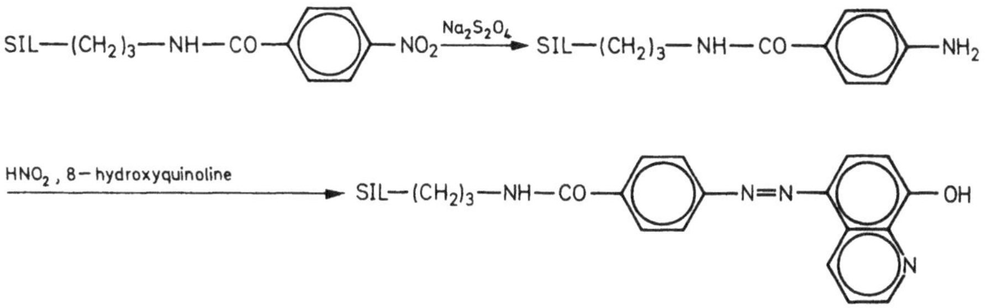

This reaction has been studied in detail by Fulcher *et al.* [643]. The red reaction product is efficient in binding Cu^{2+} ions from dilute solutions.

Kimura *et al.* [659] treated SIL—$(CH_2)_2$—NH_2 with acrylic anhydride, thus producing a grafted double bond

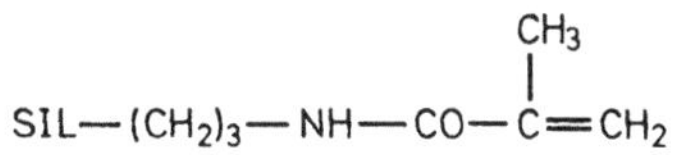

which was co-polymerized — using azoisobutyronitrile as initiator — with methacrylylaminobenzo-15-crown-5

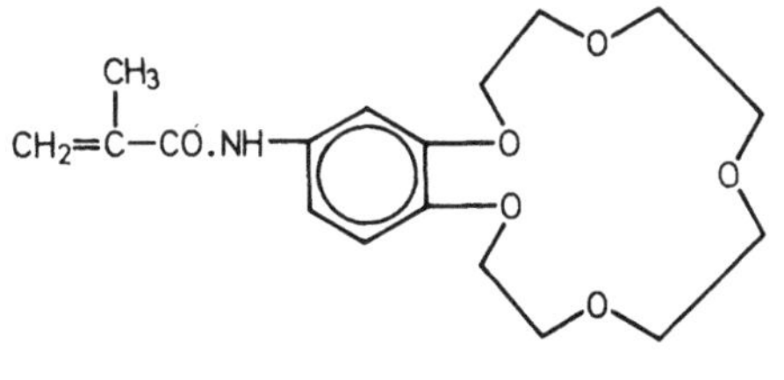

e) Immobilization of transition metal carbonyls on silica with pendant ligands

Modified silica carrying amine and, especially, phosphine groups has found great interest lately for the binding of transition metal carbonyl complexes and metal cluster complexes. The aim of this work, which is founded on earlier studies using functionalized polymers, is to immobilize complexes which have shown good catalytic activity in homogeneous dispersion for hydrogenation, hydroformylation, Fischer-Tropsch synthesis and other reactions of actual interest. Extensive reviews of these studies have recently appeared [229, 666, 667].

The surface-bound phosphine ligands are prepared either by reaction of SIL—$(CH_2)_3$—Cl groups with $KPPh_2$ or by treatment of silica with X_3Si—$(CH_2)_n$—PR_2. The latter reagent is prepared by UV irradiation of vinyl- or allyltrialkoxysilane in the presence of the phosphine HPR_2 [662].

The phosphinated silica is then reacted with the carbonyl compound. Alternatively, the carbonyl may be first reacted with the phosphine trialkoxysilane and then grafted to the silica surface [662]. Horner and Schuhmacher [668] condensed diphenylphosphine carboxylic acids PPh_2—R—COOH with amine groups on silica by the dicyclohexylcarbodiimide method.

Smith *et al.* [669] found that $Ni(CO)_4$ is anchored to the surface with displacement of CO by the phosphine ligand, yielding a light yellow nickel tricarbonyl surface complex:

A nickel dicarbonyl species anchored by two pendant phosphine ligands could be obtained only be reacting the tetracarbonyl with two phosphine-silane molecules which were subsequently condensed onto the surface. The surface nickel carbonyl species lose their CO ligands on heating *in vacuo* [669]. This decarbonylation is a general phenomenon that has been observed with many surface-bound carbonyl complexes. It occurs frequently at temperatures which are lower than the decomposition temperatures of the free carbonyls. The CO evolution is complete at 425 K for the nickel tricarbonyl species. On exposure to CO, the tricarbonyl complex is regenerated. Smith *et al.* [669] demonstrated by a few simple but convincing experiments that the zero-valent metal atoms produced by decarbonylation do not stay on the phosphine groups. The easy recarbonylation shows that they are mobile on the surface. Treatment of phosphinated silica with $[Rh(CO)_2Cl]_2$ yielded

$$\text{SIL—(CH}_2)_n\text{—PPh}_2 \rightarrow \text{RhCl(CO)}_2$$

which has been claimed to be in *cis* configuration on the basis of IR spectroscopic evidence [662]. According to Murrell [228], however, this reaction yields

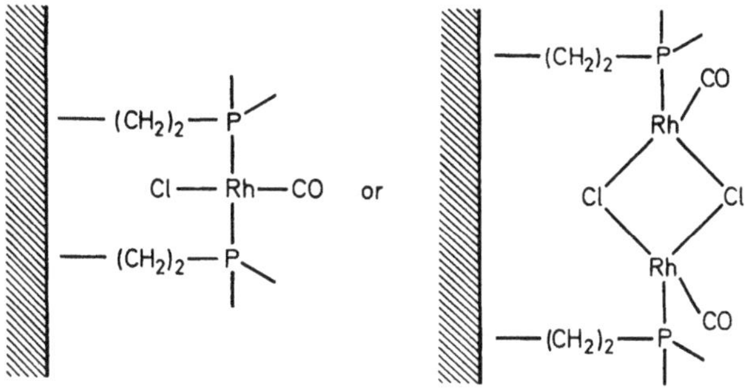

Similarly, di-μ-chloro-bis(1,5-hexadiene)-dirhodium(I) has been anchored to the surface [668], producing an active catalyst for olefin hydrogenation. Reaction of SIL—$(CH_2)_2PPh_2$ with $IrCl(CO)_2$(*p*-toluidine) in the presence of zinc and CO (under 370 kPa pressure) resulted in the formation of a surface-bound Ir_4 cluster [670, 671]

$$SIL{-}(CH_2)_2{-}PPh_2 \rightarrow Ir_4(CO)_{11}$$

Knözinger and Rumpf [648] treated SIL—$(CH_2)_3$—PPh_2 with $Rh_6(CO)_{16}$. A detailed IR spectroscopic study [672] revealed, however, that the cluster disintegrates in this reaction giving mononuclear $L_nRh^I(CO)_2$ moieties bound to the surface (*n* is unknown). Silica with amine functions, *e.g.* SIL—$(CH_2)_3$—NHC_6H_{11}, has also been used with good success in this study. On evacuation the anchored species loses reversibly one CO ligand, forming $L_nRh(CO)$. Exposure of the surface rhodium carbonyl complexes to hydrogen above 370 K resulted in reduction and aggregation of the metal [673]. The metal particles were very small but uniform in size. The $Rh^I(CO)_2$ group has also been observed by Bilhou *et al.* [674] after contacting phosphinated silica with $Rh_6(CO)_{16}$ at 353 K. In addition, rhodium aggregates of unknown nuclearity were observed which gave $Rh_6(CO)_{1x}$ on carbonylation (*x* is probably 3).

A triosmium species was anchored in the reaction of $Os_3(CO)_{11}$(MeCN) with phosphine-modified silica [675]

$$SIL{-}(CH_2)_2{-}PPh_2 \rightarrow Os_3(CO)_{11}$$

$HAuOs_3(CO)_{10}PPh_2$—$(CH_2)_n$—SIL, $H_4Ru_4(CO)_{12-x}(PPh_2{-}(CH_2)_n{-}SIL)_x$ and $H_2Os_3(CO)_9PPh_2$—$(CH_2)_n$—SIL have also been described [676].

When $Ir_4(CO)_{12}$ was reacted with phosphine-on-silica, binding was observed only in the presence of pyridine, leading to a mixture of mono- and trisubstituted Ir_4 clusters [677]. Disubstituted and, finally, trisubstituted clusters form upon evacuation of the monosubstituted cluster at 343 K.

Reaction of $Ir_4(CO)_{11}(PPh_3)$ and $Ir_4(CO)_{10}(PPh_3)_2$ with phosphinated silica at 358 K leeds to

$$Ir_4(CO)_{10}(PPh_3)PPh_2—(CH_2)_3—SIL \quad \text{and}$$

$$Ir_4(CO)_9(PPh_3)\,[PPh_2—(CH_2)_3—SIL]_2$$

respectively [670].

Castrillo *et al.* [678] succeeded in anchoring trinuclear osmium and ruthenium carbonyl clusters onto silica functionalized with SH-groups, SIL—$(CH_2)_3$—SH. The starting compounds were cyclohexadiene-triosmium-carbonyl, $Os_3(CO)_{10}(C_6H_8)$ and $Ru_3(CO)_{12}$, respectively. The structure of the osmium compound $HOs_3(CO)_{10}S—(CH_2)_3—SIL$ is similar to that observed in bonding of $Os_3(CO)_{12}$ onto bare silica surfaces (*vide infra*).

Wild *et al.* [679] prepared cyclopentadiene-modified silica, SIL—C_5H_5 and SIL—$(CH_2)_3$—C_5H_5. These surface compounds were treated with $Co_2(CO)_8$ or $Fe_2(CO)_9$ resulting in reddisch brown

$$SIL \sim C_5H_4Co(CO)_2 \text{ or } SIL \sim C_5H_4FeH(CO)_2\text{, respectively.}$$

f) Direct reaction of transition metal carbonyls with the silica surface

It has also been frequently attempted to react mononuclear or polynuclear transition metal carbonyls directly with the silica surface, that is without the vehicle of pendant phosphine ligands. Such reactions have met with great success on the surface of alumina (*vide infra*). However, all authors agree that the surface of silica is much less reactive than the surfaces of more basic oxides, *e.g.* alumina. At room temperature, the carbonyls are only physically adsorbed, and on heating decarbonylation occurs at relatively low temperatures. It was not possible to identify subcarbonyl species as were found on alumina [440, 680–682]. However, hydrogen evolution was observed on heating $Mo(CO)_6/SiO_2$, and the metal was oxidized, very likely to Mo^{4+} [440]. The light yellow color appearing at 373 K which fades at higher temperatures was taken as indication that subcarbonyls form transitionally [440]. The materials activated near 400 K were catalytically active in olefin metathesis.

Most polynuclear carbonyls are only weakly adsorbed on silica [683]. Robertson and Webb [684], however, concluded that a trinuclear cluster which is stable up to 400 K is bound on warming silica with $Ru_3(CO)_{12}$. Kuznetsov *et al.* [685] reported that this compound is only physically adsorbed at room temperature, but that new bands appear in the infrared spectrum at 358 K. The surface-bound cluster is catalytically active in hydrogenation and isomerization reactions [684]. The surface cluster contained 5 CO molecules under catalysis conditions.

The rhodium cluster carbonyls $Rh_4(CO)_{12}$ and $Rh_6(CO)_{16}$ were reported to be stable on silica surfaces up to 323 K [686], but they are easily decarbonylated by heating to higher temperatures *in vacuo*. Infrared spectroscopic observations on recarbonylation of the thermal degradation product of $Rh_6(CO)_{16}$ led to the conclusion that the Rh_6 cluster remains at least partially intact during decarbonylation [687]. In a more detailed study, Theolier *et al.* [688] found that $Rh_6(CO)_{16}$ was produced at room temperature when $Rh_4(CO)_{12}$ was adsorbed on silica from solution in hexane. This implies that rhodium(0) carbonyl fragments must be mobile on the surface. $Rh_4(CO)_{12}$ was stable on SiO_2 in a CO atmosphere, however. The complex on the surface is sensitive to oxygen, the resulting surface species was postulated to be $Rh^I(CO)_2$, perhaps as

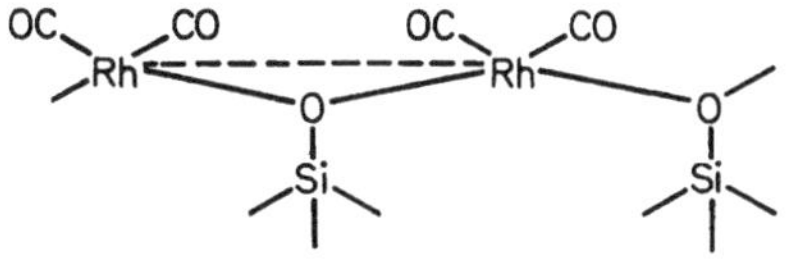

Also with $Ir_4(CO)_{12}$, only very scanty evidence for a surface-bound cluster (after heating to 373 K) has been described [689].

To sum up, although there is substantial evidence for the binding of carbonyls to the silica surface at temperatures in the region of 370 K, the resulting structures are insufficiently defined. The only well-characterized carbonyl cluster attached by chemical bonds to the silica surface is produced in the reaction with triosmium dodecacarbonyl, $Os_3(CO)_{12}$, a complex of particularly high stability. The light yellow surface complex, $HOs_3(CO)_{10}$—O—SIL, is formed in the reaction of an $Os_3(CO)_{12}$ solution in hexane with silica at room temperature [690] (other authors claim that refluxing in *n*-octane [691] or heating to 423 K [692, 693] is necessary). The structure of this surface complex follows from IR spectra and from analogy with the reaction products of $Os_3(CO)_{12}$ with alcohols, phenols, or triphenylsilanol [690–693]

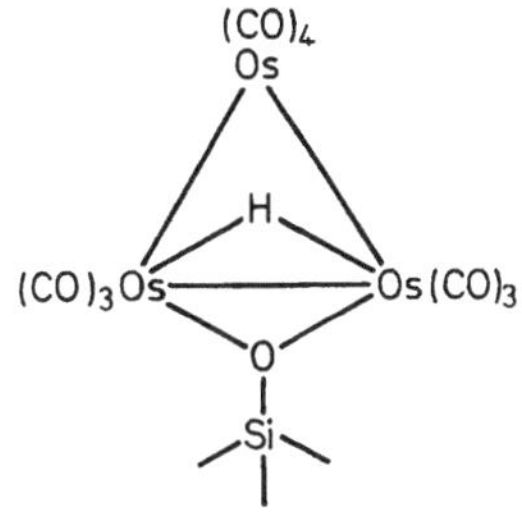

The same surface compound was obtained in the reaction of silica with the cyclohexadiene complex $Os_3(C_6H_8)(CO)_{10}$ [690]. When the triosmium surface cluster is heated to 673 K in helium or hydrogen, carbon monoxide and methane are given off [691], or carbon monoxide and hydrogen according to other authors [693], and a new surface species develops which is stable at 673 K in helium. Its infrared spectrum is similar to that of $Os(CO)_2I_2$. From this analogy, the authors conclude that the structure is similar to this polymeric dicarbonyl osmium compound which is thought to have bridging iodide ligands [690, 693]. The structure would be similar to that of the rhodium complex shown above. The trinuclear osmium surface complex is a Fischer-Tropsch catalyst with a high selectivity for methane [694].

Yermakov [250] found that π-allyl complexes of transition metals react with the surface hydroxyl groups of silica. The surface complex $[Si)—O]_2Mo(C_3H_5)_2$ has been postulated for the reaction with $Mo(C_3H_5)_4$ (see section A.1.a.iv). Similarly, Ward *et al.* [695] reacted triallyl rhodium, $Rh(\pi\text{-allyl})_3$, with silica at room temperature and found fixation of the metal and evolution of propene

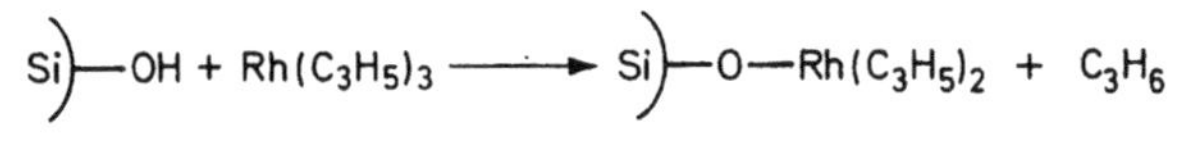

One of the allyl ligands can be replaced by hydrogen [696, 697]

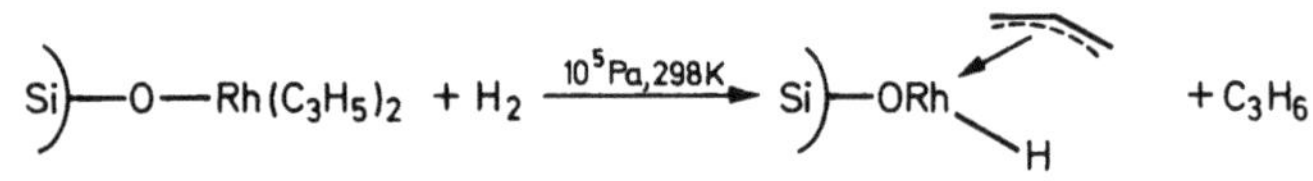

This surface complex is highly active in catalyzing hydrogenation of olefins. It can be transformed by further reactions [697] to

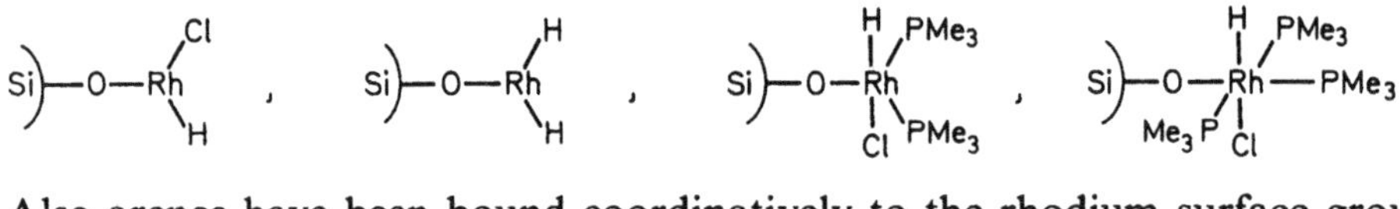

Also arenes have been bound coordinatively to the rhodium surface group [696]

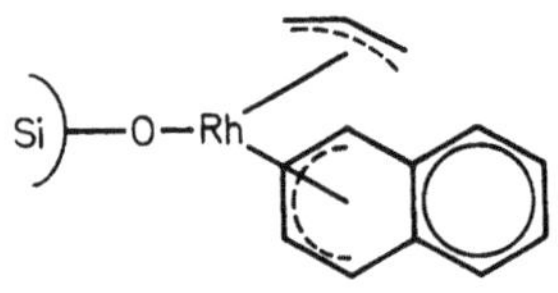

Karol *et al.* [698] found that chromocene, $CrCp_2$, was chemisorbed on hydroxylated silica at room temperature from solutions in hexane or toluene; cyclopentadiene is formed in the reaction. It was assumed that chromium is bound by one as well as by two bonds to surface oxygen:

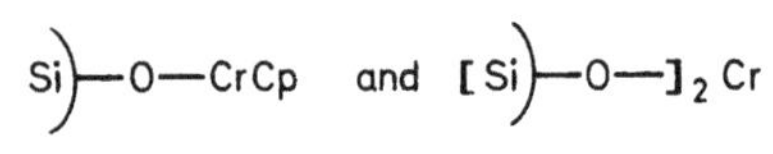

The latter species is identical to that present in Philips catalysts (see section II.A.1.a.iv, II.A.2.a.ii, and II.B.3.b.). The surface compounds obtained with chromocene were also active in polymerization of ethylene.

g) Miscellaneous methods of attaching functional groups to silica

Other more straightforward methods of attaching organic groups to the silica surface comprise the *acetylation* (or acylation) with acetyl chloride or other acyl chlorides [583, 699], acetic anhydride [583, 699, 700], or even free carboxylic acids [699, 700]. *Isocyanates* are also well suited for the binding of organic groups [249, 701]

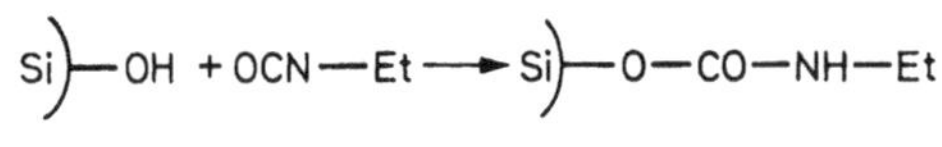

However, substantial quantities of diethyl urea were also produced, especially if the silica had not been heated to higher temperatures [249]. Use of diisocyanates, *e.g.* toluene-(2,4-)diisocyanate, has the advantage that a highly reactive group is anchored to the silica surface which can be used for grafting of other molecules [701, 702].

The reaction of *lithium organic compounds* with chlorinated silica surfaces has already been mentioned. It was found, however, that replacement of surface hydroxyl by chlorine is not necessary. Lithium organic compounds are powerful nucleophilic agents which are able to open up siloxane bonds [703]

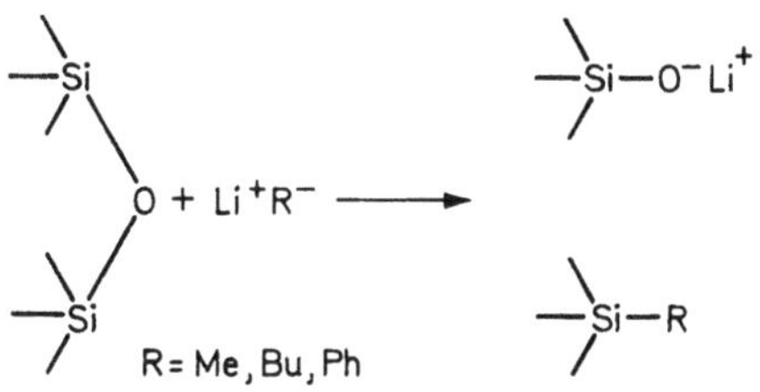

The silica is partly dissolved in this reaction, and the formation of $SiPh_4$, Ph_3SiOH, and $Ph_2Si(OH)_2$ has been observed. The disolution is halted by the accumulation of organic groups on the surface when about one monolayer of SiO_2 is removed. It has been pointed out recently [405] that the diethyl ether used as solvent reacts also with lithium phenyl, giving lithium ethylate and ethylbenzene; the lithium ethylate may also cleave siloxane bonds

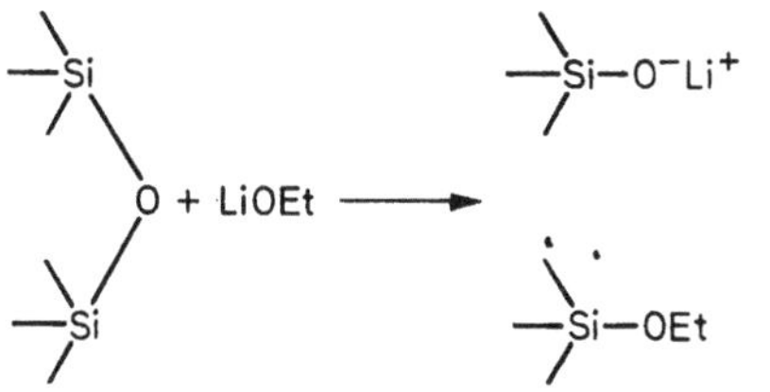

Cleavage of siloxane bonds occurs also on *grinding* of quartz. When quartz was ground in a vibration mill under organic halides or alcohols, formation of surface alkoxy groups has been observed by infrared spectroscopy [704]. Reactive surface groups by rupture of siloxane bonds on drying of a silica hydrogel has been made responsible for the irreversible binding of dyes [705]. The necessary energy is provided by the shrinking of the gel.

Silica with organic groups on the surface can also be prepared by *co-hydrolysis of organotrialkoxysilanes and tetramethoxysilane* or polyethoxysiloxane. These methods have been reviewed by Unger [227]. Partially hydrolyzed solutions of alkyltrichlorosilanes in acetonitrile have been adsorbed on silica; by addition of more water a siloxane layer was formed on the silica surface [706].

h) Anchoring of polymers onto the silica surface

The bonding of polymers to the silica surface has been repeatedly described. Parr and Grohmann [636] used silica with grafted chloromethyl groups

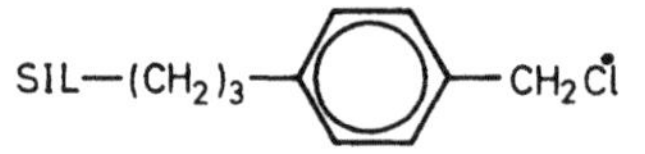

for the Merrifield synthesis of polypeptides. It was found that the splitting off of the finished polypeptide is much easier when the benzene ring is removed from the silicon atom by the propylene group. The synthesis of this group is described in detail.

Dietz *et al.* [707] used aminophenyl-modified silica as initiating agent for the polymerization of N-carboxy-α-amino acid anhydrides

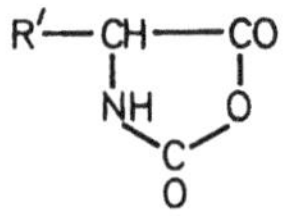

The reaction product is a polyamide grafted to the surface via the aminophenyl group. Fery *et al.* [582] produced surface diazo compounds by treating aminophenyl groups with nitrous acid as described above, and coupling, with thiophenols. The resulting thioazo compounds are thermally labile and produce free radicals on warming above room temperature

If the thermolysis is done in a solution of styrene (at 333 K), polystyrene grafted to the silica is obtained.

Papirer and Nguyen [708] grafted polystyrene to silica by treating the "strained siloxane" groups on a silica heat-treated at 1020 K with a "living polymer", *i.e.* Li^+ $^-CHPh—CH_2—(CHPh—CH_2)_n—R$. Living polymer of polystyrene [709] or of polybutadiene [710] has also been reacted with chlorinated silica. Anchoring of poly(ethyleneoxide) was achieved by anionic polymerization of ethylene oxide with potassium *tert.* butylate and hydrolyzing to $^tBu—(OCH_2CH_2)_n—OH$; this polymer was grafted onto phenylisocyanate groups on the surface [702].

Phenyldiazonium salts on the surface have also been employed to bind alkaline phosphatase in an effort to produce immobilized enzymes [711]. Other methods of grafting enzymes to the surfaces of glass or silica have been dealt with in the literature on immobilized enzymes [230, 712, 713].

3. *Modification of Metal Oxide Surfaces*

As has been pointed out in section III.A.5., the surface of oxides is normally hydroxylated, with hydroxyl groups of more or less acidic character and hydroxyl groups (hydroxide ions) of basic character present. The surface is dehydroxylated by heat-treatment, the temperature of beginning dehydroxylation varying with the oxide, *e.g.* ~375 K for alumina, 425 K for

TiO_2 *etc.* The greatest part of the hydroxyl groups was found to be removed from the TiO_2 surface at 575 K [80, 445]. However, a few isolated hydroxyl groups remain on the surface of some samples at higher temperatures [78, 714–716]. In the case of alumina, a relatively higher proportion of hydroxyl groups is retained after heat treatment to temperatures in the region of 670 K to 870 K [88, 717]. Also on SnO_2, residual OH groups could be detected after heating to 773 K [75]. The dehydroxylated surface is rehydroxylated on exposure to water. Obviously, in analogy to the silica surface, amide groups will form, in addition to hydroxyl groups, in the reaction with ammonia [236], or alkoxy groups with alcohols [718–720], and so on.

The presence of weakly acidic hydroxyl groups makes one expect that the reactions described for the hydroxylated silica surface will also work with hydroxylated oxides. The polarizing effect of the cations is the stronger, the higher their charge and the smaller their radius is. In consequence the hydroxyl groups are the more acidic and their bonding is the more covalent in nature, the higher the valency of the cations in the oxide is (see section II.A.1.a.ii). On the other hand, the presence of basic hydroxyls leads to many reactions which are not observed with silica.

The formation of salts by *ion exchange adsorption* on the Brönsted acid sites and the *hydrolytic adsorption* of multivalent metal ions will not be discussed within the scope of this article. Some aspects have been described earlier [495].

a) Surface halides

Replacement of the surface hydroxyl groups by *chlorine* can be achieved in various ways. Peri [236] found that all infrared absorptions had disappeared after treating alumina with HCl gas at 875 K. However, at lower temperatures the reaction is more complicated [721]. On treatment of TiO_2 (anatase) with anhydrous HCl at 298 K, a quantity of chloride was found on the surface after evacuation that corresponded to the original OH content [445]. It was shown by infrared spectroscopy that the OH groups were not all replaced by Cl; the water produced is retained on the surface, and species like $Ti^{4+}(H_2O)\,Cl^-$ and $H_3O^+Cl^-$ were observed [722, 723]. On outgassing, HCl is evolved, and the chloride content approaches half the original hydroxyl content at 513 K [445]. Presumably, the basic OH^- ions on the surface are substituted with Cl^- at this stage. On further heating, more HCl comes off near 625 K, resulting from the remaining OH groups. In the reaction with $SOCl_2$, HCl and SO_2 are formed which are firmly adsorbed on TiO_2 at low temperatures [445, 714]. The SO_2 content became negligible only after evacuation at 525–575 K, and the quantity of chloride on the surface corresponded to that of active hydrogen on the original surface [445, 495]. The surface of alumina has been chlorinated with CCl_4 vapor at 475 K [723]. The reaction runs very similarly with titania [714]. The reaction begins already at room temperature, and some chemisorbed phosgene, $COCl_2$, as well as carbonate is produced [714, 724]. All hydroxyls were removed at 475 K, and at 575 K reaction with lattice oxygen became dominant, leading to formation of $AlCl_3$ or $TiCl_4$. Some chlorination of alumina has also been achieved by reaction with chlorine gas at 575 K [717].

When TiO_2 (rutile) was treated with anhydrous hydrogen fluoride at 470 to 520 K, more *fluoride* was bound than corresponded to a monolayer. X-Ray diffraction showed presence of $TiOF_2$ [725]. The oxide was completely volatilized as TiF_4 at 670 K. Approximately monolayer coverage was observed in adsorption from 40% aqueous hydrofluoric acid; about a quarter of the surface fluoride could be washed off with water at ambient temperature. Infrared spectra indicated that most of the HF was bound by hydrogen bonding in this treatment.

Surface *isocyanate* species are formed in the reduction of nitric oxide with carbon monoxide on catalysts such as noble metal-on-alumina. This species was also identified by its IR absorption on a SnO_2/CuO mixture [726] or on α-Cr_2O_3 [727]. For comparison, authentic isocyanate was prepared by adsorption of HNCO vapor on the chromia. An eightfold excess of CO was found to be optimal for the formation of Cr—NCO from NO at 723 K. A weak band observed at 2150 cm^{-1} was tentatively assigned to surface cyanide, Cr—CN [727].

b) Surface esters

The weakly acidic character of a part of the hydroxyl groups on oxide surfaces allows also "esterification" with alcohols, *i.e.* replacement of OH by OR [540, 728–730]. Depending on the reaction conditions, only the acidic or all hydroxyl groups on TiO_2 were replaced by methoxyl groups on treatment with hot methanol [540]: by refluxing with methanol only half of the hydroxyl population was methylated whilst all hydroxyl groups were methylated by passing methanol vapor over the oxide at 375 K. In this case the water formed in the reaction was removed from the equilibrium; the same end can be achieved by using a Soxhlet extractor loaded with zeolite 4A, CaH_2 or other good drying agents.

Methoxyl groups on SnO_2, formed by reaction with methanol, are easily oxidized to formate above 320 K, this oxidation is complete at 463 K [731]. Surface esters have also been prepared by reaction of chlorinated alumina surfaces with solutions of sodium alcoholates [717]. Methoxyl groups are also formed in the reaction with diazomethane [3, 495, 540], and benzoxyl groups in the reaction of phenyldiazomethane [3, 495].

c) Reactions of basic surface hydroxyl groups

Carbon dioxide adds onto the basic hydroxides on the surface forming hydrogen carbonate. Infrared absorptions characteristic for HCO_3^- have been observed on MgO [732], Al_2O_3 [733], TiO_2 [78, 734], and SnO_2 [75]. The reaction

$$M^{n+}\rangle\!-\!OH^- + CO_2 \longrightarrow M^{n+}\rangle\!-\!HCO_3^-$$

is reversible [540, 735]. When the basic hydroxide ions on TiO_2 are replaced by hydrogenphosphate ions, the adsorption of CO_2 is drastically reduced, as well as the heat of adsorption [540]. Only sulfite ions, not hydrogen sulfite, were observed after reaction of SO_2 with TiO_2 (rutile) surfaces [714]. This is not surprising in view of the fact that HSO_3^- ions condense to $S_2O_5^{2-}$ on concentration of aqueous bisulfite solutions.

The basic hydroxyl groups are added onto reactive C=C double bonds, *e.g.* cyclohexene, and a surface cyclohexanolate is formed, giving rise to an infrared absorption at 1270 cm^{-1} (C—O vibration) [736]

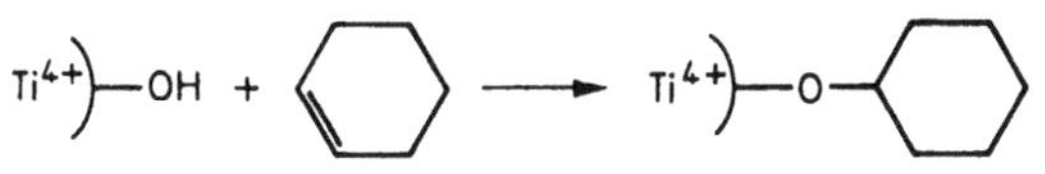

In this case, too, the quantity of CO_2 adsorbed and the heat of adsorption are reduced [736]. An analogous reaction has been described for the reaction of 1-butene with rutile at 423 K [735].

Nucleophilic attack by the basic OH groups on alumina can lead to changes in the adsorbate. Kiselev and Uvarov [737] found acetate on the surface of TiO_2 after reaction with acetone. A similar observation was made with alumina, methane was observed on treatment with acetone [738, 739]

$$CH_3COCH_3 + OH^-_{surf} \longrightarrow CH_3COO^-_{surf} + CH_4$$

This reaction of acetone has also been observed with liquid sodium hydroxide at 525 K [740].

Infrared spectra of various forms of alumina, *e.g.* γ-, η-, ϰ-Al_2O_3, treated with acetone at 423 to 525 K gave indication for the presence of CH_2 groups and of C=C bonds [741]. Obviously, the reaction is not as clear-cut as originally assumed. A 525 K sample was extracted with aqueous 1 N NaOH, and extracted with ether after acidification. There were other acids than acetic acid present in the isolated oily substance; their identification was not possible, however. The smell and insolubility in water showed that six or more carbon atoms must have been present in the acid. Propene, *iso*-butene and butane were detected in the gas phase, in addition to methane [741]. Propene and iso-butene had also been observed earlier [742].

In addition, condensation of acetone is initiated on alumina surfaces, at room temperature [743] or at reflux temperature [741]. Mesityloxide, $(CH_3)_2=$ $=CHCOCH_3$, mesitylene(1,3,5-trimethylbenzene), isophorone, $C_9H_{14}O$, and other compounds have been isolated from the alumina [741, 743]. Condensation reactions of acetone are catalyzed by acids as well as by bases.

Acetate has also been observed after adsorption of acetone or acetaldehyde on SnO_2 [731]. The authors ascribed this to oxidation of the reagents by the oxide; however, it seems very likely that the reaction was analogous to that observed on alumina with acetone, and that acetaldehyde underwent a Cannizzaro disproportionation, a typical reaction of aldehydes in alkaline media. Hexachloroacetone yields trichloroacetate on the SnO_2 surface [744].

Nitriles are converted to surface amides on alumina [72, 192] and on tin dioxide [744]

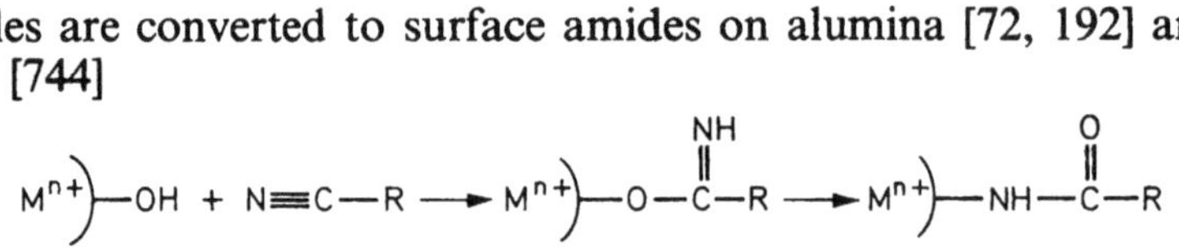

Pyridine reacts on alumina with hydrogen evolution and formation of adsorbed pyridone [72, 193].

The basic surface OH groups react with carboxylic acids producing carboxylate ions on the surface. The carboxylate ions are easily recognized by the strong infrared absorption for symmetric and asymmetric stretching vibrations [8, 9, 14, 15]. Carboxylate ions which have been first observed on alumina were found also on SnO_2 [731] and on Fe_2O_3 [745–747].

Acetylation with acetic anhydride can be used for the quantitative determination of surface hydroxyl groups [540]. In this reaction all hydroxyl groups were acetylated as with gaseous acetic acid at 393 K [540]. After reaction with carboxylic acids the IR stretching frequencies of free carboxylic groups near 1710 cm^{-1} have been observed, too [731, 745, 747]. The authors conjectured that the free acids are bonded to surface Lewis acids by their carbonyl groups or to surface oxygen by hydrogen bridges. Phthalic anhydride reacted only with the basic hydroxyls on TiO_2 and a new carbonyl group was formed in the reaction; twice as many methoxyl groups were formed on methylation with diazomethane than on the original TiO_2 [748]

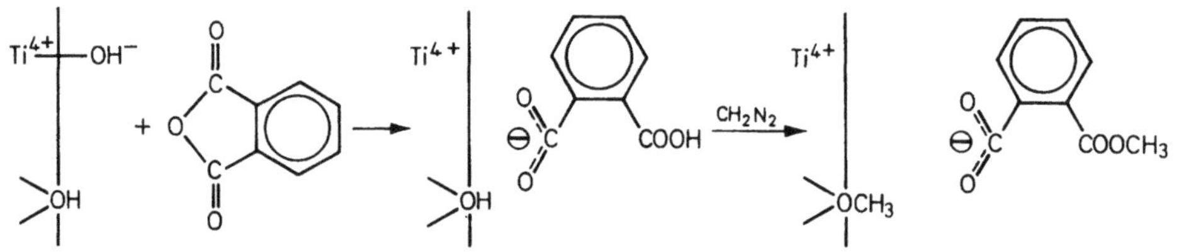

The reaction of carboxyl groups or their chlorides with oxide surfaces, namely SnO_2 surfaces, has been utilized in the attempt to modify electrodes. Fujihira *et al.* [749] found that Rhodamin B which was attached to the SnO_2 surface by an ester bond (condensation by the dicyclohexylcarbodiimide method) was photoelectrochemically more efficient than the same molecule grafted by an amide bond to the silanized surface. The same holds for tetra-(*p*-chlorocarbonylphenyl)-porphyrine, which reacts — very likely — with two of its —COCl groups [750]. Also in this case, the sensitizer was more efficient when it was bound immediately on the surface without a propylene linking group. However, other authors were disappointed by the low solvolytic stability of organic moieties anchored in this fashion onto SnO_2 electrodes [751, 752].

Reaction of isocyanates with isolated hydroxyl groups on SnO_2 at 320 K produces an urethane [753]

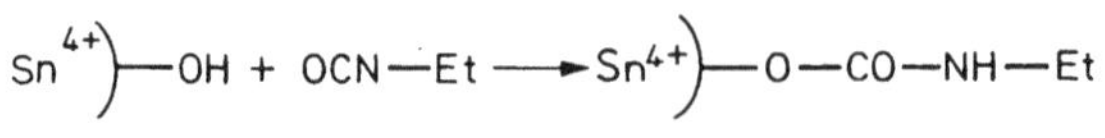

With neighboring hydroxyl groups, however, formation of a di-substituted urea occured which is bound to the surface in "carboxylate" form (more precisely: in the diolate form):

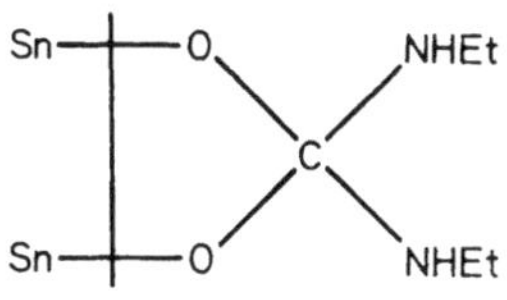

The binding of isocyanates is reversed at higher temperatures; they are less stable on SnO_2 than on silica. Phenyl isocyanate is already desorbed at 320 K [753].

d) Condensation with chloro or alkoxy silanes

The reaction of reactive chlorides with oxide surfaces has been employed repeatedly. The general rules and observations described for the reaction with silica are valid also here. Vidal *et al.* [717] found that $TiCl_4$ reacts with two hydroxyl groups on the surface, the ratio of Cl:Ti on the surface after outgassing was very near to 2. Fox *et al.* [751] condensed $SiCl_4$ onto SnO_2 surfaces and used the remaining Si—Cl bonds for grafting with lithium arenes.

A large amount of work has been done in the last years in modifying SnO_2 or TiO_2 electrodes via reaction with trichloro- or trialkoxy silanes. Mostly, SnO_2 films on glass were used, but also single crystals of rutile (TiO_2). The literature has been reviewed extensively [565, 754]. The problems encountered in this type of reaction have been dealt with in the description of the silanization of silica surfaces (see section III.B.2.c.). There is one notable difference, however: the hydrogen chloride liberated in the condensation is strongly adsorbed on metal oxide surfaces, not on silica [755].

Harrison and Thornton [248] treated SnO_2 with Me_3SiCl and with Me_2SiCl_2. The hydroxyl content was reduced to zero when Me_3SiCl was employed at 470 K. SnO_2 was attacked by Me_2SiCl_2 above 533 K, and $SnCl_4$ as well as polymeric silicone, $(Me_2SiO)_n$, was formed. Some Si—Cl bonds were retained after reaction with Me_2SiCl_2 at lower temperatures. They could be reacted with acetic acid giving rise to $—SiMe_2—O—COCH_3$ groups. The hydrogen chloride formed as condensation product was, at least partially, retained after the reaction; ammonium chloride was found on exposure to ammonia. Adsorption of CO_2 or CO on SnO_2 was inhibited by silanization with Me_3SiCl.

It is very likely, from the analogy with the reaction with silica and the observations in the reaction with oxidized carbon surfaces [558], that — in average — only two of the three active functions in $X_3Si—R$ are used for the bonding [756]. Vapour phase silanization by the method described by Haller [642] produces smooth, reproducible surface films whilst partially polymerized coatings were obtained by grafting from solution — leading to thick, opaque patches on SnO_2 covered glass [751]. The electrochemical behavior of the electrodes is influenced by the technique of surface treatment.

Finklea and Murray [757] found that immersion of TiO_2 in solutions of the silanes for five minutes at room temperature is sufficient for formation of a monolayer. The pH dependence of the flatband potential indicated that a substantial part of the surface hydroxyl groups were not affected. Most authors employed higher reaction temperatures. However, often more than a monolayer was bound to the oxide surface, which might be caused by cross-linking due to incomplete removal of free water. In one case, 83 $\mu mol\ cm^{-2}$ of the silane was bound, corresponding to 5000 molecules per nm^2. The authors [752] of this report had synthesized a tris(bipyridyl)-ruthenium(II) complex which carried a $—(CH_2)_2—SiCl_3$ group attached to one of the bipyridyl molecules. Wrighton *et al.* [758], too, obtained very thick layers using (1,1′-ferrocenediyl)dichlorosilane. This was attributed to oligomerization by cross-linking. However, it seems not improbable that micelle formation on the surface was contributing to the unexpectedly high surface coverage. Only a small portion of the ruthenium complexes on the surface, corresponding to 1.25 nm^{-2}, was photoelectrochemically active [752]. The adsorbed quantity of ruthenium complex decreased to 10% of its original value in the course of the measurements.

XPS measurements [756] indicated that only a fraction of $—(CH_2)_3NH_2$ or $—(CH_2)_3NH(CH_2)_2NH_2$ groups on silanized oxide surfaces reacted with p-nitrobenzoyl chloride. The spectra gave evidence, furthermore, that always a part of the amine functions was present as ammonium ions, even after treatment with bases. On the other hand, not all amines were converted to ammonium salts with dilute hydrochloric acid. Apparently, the organic layer on the surface was very tangled and disordered, and many functions were inaccessible from the outside. Moses *et al.* [756] concluded that ring structures formed by hydrogen bonding to silanol groups:

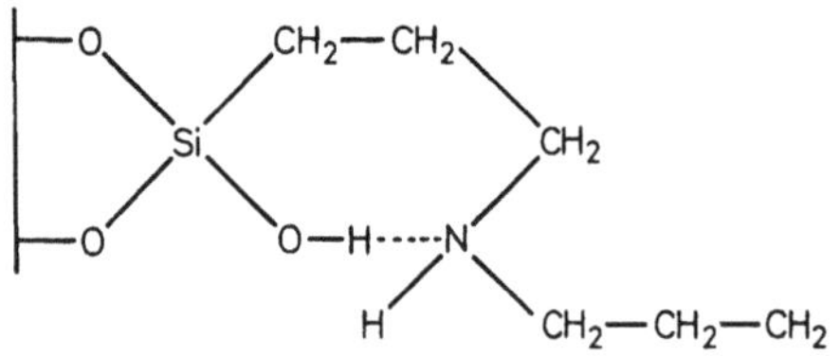

The surface density of grafted chains as determined by White and Murray [759], agreed with the ring model for PtO-on-Pt (1.6 nm^{-2}), but on SnO_2 a packing of 4.5 amine chains per nm^2 was calculated. Employing XPS methods, Willman *et al.* [760] determined the average spacing between the dangling amino groups by condensation with the chlorides of dicarboxylic acids. Oxalyl chloride, Cl—CO—CO—Cl, reacted with one —COCl function only, whilst 75% bifunctional amide formation was observed when the two —COCl groups were separated by at least three $—CH_2—$ units. The separation of the amino groups was estimated to be 400 to 800 pm.

Most researchers agree that the silane coating on oxides is quite stable towards hydrolysis. Moses *et al.* [761] found that a surface film of 3-aminopropyl silane was not destroyed by 1 M mineral acids. It was reduced by 50% after 2–3 hours in hot concentrated hydrochloric acid. Rapid hydrolysis occured in 0.1 M NaOH at room temperature. Similarly, Osa and Fujihira [762] observed good stability, except against strong acids or bases. Hawn and Armstrong [763] could not remove erythrosin bonded via silane groups by prolonged Soxhlet extraction with water, ethanol or acetone. Tomkievicz [764] found an unchanged XPS pattern for a $\equiv Si-(CH_2)_2-PO(OH)_2$ coating on TiO_2 after 90 hours in water; the surface groups were removed by hot 10 N NaOH. However, 50% of iron tetraphenylporphyrine tetracarboxylic acid which had been condensed onto amino silane on the surface was removed by one week's extraction with DMF [765]. Finklea and Murray reported [757] that the organosilane on the surface was not oxidized by holes produced in the valence band by UV irradiation.

Among the groups seccessfully grafted to oxide surfaces are

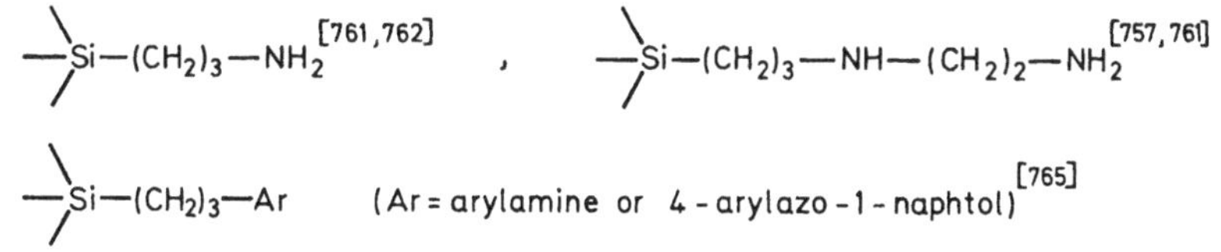

An anion exchanger was produced by reaction of β-trichlorosilyl-2-ethyl-pyridine with SnO_2 films or with a TiO_2 layer on titanium metal and subsequent refluxing with methyl iodide [766]

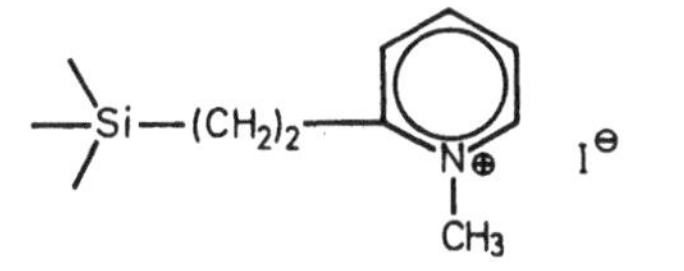

Again, as on silica, amine functions were used for anchoring further groups. Carboxylic acids, *e.g.* Rhodamin B [762], Erythrosin (tetraiodo-fluoresceine) [763], or iron(II)tetraphenylporphyrine-tetracarboxylic acid [767], have been condensed onto the amine functions by the DCC method (see section III.B.2.d.). Erythrosin reacts very easily with $-Si-(CH_2)_2-SH$, the sulfur substitutes for one of the iodine atoms [763]. Firth and Miller [768] grafted (—)camphoric anhydride onto a 3-(aminopropyl)-silane-modified SnO_2 electrode and found that a slight excess of the (—)enantiomer of

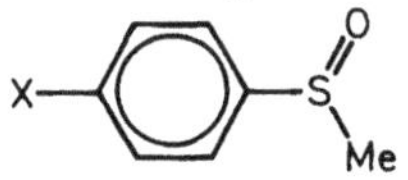

was produced by electrochemical oxidation of the corresponding thiophenol ($X=NO_2$, CH_3).

Cyanuric chloride has been used as a linking agent for the fixation of functional organic moieties also onto oxides [751, 765, 769]. The anchoring is fairly easily destroyed by hydrolysis, unfortunately [769].

Most metals are superficially oxidized on exposure to oxygen or air. This surface film is often dense and adherent, and it protects the metal from corrosion. The oxide film can often be thickened by oxidative treatment of the metal. It has been found that grafting of other groups by the methods described above is also possible on these oxide layers [766, 769].

e) Reaction of transition metal carbonyls with oxide surfaces

Within the studies of the reaction of transition metal carbonyls with solid supports (see reviews [229, 666, 667]) it was discovered that many of these carbonyls are firmly bound on metal oxides, in particular alumina, without a complex-forming pendant group, such as a phosphine group. Some carbonyl groups are lost in the binding, and surface species with fewer CO ligands on the transition metal are formed, which are called subcarbonyls. These subcarbonyls are relatively stable on the surface of alumina or magnesia, quite in contrast to the behavior on silica surfaces. Interest in such surface compounds was intensified by their catalytic activity in the metathesis of olefins or reduction of carbon monoxide to methane.

The difference in behavior of silica and alumina towards carbonyls arises from the presence of basic hydroxide ions on the surface of the latter. Howe *et al.* [680] concluded that stable subcarbonyls can be observed only on surfaces with electron-donating hydroxyl groups, that is surfaces which produce anion radicals with strong electron acceptors. Hugues *et al.* [770] considered the reaction of iron carbonyls with alumina to be analogous to the well-known base reaction of carbonyls [771, 772], which is caused by nucleophilic attack of OH^- ions in aqueous alkaline solutions, *e.g.*

$$Fe(CO)_5 + OH^- \longrightarrow [HFe(CO)_4]^- + CO_2$$

Formation of subcarbonyls was usually observed at temperatures above room temperature, *e.g.* 320 to 373 K. Carbon dioxide is firmly adsorbed on alumina [8, 9], and it is not astonishing that no gas phase CO_2 has been detected. IR bands of surface carbonates have been observed, however [680, 770]. Significant quantities of CO_2 are evolved on further heating. Formation of metal-hydrogen bonds has been proven in only one case: $[HFe_3(CO)_{11}]^-$ ions formed in the reaction of alumina with $Fe(CO)_5$ or $Fe_3(CO)_{12}$, even at room temperature [770]. Many authors described evolution of hydrogen in the course of thermal treatment. At least some of the hydrogen might arise from the reaction of protonic hydrogen (from acidic surface hydroxyl groups) with metal hydride, as has been suggested by Smith *et al.* [773].

Molybdenum hexacarbonyl is only physisorbed on alumina at room temperature, but changes in the IR spectrum were observed already at 318 K [680]. A $Mo(CO)_5$ species formed in a closed system at 373 K under a CO pressure of *ca.* 7 kPa [774], but other species, *i.e.* $Mo(CO)_{4\,ads}$ or $Mo(CO)_{3\,ads}$ were also isolated [774]. The reaction is reversible in CO at 373 K. The golden yellow tricarbonyl species is especially stable, it is the endproduct of decarbonylation at 373 K [252, 775]: further reaction, which is irreversible, occurs above 473 K. Formation of surface-bonded $W(CO)_5$ has been reported by Bilhou *et al.* [776].

Infrared spectroscopic studies by Kazusaka and Howe [253] gave evidence that also binuclear species are formed; see the scheme in section II.A.1.a.iv and Table 11. Analogous species were formed with $W(CO)_6$, whilst only mononuclear species were formed by $Cr(CO)_6$ [254]. Brenner and Burwell

[252] surmised that surface hydroxyl groups or oxide ions fill the vacant coordination sites on the molybdenum atoms, such as in $(\sigma\text{-OH})_3Mo(CO)_3$ on a fully hydroxylated surface. Evidence for coordination of terminal and bridging CO ligands to surface Lewis acid sites has been discussed by Kazusaka and Howe [253].

The hydroxyl content of the alumina surface has been found to be important for the course of reaction on heating from 473 to 773 K [252, 777]. If partially hydroxylated alumina was used (pre-treatment at 723 K) *ca.* 70% of the molybdenum was oxidized to Mo(II), and hydrogen was formed at 543 K

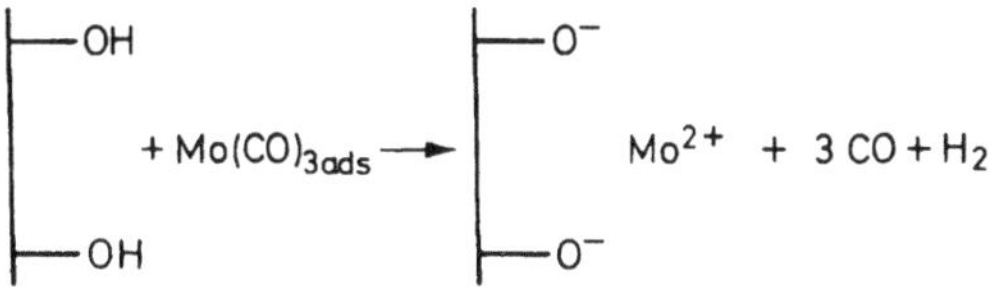

Further oxidation to Mo(IV) was observed at 773 K, methane and small quantities of CO_2 were given off besides hydrogen. If, however, dehydroxylated alumina (pre-treatment at 1220 K) is used, mainly metallic molybdenum is formed at 1220 K in a helium atmosphere. It is probably the oxidized species without CO ligands which is active in catalysis of olefin metathesis [680, 778].

The subcarbonyl species are very reactive. Approximately 90% of the bound molybdenum on a sample activated at 373 K was volatilized as $Mo(CO)_6$ with carbon monoxide at 373 K [774]. Oxygen was reported to displace one CO ligand at room temperature in the surface $Mo(CO)_3$ complex giving $Mo(CO)_2(O_2)_{ads}$; at 358 K $Mo(CO)(O_2)_{2\,ads}$ was formed [252]. This finding has been disputed by other authors [253]. Nitric oxide displaces CO, too, and various metal nitrosyl species are formed on alumina with preadsorbed $Cr(CO)_6$, $Mo(CO)_6$, or $W(CO)_6$ [266]. The metals are oxidized, and N_2O was found as reduction product. The IR and ESR spectra speak for the formation of such species as $Cr^{2+}(NO)$ (probably $Cr^{+}\text{—}NO^{+}$), $Cr^{3+}(NO)$, $Cr^{3+}(NO)_2$, $Mo^{4+}(NO)_2$. It is interesting that the IR spectra agreed very well with those observed on NO adsorption on reduced CrO_3/Al_2O_3 or MoO_3/Al_2O_3 catalysts (see section II.A.1.a.iv).

Analogous surface subcarbonyls were found after reaction of $Mo(CO)_6$ with magnesium oxide [680, 778].

Similar findings as with $Mo(CO)_6$ were obtained when $Ni(CO)_4$ was adsorbed on partially hydroxylated alumina at 300 to 323 K a yellow surface complex, probably $Ni(CO)_{3\,ads}$, formed; it lost more CO on further warming in a He stream [779]. The surface nickel carbonyl complex is very sensitive to oxygen, nickel oxide and carbon dioxide are formed [780].

Subcarbonyl species with zero-valent iron were generated on treatment of alumina with $Fe(CO)_5$, $Fe_2(CO)_9$ or $Fe_3(CO)_{12}$ [781]. Some loss of CO was observed already at room temperature with the di- and trinuclear iron carbonyls. Two CO molecules were lost per cluster when a solution of $Fe_3(CO)_{12}$ (in pentane) reacted at 348 K. Oxidation of iron began at 423 K

[781]; evolution of CO, H_2, CH_4 and CO_2 was observed in TPDE (see section II.A.5.) up to 875 K. The oxidation number of iron was determined by oxygen titration, it was near 2 [781]. Hugues *et al.* [770], however, reported that a deep-red surface compound was formed on alumina or magnesia by adsorption (at room temperature) of green $Fe_3(CO)_{12}$ from solution in hexane. IR spectra indicated the presence of $[HFe_3(CO)_{11}]^-$ ions; this ion could be isolated as its tetraethyl ammonium salt. The generation of surface carbonate during chemisorption was taken as evidence for a base reaction, as explained above. The same species $[HFe_3(CO)_{10}]^-$ was formed, albeit distinctly slower, in the reaction with $Fe(CO)_5$ [770]. Similar results were obtained with ZnO or La_2O_3, but not with silica or silica-alumina. McVicker and Vannice [782] impregnated alumina (pre-treated at 625 K) with solutions of iron, ruthenium or iridium carbonylates, *e.g.* $K_2[Fe(CO)_4] \cdot 2\,THF$, $K[Fe(CO)_2Cp]$, $K_2[Ru_3(CO)_{12}]$, $K[Ir(CO)_4]$. The IR spectra of the preparations showed unchanged complexes. Fischer-Tropsch catalysts with a high selectivity for C_2 to C_5 olefins were derived from these materials.

Anderson *et al.* [683] reported that $Co_4(CO)_{12}$, $Co_3Rh(CO)_{12}$ or $Co_2Rh_2(CO)_{12}$, are very readily adsorbed on hydroxylated alumina at room temperature from solutions in hexane, cyclohexane or chloroform. The infrared spectra indicate disappearance of the bridging carbonyls for $Co_2Rh_2(CO)_{12}$ during adsorption. The terminal CO groups are gradually lost in the course of outgassing above 300 K as well as on standing in air. The authors discuss that possibly initial adsorption occurs by binding of carbonyl ligands to Lewis acid sites on the surface (the oxygen of bridging carbonyls is more basic than that of terminal carbonyl groups). $Ir_4(CO)_{12}$ which has terminal carbonyl groups only, is initially only physisorbed, but loses CO progressively on heating to 320–620 K [683]. This observation was confirmed by Howe [689] who conjectured that surface hydroxyl groups displace CO as ligands.

Smith *et al.* [783] studied the gas evolution on heating of adsorbed cluster carbonyls ($Ru_3(CO)_{12}$, $Os_3(CO)_{12}$, $Os_6(CO)_{18}$, $Rh_4(CO)_{12}$, $Rh_6(CO)_{16}$, $Ir_4(CO)_{12}$) adsorbed on hydroxylated alumina. Various hydrocarbons, methane to propane and ethylene to butenes, were produced on heating, beginning at temperatures as low as 425 K. When the temperature reached 523 K, only methane was liberated. Also copious amounts of CO_2 were observed. When a deuterated alumina surface was used, isotopic substitution in the methane was noticed, giving clear evidence that the surface hydroxyl groups participate in the reaction. The yield on CO_2, CH_4 and higher hydrocarbons increased significantly when the heating was performed in a CO atmosphere [783].

Kuznetsov *et al.* [685] discovered that the binding of $Ru_3(CO)_{12}$, $Ru_4(CO)_{12}$, or $Ru_6C(CO)_{17}$ is strongly influenced by the pre-treatment of the alumina. Maxima in the $Ru_3(CO)_{12}$ uptake were observed for pre-treatment temperatures of 475 and 900 K. This was explained by different adsorption mechanisms; at lower temperatures the reaction is by a base reaction involving basic surface hydroxyls. The decrease in uptake above 475 K is therefore a consequence of dehydroxylation. However, the Lewis acidity of alumina increases with increasing temperature, peaking near 875 K, and this is thought to give rise to the second adsorption maximum; in this case a Lewis acid base interaction is assumed. A down-scale shift of the IR frequencies for $Ru_3(CO)_{12}$ on alumina outgassed below 525 K is taken as indication for a negatively charged surface cluster. This down-scale shift was observed for $H_4[Ru_4(CO)_{12}]$ and $Ru_6C(CO)_{17}$ also on alumina dehydroxylated at 925 K. Changes in the infrared spectrum indicative of formation of $[Ru(CO)_2X_2]_n$ species were observed on outgassing at 373 to 440 K, or on standing in air at ~295 K of the adsorbate of $Ru_3(CO)_{12}$ [685] (the value of *n* is not known). Anderson *et al.* [683] found no adsorption of $Ru_3(CO)_{12}$ on alumina dehydroxylated at 630 K.

The adsorption of the hexarhodium cluster carbonyl, $Rh_6(CO)_{16}$, on alumina has been the subject of several detailed studies [686, 773, 784, 785]. $Rh_6(CO)_{16}$ is obviously only physisorbed on alumina pretreated at 775 K when the complex is adsorbed under exclusion of air [773]. The adsorbed cluster is easily decarbonylated already at room temperature in air. The presence of water vapour is apparently necessary for decarbonylation [773]. Oxygen is taken up during decarbonylation [784]. The oxidized species is very easily recarbonylated under carbon monoxide provided that water vapor is present [773, 784]. The capability for reversible carbonylation/decarbonylation is lost at 473 K [784]. XPS data indicate that finely dispersed rhodium metal is formed [785]. The $Rh_6(CO)_{16}$ cluster is less stable on hydroxylated alumina (outgassed at 298 K) [773]. The first product of decarbonylation contains still residual CO ligands, giving rise to a sharp doublet in the 2000 cm^{-1} infrared absorption region. The two laboratories disagree, however, on the nature of the oxidized species. One group [784] claims that the Rh_6 cluster stays intact during oxygenation whilst the other group [773] interprets their data with the assumption of dissociation of the cluster to

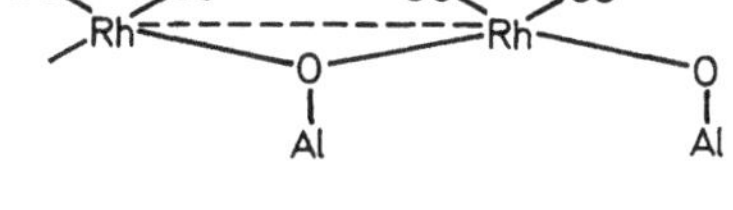

Formation of hydrogen had been observed in the reaction with moist oxygen. The easy reformation of $Rh_6(CO)_{16}$ implies that the rhodium atoms stay in close contact. There was no evidence of bonding of CO ligands to exposed Al^{3+} ions (Lewis acids).

Carbon dioxide is formed during decarbonylation in air, as well as during recarbonylation. The reaction has been studied with isotope-labelled oxygen and carbon monoxide.

Complete loss of CO ligands is observed on prolonged standing in air, but recarbonylation is not inhibited. The following reaction scheme has been postulated for the stepwise oxidation and for recarbonylation [784]

$$Rh_6(CO)_{16} \underset{CO}{\overset{O_2}{\rightleftharpoons}} Rh_6(CO)_{12}O_{12} \underset{+CO}{\overset{-CO}{\rightleftharpoons}} Rh_6O_{12}$$

The species Rh_6O_{12} was interpreted on the basis of XPS measurements to be $Rh_6^+(O_2^-)_6$ [785]. Smith *et al.* [686] concluded that the bridging carbonyls in the cluster are preferentially displaced by oxygen. Reduction of the oxidized species with hydrogen under mild conditions lead to rhodium metal which reconverts easily to $Rh_6(CO)_{16}$ [773]. Rh_6^0 clusters were assumed on the surface because of the easy recarbonylation [784, 785]. $Rh_4(CO)_{12}$ or $[Rh(CO)_2Cl]_2$ adsorbed on alumina are easily decarbonylated, and $Rh_6(CO)_{16}$ is formed on subsequent exposure to CO [686].

The reaction of alumina with $Os_3(CO)_{12}$ has been described in several nearly simultaneous publications [690–692, 786]. The structure of the surface complex has been elucidated unambigously as

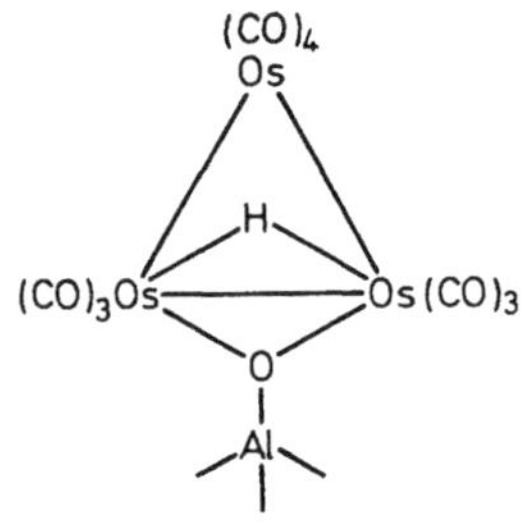

It is analogous to the structure obtained on silica, and it may be inferred that it is the more acidic hydroxyl group on the oxide surface which reacts with the cluster. The Os—Os vibrations have been observed in the Raman spectrum [787]. The same reaction with surface hydroxyl groups has been observed with TiO_2 [690, 691], ZrO_2 [690], ZnO [691], MgO [788]. The reports on the necessary reaction temperature differ somewhat: room temperature [690], 398 K [691, 786], 423 K [692].

The cluster disintegrates on heating at approximately 475 K [692], and two mononuclear species are formed, the IR spectra of which are similar to those of $[Os(CO)_3X_2]_2$ and $[Os(CO)_2X_2]_n$ (X = Cl, Br, I) [691, 692, 786, 787]. These surface species can also be prepared by direct interaction with the mononuclear carbonyl halides [692]. The halogen is replaced by surface oxygen atoms in all likelihood. XPS measurements confirm the oxidation number +2 for osmium in the surface complex [786, 789]. Knözinger *et al.* [789] obtained some evidence that the three osmium atoms of each precursor complex are still in relatively close neighbourhood in the $Os^{II}(CO)_2$ surface complex. This finding has some implication for the role of these materials in catalytic hydrogenation of CO. In the thermal degradation of alumina-supported osmium carbonyl, CO_2, CH_4 and C_2–C_4 hydrocarbons have been observed [692, 783, 786]. The mononuclear osmium species formed in the thermolysis of the $HOs_3(CO)_{10}O \cdot M$ cluster on MgO is apparently somewhat different [788].

In summary, transition metal carbonyl and cluster carbonyl complexes are bound with loss of CO ligands on the surface of oxides such as alumina. The surface hydroxyl groups are involved in this reaction. The structure of the surface complexes is not known in sufficient detail, however. The experimental methods used in these studies involved mainly infrared spectroscopy and qualitative and quantitative analyses of the gases evolved. Few researchers were concerned with a quantitative correlation of available surface area or surface hydroxyl groups and fixed metal entities. In several experiments, chloroform has been used as a solvent in contacting carbonyl solutions with the oxide. This procedure seems contestable, however, since many metal carbonyls are known to be reactive towards chlorinated hydrocarbons [790]. In addition, the surface of alumina — and presumably other metal oxides — tends to react with carbon tetrachloride already at room temperature (see section III.B.3.a); the inertness towards chloroform has never been tested.

f) Surface hydrides and surface organic compounds

The dehydroxylated surfaces of the more basic oxides, ZnO and MgO, show characteristic behavior towards hydrogen and towards hydrocarbons with slightly acidic hydrogen. Dent and Kokes [791] found that hydrogen is chemisorbed on zinc oxide (outgassed at 670 K) producing two types of chemisorbed hydrogen. Type I is formed rapidly, reversibly, and it gives rise to two sharp IR peaks, characteristic for O—H and Zn—H stretching modes. This species is active in catalytic hydrogenation. This work has been summarized in a review [210]. The nature of type II hydrogen, which is adsorbed irreversibly and over a longer period, was explained by Boccuzzi *et al.* [792]. The authors suggested that extremely broad IR bands are caused by bridging hydrogen on oxygen and on zinc:

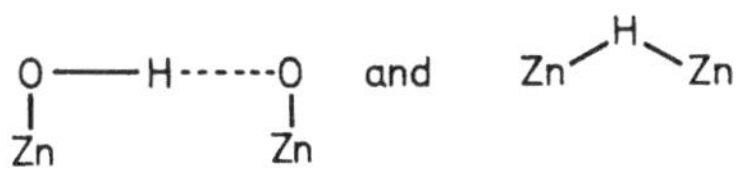

The position of the IR absorption is shifted when D_2 is chemisorbed. Chemisorption by heterolytic dissociation of hydrogen on strongly polar magnesium oxide has also been described [793].

Toluene is slowly chemisorbed on dehydroxylated zinc oxide, giving rise to O—H and Zn—CH_2Ph groups [794]. Hydroxyl groups are also formed in the irreversible adsorption of acetylene on zinc oxide [210, 795]. This is also the case, when the surface is contacted with propene. The organic moiety is bound as a π-allyl complex, however, in this reaction [210, 796].

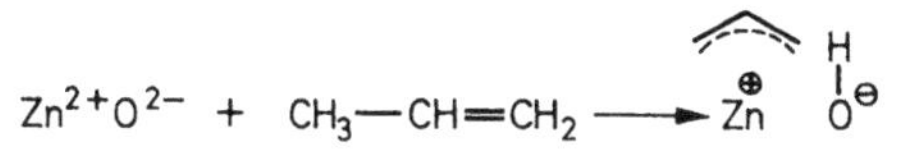

Identical surface species are formed with CH_3—CH=CD_2 and with CD_3—CH=CH_2. The shift for the C=C absorption is somewhat smaller than in transition metal π-allyl complexes. Similar complexes are formed in the adsorption of 1-butene [797]. Ethylene is weakly adsorbed by π-bonding to surface sites [210, 798]. Studies of the adsorption of propyne (methylacetylene) indicated that the reaction is analogous to that of propene, and that the adsorbed species is in equilibrium with the allene anion [210, 795, 799]

$$^{\ominus}CH_2{-}C{\equiv}CH \rightleftharpoons CH_2{=}C{=}CH^{\ominus}$$

The analogous reaction has also been observed with 3-phenyl-1-propyne [800].

g) Anchoring of polymers onto metal oxide surfaces

Polymers have been grafted to metal oxide surfaces, too. An approach used by Nollen *et al.* [801] was to adsorb acrylic or methacrylic acid on TiO_2 from solution in non-polar solvents. Then styrene was polymerized onto the anchoring groups employing benzoyl peroxide as a starter. This reaction has also been untilized for grafting polystyrene onto Cr_2O_3, α-Fe_2O_3, FeO(OH), or Pb_3O_4 [801]. Copolymerization with butadiene has also been achieved. Vidal *et al.* [802] found that polystyrene was grafted to the thin Al_2O_3 layer on aluminium when it was irradiated with a low-pressure mercury lamp under a solution of polystyrene in cyclohexane. The binding was ascribed to the formation of free radicals in the oxide surface. However, it has been found that isobutyronitrile radicals, produced by thermolysis of azo-isobutyronitrile, are adsorbed on TiO_2 [495]. 2,4-dichlorophenyl radicals (from 2,4,2′,4′-tetrachlorodibenzoyl peroxide) and diphenylamine radicals (from tetraphenylhydrazine) are also fixed on the surface [495].

p-aminobenzoic acid adsorbed onto aluminium oxide hydroxide (boehmite) has been diazotized, the enzyme urease was irreversibly attached to the alumina support [803]. Imino-type groups have been produced on alumina or TiO_2 surfaces with cyanogen bromide, which were used for the fixation of enzymes [804]. Many more ways of immobilizing enzymes on oxides are feasible [230].

h) Surface species related to functional groups

Tauster *et al.* [805, 806] found *strong metal-support interactions* (SMSI) when finely dispersed platinum metals on TiO_2, V_2O_3, Nb_2O_5 or MnO were reduced with hydrogen at 775 K. The bound platinum lost its ability to chemisorb hydrogen or carbon monoxide. A similar observation had been described earlier by Ioffe *et. al* [807] after deposition of platinum from bis-π-methallyl platinum on low-valent tungsten ions on a silica surface. Formation of bonds between reduced metal atoms in the oxide surface and the platinum metal was considered [805–807]. This would place such preparations near to surface groups.

Many oxides, *e.g.* ZnO or TiO_2, are *n*-conducting as a consequence of non-stoichiometry. On such oxides with free electrons in the conduction band, *oxygen* is chemisorbed. ESR gave undisputed evidence for the formation of O_2^- [808–810] and O^- ions [808], the latter being preferred at higher temperatures [811–812]. However, the concentration of these anion radicals is so small as compared to the functional groups discussed so far that they are not included in full discussion, even through they play an important role in many catalytic reactions and photo-assisted oxidations.

The surfaces of other solids contain functional groups as well. There are $H_2PO_4^-$ ions on the surface of hydroxyl apatite, or HCO_3^- ions on the surface of calcium carbonate. The chemistry of such surfaces is, in principle, quite similar to the chemistry of silica or metal oxide surfaces, and most of the chemical reactions described in this chapter will be applicable in these cases, too.

References

1. Boehm, H. P., Diehl, E., Heck, W., Sappok, R.: Angew. Chem. **76**, 742 (1964); Angew. Chem. Intern. Ed. **3**, 669 (1964)
2. Johnson, B. F. G. (ed.): Transition Metal Clusters, Chichester, New York, Brisbane, Toronto, J. Wiley & Sons 1980
3. Boehm, H. P.: Adv. Catal. **16**, 179 (1966)
4. Eischens, R. P., Pliskin, W. A.: Adv. Catal. **10**, 1 (1958)
5. Crawford, V.: Quart. Rev. **14**, 378 (1960)
6. Sheppard, N.: Pure Appl. Chem. **4**, 71 (1962)
7. Kiselev, A. V., Lygin, V. I.: Russ. Chem. Rev. **31**, 175 (1962)
8. Little, L. H.: Infrared Spectroscopy of Adsorbed Species, New York: Academic Press 1966
9. Hair, M. L.: Infrared Spectroscopy in Surface Chemistry, New York, Dekker 1967
10. Amberg, C. H.: Infrared Spectroscopy and the Solid-Gas Interface. In: The Solid-Gas Interface Flood, E. A. (ed.), New York: Dekker 1967, pp. 869–893
11. Basila, M. R.: Appl. Spectrosc. Rev. **1**, 289 (1968)
12. Blyholder, G.: Spectra of Adsorbed Species. In: Experimental Methods in Catalytic Research (Anderson, R. B., ed.), New York, London: Academic Press 1968, pp. 323–360
13. Knözinger, H.: Acta Cient. Venez. **24**, Supl. 2, 76 (1973)
14. Kiselev, A. V., Lygin, V. I.: Infrared Spectra of Surface Compounds, New York, Toronto: J. Wiley, 1975
15. Pritchard, J., Catterick, T.: Infrared Spectroscopy. In: Experimental Methods in Catalytic Research, Anderson, R. B., Dawson, P. T. (eds.). New York, San Francisco, London: Academic Press 1976, pp. 281–318
16. Delgass, W. N., Haller, G. L., Kellerman, R., Lunsford, J. H., Spectroscopy in Heterogeneous Catalysis, New York, San Francisco, London: Academic Press 1979
17. Treibmann, D.: Wiss. Z. Techn. Univ. Dresden, **26**, 1043 (1977)
18. Wright, J. C.: Appl. Spectrosc. **34**, 151 (1980)
19. Mantz, A. W.: Appl. Spectrosc. **30**, 459 (1976)
20. Garrison, A. A., Crocombe, R. A., Mamantov, G., de Haseth, J. A.: Appl. Spectrosc. **34**, 399 (1980)
21. Kortüm, G.: Reflexionsspektroskopie. Berlin: Springer 1969
22. Wendlandt, W. W., Hecht, H. G.: Reflectance Spectroscopy. New York: Interscience Publ. 1966
23. Wendlandt, W. W. (ed.): Modern Aspects of Reflectance Spectroscopy. New York: Plenum Press 1968
24. Klier, K.: Catal. Rev. **1**, 207 (1967)
25. Resing, H. A.: Advan. Mol. Relaxation Processes **1**, 109 (1967/68)
26. Fripiat, J. J.: Catal. Rev. **5**, 269 (1971)
27. Pfeifer, H.: Nuclear Magnetic Resonance and Relaxation of Molecules Adsorbed on Solids. In: NMR Basic Principles and Progress. Diehl, P., Fluck, E., Kosfeld, R. (eds.), Vol. 2. Berlin, Heidelberg, New York 1972, pp. 53–153
28. Derouane, E. G., Fraissard, J., Fripiat, J. J., Stone, W. E. E.: Catal. Rev. **7**, 121 (1973)
29. O'Reilly, D. E.: Advan. Catal. **12**, 31 (1960)
30. Kokes, R. J.: Electron Spin Resonance Spectroscopy. In: Experimental Methods in Catalytic Research. Anderson, R. B. (ed.). New York: Academic Press 1968, pp. 436–475

31. Lunsford, J. H.: Advan. Catal. **22**, 265 (1972)
32. Thomas, J. M., Lambert, R. M. (eds.): Characterization of Catalysts. Chichester, New York, Brisbane, Toronto: J. Wiley & Sons 1980
33. Kane, P. F., Larrabee, G. R. (eds.): Characterization of Solid Surfaces. New York, London: Plenum Press 1974
34. Czanderna, A. W. (ed.): Methods of Surface Analysis. Amsterdam, Oxford, New York: Elservier 1975
35. Poate, J. M., Buck, T. M.: Ion scattering spectroscopy. In: Experimental Methods in Catalytic Research. Anderson, R. B., Dawson, P. T. (eds.). New York, San Francisco, London: Academic Press 1976. Vol. 3, pp. 175–210
36. Nelson, G. C.: J. Colloid Interface Sci. **55**, 289 (1976)
37. Taglauer, E., Heiland, W.: Appl. Phys. **9**, 261 (1976)
38. Cvetanović, R. J., Amenomiya, Y.: Advan. Catal. **17**, 103 (1967)
39. Smůtek, M., Černý, S., Buzek, F.: Advan. Catal. **24**, 343 (1975)
40. King, D. A., Surface Sci. **47**, 384 (1975)
41. Rochester, C. H.: Adv. Colloid Interface Sci. **12**, 43 (1980) and Progr. Colloid Polym. Sci. **67**, 7 (1980)
42. Hobert, H., Marx, G., Fink, P., Schulz, H.-R.: Wiss. Z. Friedrich-Schiller-Univ. Jena, Math. Nat. R. **20**, 123 (1971)
43. Gallei, E., Schadow, E.: Rev. Sci. Instrum. **45**, 1504 (1974)
44. Low, M. J. D.: J. Chem. Educ. **47**, A163, A255, A349, A415 (1970)
45. Bates, J. B.: Science **191**, 31 (1976)
46. Bouwman, R., Freriks, J. L. C.: Appl. Surface Sci. **4**, 11 (1980)
47. Anderson, Jr., J. H., Wickersheim, K. A.: Surface Sci. **2**, 252 (1964)
48. Boccuzzi, F., Coluccia, S., Ghiotti, G., Morterra, C., Zecchina, A.: J. Phys. Chem. **82**, 1298 (1978)
49. Knözinger, H., Ress, E.: Z. Phys. Chem. (Frankfurt) **59**, 49 (1968)
50. Borello, E., Della Gatta, G., Fubini, B., Morterra, C., Venturello, G.: J. Catal. **35**, 1 (1974)
51. Yates, D. J. C.: J. Phys. Chem. **65**, 746 (1961)
52. Knözinger, H.: Hydrogen Bonds in Systems of Adsorbed Molecules. In: The Hydrogen Bond. Schuster, P., Zundel, G., Sandorfy, C. (eds.). Amsterdam, New York, Oxford: North-Holland Publ. Comp. 1976, Vol. III, pp. 1263–1364.
53. Kiselev, A. V.: Disc. Faraday Soc. **52**, 14 (1971)
54. Zettlemoyer, A. C., Klier, K.: Disc. Faraday Soc. **52**, 46 (1971)
55. Klier, K.: J. Chem. Phys. **58**, 737 (1973)
56. McDonald, R. S.: J. Phys. Chem. **62**, 1168 (1958)
57. Sidorov, A. N.: Optika Spektrosk. **8**, 806 (1960)
58. Armistead, C. G., Tyler, A. J., Hambleton, F. H., Mitchell, S. A., Hockey, J. A.: J. Phys. Chem. **73**, 3947 (1969)
59. Kubelkovà, L., Schürer, P., Jirů, P.: Surface Sci. **18**, 245 (1969)
60. Davydov, V. Ya., Zhuravlev, L. T., Kiselev, A. V.: Zh. Fiz. Khim. **38**, 2047 (1964)
61. van Cauwelaert, F. H., Vermoortele, F., Uytterhoeven, J. B.: Disc. Faraday Soc. **52**, 66 (1971)
62. Tsyganenko, A. A., Filimonov, V. N.: J. Mol. Structure **19**, 579 (1973)
63. Tretyakov, N. E., Filimonov, V. N.: Kinet. Katal. **11**, 815 (1970)
64. Anderson, P. J., Horlock, R. F., Oliver, J. F.: Trans. Faraday Soc. **61**, 2754 (1965)
65. Tench, A. J.: J. Chem. Soc. Faraday I **68**, 197 (1972)
66. Takezawa, N.: Bull. Chem. Soc. Japan **44**, 3177 (1971)
67. Takezawa, N., Miyahara, K., Toyoshima, I.: J. Res. Inst. Catalysis, Hokkaido Univ. **19**, 56 (1971)
68. Faure, M.: Bull. Soc. Chim. France 69 (1970)
69. Eley, D. D., Kiwanuka, G. M., Rochester, C. N., JCS Faraday I **72**, 876 (1976)
70. Low, M. J. D., Takezawa, N., Goodsel, A. J.: J. Colloid Interface Sci. **37**, 422 (1971)
71. Iizuka, T., Hattori, H., Ohno, Y., Sohna, J., Tanabe, K.: J. Catal. **22**, 130 (1971)
72. Knözinger, H.: Advan. Catal. **25**, 184 (1976)
73. Knözinger, H., Ratnasamy, P.: Catal. Rev. Sci. Eng. **17**, 31 (1978)

74. Low, M. J. D., Maddison, N., Ramamurthy, P.: Surface Sci. **13**, 238 (1969)
75. Thornton, E. W., Harrison, P. G.: J. Chem. Soc. Faraday I **71**, 461 (1975)
76. Lewis, K. E., Parfitt, G. D.: Trans. Faraday Soc. **62**, 204 (1966)
77. Criado, J. M., Dominguez, J., Gonzalez, F., Munuera, G., Trillo, J. M.: Proc. Int. Congr. Catal., 4th, Moscow, 1968, **1**, 504 (1971)
78. Parkyns, N. D.: Surface Properties of Metal Oxides. In: Chemisorption and Catalysis. Hepple, P. (ed.). Amsterdam, London, New York: Elsevier Publ. Comp. 1970, pp. 150–172
79. Primet, M., Pichat, P., Mathieu, M. V.: C. R. Acad. Sci., Ser. B **267**, 799 (1968)
80. Primet, M., Pichat, P., Mathieu, M. V.: J. Phys. Chem. **75**, 1216 (1971)
81. Jackson, P., Parfitt, G. D.: Trans. Faraday Soc. **67**, 2469 (1971)
82. Zecchina, A., Coluccia, S., Guglielminotti, E., Ghiotti, G.: J. Phys. Chem. **75**, 2774 (1971)
83. Atherton, K., Newbold, G., Hockey, J. A.: Disc. Faraday Soc. **52**, 33 (1971)
84. Mattmann, G., Oswald, H. R., Schweizer, F.: Helv. Chim. Acta **55**, 1249 (1972)
85. Tretyakov, N., Pozdnyakov, E., Oranskaya, O. M., Filimonov, V. N.: Russ. J. Phys. Chem. **44**, 596 (1970)
86. Agron, P. A., Fuller, Jr., E. L., Holmes, H. F.: J. Colloid Interface Sci. **52**, 553 (1975)
87. Fuller, E. L., Holmes, H. F., Gammage, R. B.: J. Colloid Interface Sci. **33**, 623 (1970)
88. Peri, J. B.: J. Phys. Chem. **69**, 220 (1965)
89. Munuera, G., Moreno, F., Gonzalez, F.: Proc. Int. Symp. Reactivity Solids, 7th, Bristol, p. 681 (1972)
90. Munuera, G., Stone, F. S.: Disc. Faraday Soc. **52**, 205 (1971)
91. Jones, P., Hockey, J. A.: Trans. Faraday Soc. **67**, 2679 (1971)
92. Dent, A. L., Kokes, R. J.: J. Phys. Chem. **73**, 3781 (1969)
93. Krebs, H.: Fundamentals in Inorganic Crystal Chemistry, London: McGraw-Hill 1968
94. Platonov, V. V., Tretyakov, N. E., Filimonov, V. N.: Usp. Fotoniki **2**, 92 (1971)
95. Peri, J. B., Hannan, R. B.: J. Phys. Chem. **64**, 1526 (1960)
96. Peri, J. B.: J. Phys. Chem. **69**, 211 (1965)
97. Dunken, H., Fink, P.: Z. Chem. **6**, 194 (1966)
98. Carter, J. L., Lucchesi, P. J., Corneil, P., Yates, D. J. C., Sinfelt, J. H.: J. Phys. Chem. **69**, 3070 (1965)
99. Knözinger, H., Stolz, H.: Fortschr. Kolloide Polym. **55**, 16 (1971)
100. Meye, W.: Thesis, University of Munich, 1972
101. Lippens, B. C.: Thesis, Delft, 1961
102. Lippens, B. C., Steggerda, J. J.: Active Alumina. In: Physical and Chemical Aspects of Adsorbents and Catalysts. Linsen, B. G. (ed.). New York: Academic Press, 1970, pp. 171–211
103. Pauling, L.: The Nature of the Chemical Bond, Ithaca, New York: Cornell University Press, 1960, p. 548
104. Jones, L. H.: J. Chem. Phys. **22**, 217 (1954)
105. Zecchina, A.: Disc. Faraday Soc. **52**, 89 (1971)
106. Borello, E., Cimino, A., Ghiotti, G., LoJacono, M., Schiavello, M., Zecchina, A.: Disc. Faraday Soc. **52**, 149 (1971)
107. Morterra, C., Ghiotti, G., Garrone, E., Boccuzzi, F.: J. Chem. Soc. Faraday I **72**, 2722 (1976)
108. Peri, J. B.: Disc. Faraday Soc. **52**, 55 (1971)
109. Morterra, C., Ghiotti, G., Boccuzzi, F., Coluccia, S.: J. Catal. **51**, 299 (1978)
110. Smirnov, E. P., Tsyganenko, A. A.: React. Kinet. Catal. Letters **7**, 425 (1977)
111. Smirnov, E. P., Filimonov, V. N.: React. Kinet. Catal. Letters **7**, 431 (1977)
112. Dunken, H., Hobert, H.: Z. Phys. Chem. (Leipzig) **260**, 913 (1979)
113. Hair, M. L., Hertl, W.: J. Phys. Chem. **73**, 2372 (1969)
114. Hockey, J. A.: J. Phys. Chem. **74**, 2570 (1970)
115. van Cauwelaert, F. H., Jacobs, P. A., Uytterhoeven, J. B.: J. Phys. Chem. **76**, 1434 (1972)
116. Morrow, B. A., Cody, I. A.: J. Phys. Chem. **77**, 1465 (1973)
117. Klier, K., Shen, J. H., Zettlemoyer, A. C.: J. Phys. Chem. **77**, 1458 (1973)

118. van Cauwelaert, F. H., Jacobs, P. A., Uytterhoeven, J. B.: J. Phys. Chem. **77**, 1470 (1973)
119. Ryason, P. R., Russell, B. G.: J. Phys. Chem. **79**, 1276 (1975)
120. van Rosmalen, A. J., Mol, J. C.: J. Phys. Chem. **82**, 2748 (1978)
121. Peri, J. B.: J. Phys. Chem. **70**, 2937 (1966)
122. Ghiotti, G., Garrone, E., Morterra, C., Boccuzzi, F.: J. Phys. Chem. **83**, 2863 (1979)
123. Peri, J. B., Hensley, Jr., A. L.: J. Phys. Chem. **70**, 3168 (1966)
123a. Fink, P., Plotzki, I.: Wiss. Z. Friedrich Schiller Univ. Jena, Math.-Naturwiss. R. **29**, 809 (1981); and Fink, P., Müller, B.: ibid. **30**, 589 (1981)
124. Chackalackal, S. M., Stafford, F. E.: J. Amer. Chem. Soc. **88**, 4815, 4823 (1966)
125. McGraw, G. E., Bernitt, B. L., Hisatsune, I. C.: J. Chem. Phys. **42**, 237 (1965)
126. Giguere, P. A., Savoie, R.: Can. J. Chem. **40**, 495 (1962)
127. Pak, V. N.: Zh. Fiz. Khim. **48**, 1631, 2338 (1974)
128. Lygin, V. I., Smolikov, V. V.: Zh. Fiz. Khim. **48**, 987, 1003 (1974)
129. Kiselev, A. V., Lygin, V. I., Smolikov, N. V.: Zh. Fiz. Khim. **49**, 1026 (1975)
130. Fink, P., Camara, B., Welz, E., Ty Pham dinh: Z. Chem. **11**, 473 (1971)
131. Low, M. J. D., Ramasubramanian, N.: J. Phys. Chem. **70**, 2740 (1966)
132. Camara, B., Dunken, H., Fink, P.: Z. Chem. **8**, 155 (1968)
133. Low, M. J. D., Harano, Y.: J. Research. Inst. Catal., Hokkaido University, **16**, 271 (1968)
134. Hair, M. L., Hertl, W.: J. Phys. Chem. **74**, 91 (1970)
135. Hair, M. L., Hertl, W.: J. Phys. Chem. **77**, 1965 (1973)
136. Low, M. J. D., Ramamurthy, P.: Chem. Commun. 609 (1967)
137. Bourne, K. H., Cannings, F. R., Pitkethly, R. C.: J. Phys. Chem. **74**, 2197 (1970)
138. Basila, M. R.: J. Phys. Chem. **66**, 2223 (1966)
139. Peri, J. B.: J. Phys. Chem. **70**, 3168 (1966)
140. Bremer, H., Steinberg, K.-H., Tran-Khac Chuong: Z. anorg. allg. Chem. **400**, 115 (1973)
141. Rouxhet, P. G., Sempels, R. E.: JCS Faraday I, **70**, 2021 (1974)
142. Peri, J. B.: J. Catal. **41**, 227 (1976)
143. Ward, J. W.: Infrared Studies of Zeolite Surface and Surface Reactions. In: Zeolite Chemistry and Catalysis. Rabo, J. A. (ed.). Washington: American Chemical Society, 1976, pp. 118–284
144. Ward, J. W.: Adv. Chem. Ser. **101**, 380 (1971)
145. Uytterhoeven, J., Christner, L. G., Hall, W. K.: J. Phys. Chem. **69**, 2117 (1965)
146. Karge, H. G.: Z. Phys. Chem. (Frankfurt) **95**, 241 (1975)
147. Jacobs, P., Uytterhoeven, J. B.: J. Catal. **22**, 193 (1971)
148. Kerr, G. T.: Adv. Chem. Ser. **121**, 219 (1973)
149. Breck, D. W., Skeels, G. W.: Proc. Int. Congr. Catal., 6th, London, 1976, **2**, 645 (1977)
150. Karge, H. G.: Infrared Spectroscopic and Catalytic Studies on Mordenite-like Zeolites. In: Molecular Sieves II. Katzer, J. R. (ed.). Washington: American Chemical Society, 1977, pp. 584–595
151. Fransen, T., van der Meer, O., Mars, P.: J. Catal. **42**, 79 (1976)
152. Ratnasamy, P., Knözinger, H.: J. Catal. **54**, 155 (1978)
153. Topsøe, N.: J. Catal. **64**, 235 (1980)
154. Gil, F. J., García Fierro, J. L., Lopez Agudo, A.: Z. Phys. Chem. (Frankfurt) **123**, 115 (1980)
155. Millman, W. S., Crespin, M., Cirillo, Jr., A. C., Abdo, S., Hall, W. K.: J. Catal. **60**, 404 (1979)
156. Hierl, R., Knözinger, H., Urbach, H.-P.: J. Catal. **69**, 475 (1981)
157. Wirzing, G.: Naturwissenschaften **51**, 211 (1964)
158. Erkelens, J., Linsen, B. G.: J. Colloid Interface Sci. **29**, 464 (1969)
159. Borello, E., Zecchina, A., Morterra, C.: J. Phys. Chem. **71**, 2938 (1967)
160. Curthoys, G., Davydov, V. Ya., Kiselev, A. V., Kiselev, S. A., Kuznetsov, B. V.: J. Colloid Interface Sci. **48**, 58 (1974)
161. Madeley, J. D., Richmond, R. C.: Z. Anorg. Allg. Chem. **389**, 92 (1972)
162. Davydov, V. Ya., Kiselev, A. V., Kiselev, S. A., Polotnyuk, V., O.-V.: J. Colloid Interface Sci. **74**, 378 (1980)
163. Winde, H., Fink, P., Köhler, A.: Z. Chem. **17**, 41 (1977)

164. Arnett, E. M., Mitchell, E. J.: J. Amer. Chem. Soc. **93**, 4052 (1971)
165. White, J. L., Jelli, A. N., André, J. M., Fripiat, J. J.: Trans. Faraday Soc. **63**, 461 (1967)
166. Noller, H., Mayerböck, B., Zundel, G.: Surface Sci. **33**, 82 (1972)
167. Basila, M. R.: J. Chem. Phys. **35**, 1151 (1961)
168. Hertl, W., Hair, M. L.: J. Phys. Chem. **72**, 4676 (1968)
169. Cusumano, J. A., Low, M. J. D.: J. Catal. **23**, 214 (1971)
170. van Cauwelaert, F. H., van Assche, J. B., Uytterhoeven, J. B.: J. Phys. Chem. **74**, 4329 (1970)
171. Knözinger, H.: Surface Sci. **41**, 339 (1974)
172. Galkin, G. A., Kiselev, A. V., Lygin, V. I.: Trans. Faraday Soc. **60**, 256 (1964)
173. Kiselev, A. V.: Surface Sci. **3**, 292 (1965)
174. Davydov, V. Ya., Kiselev, A. V., Kuznetsov, B. V.: Russ. J. Phys. Chem. **44**, 1 (1970)
175. Horill, P., Noller, H.: Z. Phys. Chem. (Frankfurt) **100**, 155 (1976)
176. Gutmann, V.: Chemische Funktionslehre, Wien and New York: Springer-Verlag 1971
177. Bellamy, L. J., Hallam, H. E., Williams, R. L.: Trans. Faraday Soc. **54**, 1120 (1958)
178. Bellamy, L. J., Pace, R. J.: Spectrochim. Acta **25A**, 319 (1969)
179. Schindler, P., Kamber, H. R.: Helv. Chim. Acta **51**, 1781 (1968)
180. Tretyakov, N. E., Filimonov, V. N.: Kinet. Katal. **13**, 815 (1972)
181. Fink, P., Pohle, W., Köhler, A.: Z. Chem. **12**, 117 (1972)
182. Borisova, M. S., Dzisko, V. A., Ignatyeva, L. A., Timofejeva, L. N.: Kinet. Katal. **3**, 461 (1961)
183. Sempels, R. E., Rouxhet, P. G.: J. Colloid Interface Sci. **55**, 263 (1976)
184. Scokart, P. O., Selim, S. A., Damon, J. P., Rouxhet, P. G.: J. Colloid Interface Sci. **70**, 209 (1979)
184a. Lercher, J. A., Noller, H.: J. Catal. **77**, 152 (1982)
184b. Sanderson, R. T.: Chemical Bonds and Bond Energy, New York: Academic Press, 1971
185. Kazansky, V. B.: Proc. Int. Congr. Catal., 6th, London, 1976, **1**, 50 (1977)
185a. Kazansky, V. B., Gritskov, A. M., Andreev, V. M., Zhidomirov, G. M.: J. Mol. Catal. **4**, 135 (1978)
185b. Kazansky, V. B.: Zh. Khim. Fiz. **56**, 318 (1982)
185c. Kustov, L. M., Borovkov, V. Yu., Kazansky, V. B.: J. Catal. **72**, 149 (1981)
186. Forni, L.: Catal. Rev. **8**, 65 (1974)
187. Tanabe, K.: Solid Acids and Bases, Tokyo, New York, London: Kodansha, Academic Press, 1970
188. Rochester, C. H., Scurrell, M. S.: Infrared Studies of Species Adsorbed on Oxide Surfaces. In: Surface and Defect Properties of Solids, Vol. 2. Roberts, M. W., Thomas, J. M. (eds.). London: The Chemical Society 1973, pp. 114–139
189. Rochester, C. H., Geok Hua Yong: J. Chem. Soc. Faraday I, **76**, 1158 (1980)
190. Pohle, W., Fink, P.: Z. Phys. Chem. (Frankfurt) **109**, 205 (1978)
191. Scokart, P. O., Rouxhet, P. G.: J. Chem. Soc. Faraday I, **76**, 1476 (1980)
192. Krietenbrink, H., Knözinger, H.: Z. Phys. Chem. (Frankfurt) **102**, 43 (1976)
193. Knözinger, H., Krietenbrink, H., Müller, H. D., Schulz, W.: Proc. Int. Congr. Catal., 6th., London, 1976, **1**, 183 (1977)
194. Morrow, B. A., Cody, A.: J. Phys. Chem. **79**, 761 (1975)
195. Clark-Monks, C., Ellis, B.: J. Colloid Interface Sci. **44**, 37 (1973)
196. Morrow, B. A., Devi, A.: J. Chem. Soc. Faraday I **68**, 403 (1972)
197. Morrow, B. A., Cody, I. A.: J. Phys. Chem. **77**, 1465 (1973)
198. Morrow, B. A., Ramamurthy, P.: J. Phys. Chem. **77**, 3052 (1973)
199. Morrow, B. A.: J. Chem. Soc. Faraday I **70**, 1527 (1974)
200. Morrow, B. A., Cody, I. A.: J. Chem. Soc. Faraday I **71**, 1021 (1975)
201. Morrow, B. A., Cody, I. A., Lee, L. S. M.: J. Phys. Chem. **79**, 2405 (1975)
202. Morrow, B. A., Cody, I. A.: J. Phys. Chem. **80**, 1995 (1976)
203. Morrow, B. A., Cody, I. A.: J. Phys. Chem. **80**, 1998 (1976)
204. Morrow, B. A., Cody, I. A., Lee, L. S. M.: J. Phys. Chem. **80**, 2761 (1976)
205. Fink, P., Camara, B.: Z. Chem. **12**, 35 (1972)

206. Knözinger, H., Stählin, W.: Progr. Colloid & Polymer Sci. **67**, 33 (1980)
207. Morterra, C.: Proc. Int. Congr. Catal., 6th, London, 1976, **1**, 194 (1977)
208. Davydov, A. A., Kuznetsova, A. S., Shchekochikhin, Yu. M.: Russ. J. Phys. Chem. **47**, 1408 (1973)
209. Morterra, C., Chiorino, A., Ghiotti, G., Garrone, E.: J. Chem. Soc. Faraday I **75**, 271 (1979)
210. Kokes, R. J., Dent, A. L.: Advan. Catal. **22**, 1 (1972)
211. Boccuzzi, F., Borello, E., Zecchina, A., Bossi, A., Camia, M.: J. Catal. **51**, 150 (1978)
212. Coluccia, S., Boccuzzi, F., Ghiotti, G., Mirra, C.: Z. Phys. Chem. (Frankfurt) **121**, 141 (1980)
213. Lavalley, J.-C., Caillod, J.: J. Chim. Phys. **77**, 373, 379 (1980)
214. Knözinger, H., Stolz, H.: Ber. Bunsenges. Phys. Chem. **75**, 1055 (1971)
215. Benesi, H. A.: J. Catal. **28**, 176 (1973)
216. Jacobs, P. A., Heylen, C. F.: J. Catal. **34**, 267 (1974)
217. Matulewicz, E. R. A., Kerkhof, F. R. J. M., Moulijn, J. A., Reitsma, H. J.: J. Colloid Interface Sci. **77**, 110 (1980)
218. Dewing, J., Monks, G. T., Youll, B.: J. Catal. **44**, 226 (1976)
219. Knözinger, H., Krietenbrink, H., Ratnasamy, P.: J. Catal. **48**, 436 (1977)
220. Pichat, P., Mathieu, M. V., Imelik, B.: Bull. Soc. Chim. Fr. 2611 (1969)
221. Hughes, T. R., White, H. M.: J. Phys. Chem. **71**, 2192 (1967)
222. Basila, M. R., Kantner, T. R.: J. Phys. Chem. **70**, 1681 (1966)
223. Ward, J. W., Hansford, R. C.: J. Catal. **13**, 154 (1969)
224. Bielanski, A., Datka, J.: Bull. Acad. Polon. Sci. **22**, 341 (1974)
225. Morrow, B. A.: J. Chem. Soc. Faraday I **70**, 1527 (1974)
226. Azrak, R. G., Angell, C. L.: J. Phys. Chem. **77**, 3048 (1973)
227. Unger, K. K., Porous Silica, Amsterdam, Oxford, New York: Elsevier Scientific Publ. Comp. 1979
228. Murrell, L. L.: Immobilization of Transition Metals: Complex Catalysts on Inorganic Supports. In: Advanced Materials in Catalysis. Burton, J. J., Garten, R. L. (eds.). New York, San Francisco, London: Academic Press 1977, pp. 235–265
229. Gates, B. C., Lieto, J.: Chemtech. **10**, 195, 248 (1980)
230. Royer, G. P.: Catal. Rev.-Sci. Eng. **22**, 29 (1980)
231. Knözinger, H., Ruhmann, W.: unpublished
232. Chwan-Hwa Chiang, Koenig, J. L.: J. Colloid Interface Sci. **83**, 361 (1981)
233. Fink, P., Marx, G., Meyer, K.: Z. Chem. **13**, 314 (1973)
234. Marx, G., Fink, P.: Wissenschaftl. Z. Friedrich-Schiller-Univ. Jena, Math.-Nat. R. **26**, 743 (1977)
235. Tsyganenko, A. A., Pozdnyakov, D. V., Filimonov, V. N.: J. Mol. Struct. **29**, 299 (1975)
236. Peri, J. B.: J. Phys. Chem. **69**, 231 (1965)
237. Dunken, H., Fink, P.: Z. Chem. **5**, 432 (1965)
238. Dunken, H., Fink, P., Pilz, E.: Chem. Techn. **18**, 490 (1966)
239. Dunken, H., Fink, P.: Acta Chim. Acad. Sci. Hung. **53**, 179 (1967)
240. Parfitt, G. D., Ramsbotham, J., Rochester, C. H.: Trans. Faraday Soc. **67**, 841 (1971)
241. Cant, N. W., Little, L. H.: Canad. J. Chem. **46**, 1373 (1968)
242. Hair, M. L.: J. Colloid Interface Sci. **60**, 154 (1977)
243. Rhee, K. H., Basila, M. R.: J. Catal. **10**, 243 (1968)
244. Fripiat, J. J., van Tongelen, M.: J. Catal. **5**, 158 (1966)
245. Armistead, C. G., Tyler, A. J., Hambleton, F. H., Mitchell, S. A., Hockey, J. A.: J. Phys. Chem. **73**, 3947 (1969)
246. Hair, M. L., Hertl, W.: J. Phys. Chem. **73**, 2372 (1969)
247. Bermudez, V. M.: J. Phys. Chem. **75**, 3249 (1971)
248. Harrison, P. G., Thornton, E. W.: J. Chem. Soc. Faraday I **72**, 1310 (1976)
249. Eley, D. D., Kiwanuka, G. M., Rochester, C. H.: J. Chem. Soc. Faraday I **69**, 2062 (1973)
250. Yermakov, Yu. I.: Catal. Rev.-Sci. Eng. **13**, 77 (1976)
251. Schwartz, J., Ward, M. D.: J. Molecular Catal. **8**, 465 (1980)
252. Brenner, A., Burwell, R. L.: J. Catal. **52**, 353 (1978)

253. Kazuska, A., Howe, R. F.: J. Molecular Catal. **9**, 183 (1980)
254. Kazuska, A., Howe, R. F.: J. Molecular Catal. **9**, 199 (1980)
255. Krauss, H. L., Weber, F., Mövik, N.: Z. Anorg. Allg. Chem. **338**, 121 (1965)
256. Krauss, H. L., Stach, H.: Inorg. Nucl. Chem. Letters **4**, 393 (1968); Z. Anorg. Chem. **366**, 34, 280 (1969)
257. Krauss, H. L.: Proc. Int. Congr. Catal., 5th, Palm Beach, 1972, **1**, 207 (1973)
258. Beck, D. D., Lunsford, J. H.: J. Catal. **68**, 121 (1981)
259. Zecchina, A., Garrone, E., Ghiotti, G., Morterra, C., Borello, E.: J. Phys. Chem. **79**, 966 (1975)
260. Hierl, G., Krauss, H. L.: Z. anorg. allg. Chem. **415**, 57 (1975)
261. Fubini, B., Ghiotti, G., Stradella, L., Garrone, E., Morterra, C.: J. Catal. **66**, 200 (1980)
262. McDaniel, M. P.: J. Catal. **67**, 71 (1981)
263. Zecchina, A., Garrone, E., Ghiotti, G., Caluccia, S.: J. Phys. Chem. **79**, 972 (1975)
264. Zecchina, A., Garrone, E., Morterra, C., Collucia, S.: J. Phys. Chem. **79**, 978 (1975)
265. Garrone, E., Ghiotti, G., Coluccia, S., Zecchina, A.: J. Phys. Chem. **79**, 984 (1975)
266. Kazusaka, A., Howe, R. F.: J. Catal. **63**, 447 (1980)
267. Peri, J. B.: J. Phys. Chem. **78**, 588 (1974)
268. Millman, W. S., Hall, W. K.: J. Phys. Chem. **83**, 427 (1979)
269. Jones, C. E., Klier, K.: Ann. Rev. Materials Sci. **2**, 1 (1972)
270. Cardona, M.: Modulation Spectroscopy. In: Solid State Physics, Suppl. 2. Seitz, F., Turnbull, D., Ehrenreich, H. (eds.). New York: Academic Press 1969
271. Harrick, N. J.: International Reflection Spectroscopy, New York: Interscience Publ., 1967
272. Kunath, D., Reklat, A.: Z. Chem. **11**, 361 (1971)
273. Mattson, J. S., Mark, H. B.: Activated Carbon, New York: M. Dekker, Inc. 1971, pp. 87–117
274. Haller, G. L., Rice, R. W., Wan, Z. C.: Catal. Rev.-Sci. Eng. **13**, 259 (1976)
275. Mattson, J. S., Mark, Jr. H. B., Weber, W. J.: Anal. Chem. **41**, 355 (1969)
276. Zawadzki, J.: Carbon **16**, 491 (1978)
277. Mattson, J. S., Mark, Jr., H. B.: J. Colloid Interface Sci. **31**, 131 (1969)
278. Haller, G. L., Rice, R. W.: J. Phys. Chem. **74**, 4386 (1970)
279. Rice, R. W., Haller, G. L.: Proc. Int. Congr. Catal., 5th, Palm Beach, 1972, **1**, 317 (1973)
280. Mark, Jr., H. B., Randall, E. N.: Symp. Faraday Soc. **4**, 157 (1970)
271. Kortüm, G., Delfs, H.: Spectrochim. Acta **20**, 405 (1964)
282. Maulhardt, H., Kunath, D., Kriegsmann, H.: Jenaer Rdsch. **23**, 242 (1978)
283. Schönemann, M., Eckermann, Th., Kunath, D., Kriegsmann, H.: Exp. Techn. Phys. **20**, 43 (1972)
284. Willey, R. D.: Appl. Spectrosc. **30**, 593 (1976)
285. Fuller, M. P., Griffiths, P. R.: Anal. Chem. **50**, 1906 (1978)
286. Maulhardt, H., Kunath, D.: Appl. Spectrosc. **34**, 383 (1980)
287. Egerton, T. A., Hardin, A. H.: Catal. Rev.-Sci. Eng. **11**, 1 (1975)
288. Cooney, R. P., Curthoys, G., Nguyen, T. T.: Advan. Catal. **24**, 293 (1975)
289. Takenaka, T.: Adv. Coll. Interface Sci. **11**, 291 (1979)
290. Hendra, P. J., Fleischmann, M.: Raman Spectroscopy at Surfaces. In: Topics in Surface Chemistry. Kay, E., Bagus, P. S. (eds.). New York: Plenum Press 1978
291. Morrow, B. A.: Raman Spectroscopic Studies of Surface Species. In: ACS Symposium Series "Vibrational Spectroscopies of Adsorbed Species on Catalysts", **137**, 119 (1980)
292. Brown, F. R., Makovsky, L. E., Rhee, K. H.: Appl. Spectrosc. **31**, 563 (1977)
293. Cheng, C. P., Ludowise, J. D., Schrader, G. L.: Appl. Spectrosc. **34**, 146 (1980)
294. Jeziorowski, H., Knözinger, H., Vielhaber, B.: unpublished
295. Freeman, J. J., Heaviside, J., Hendra, P. J., Prior, J., Reid, E. S.: Appl. Spectrosc. **35**, 196 (1981)
296. Heaviside, J., Hendra, P. J., Paul, S. O., Freeman, J. J., Friedman, R. M.: Appl. Spectrosc. **35**, 220 (1981)
297. Saperstein, D. D., Rein, A. J.: J. Phys. Chem. **81**, 2134 (1977)

298. Jeziorowski, H., Knözinger, H.: Chem. Phys. Letters **51**, 519 (1977)
299. Jeziorowski, H., Knözinger, H.: J. Phys. Chem. **83**, 1166 (1979)
300. Jeziorowski, H., Knözinger, H.: Chem. Phys. Letters **42**, 162 (1976)
301. Galeener, F. L.: Chem. Phys. Letters **48**, 7 (1977)
302. Brückner, S., Jeziorowski, H., Knözinger, H.: Proc. Int. Raman Conf., Bordeaux 1982, p. 253
303. Kiefer, W.: Advan. Infrared Raman Spectrosc. **3**, 1 (1977)
304. Krasser, W., Ervens, H., Fadini, A., Renouprez, A. J.: Raman Spectrosc. **9**, 80 (1980)
305. Murray, C. A., Greytak, T. J.: J. Chem. Phys. **71**, 3355 (1979)
306. Pernoll, I., Maier, U., Janoschek, R., Zundel, G.: J. Chem. Soc. Faraday II **71**, 210 (1975)
307. Morrow, B. A., Hardin, A. H.: J. Phys. Chem. **83**, 3155 (1979)
308. Morrow, B. A.: J. Phys. Chem. **81**, 2663 (1977)
309. Chan, S. S., personal communication
310. Jeziorowski, H., Knözinger, H.: J. Phys. Chem. **83**, 1166 (1979)
311. Knözinger, H., Jeziorowski, H.: J. Phys. Chem. **82**, 2002 (1978)
312. Jeziorowski, H., Knözinger, H., Grange, P., Gajardo, P.: J. Phys. Chem. **84**, 1825 (1980)
313. Jeziorowski, H., Knözinger, H.: Appl. Surface Sci. **5**, 335 (1980)
314. Knözinger, H., Jeziorowski, H., Taglauer, E.: Proc. Int. Congr. Catal., 7th, Tokyo, 1980, **1**, 604 (1981)
315. Brown, F. R., Makovsky, L. E., Rhee, K. H.: J. Catal. **50**, 162 and 385 (1977)
316. Medema, J., van Stam, C., de Beer, V. H. J., Konings, A. J. A., Koningsberger, D. C.: J. Catal. **53**, 386 (1978)
317. Payen, E., Barbillat, J., Grimblot, J., Bonnelle, J. P.: Spectrosc. Letters **11**, 997 (1978)
318. Cheng, C. P., Schrader, G. L.: J. Catal. **60**, 276 (1979)
319. Li Wang, Hall, W. K.: J. Catal. **66**, 251 (1980)
320. Sombret, B., Dhamlincourt, P., Wallart, F., Muller, A. C., Bouquet, M., Grosmangin, J.: J. Raman Spectrosc. **9**, 291 (1980)
321. de Vries, S. L. K. F., Pott, G. T.: Rec. Trav. Chim. Pays-Bas **96**, M 115 (1977)
322. Thomas, R., Moulijn, J. A., Kerkhof, F. P. J. M.: Rec. Trav. Chim. Pays-Bas **96**, M 114 (1977)
323. Kerkhof, F. P. J. M., Moulijn, J. A., Thomas, R.: J. Catal. **56**, 279 (1979)
324. Roozeboom, F., Medema, J., Gellings, P. J.: Z. Phys. Chem. (Frankfurt) **111**, 215 (1978)
325. Roozeboom, F., Mittelmeijer-Hazeleger, M. C., Moulijn, J. A., Medema, J., de Beer, V. H. J., Gellings, P. J.: J. Phys. Chem. **84**, 2783 (1980)
326. Pao, Y. H.: Optoacoustic Spectroscopy and Detection, New York: Academic Press 1977
327. Rosencwaig, A.: Photoacoustics and Photoacoustic Spectroscopy, Chichester: J. Wiley & Sons 1980
328. Low, M. J. D., Parodi, G. A.: Appl. Spectrosc. **34**, 76 (1980)
328a. Riseman, S. M., Massoth, F. E., Dhar, G. M., Eyring, E. M.: J. Phys. Chem. **86**, 1760 (1982)
329. Vidrine, D. W.: Appl. Spectrosc. **34**, 314 (1980)
330. Rockley, M. G.: Chem. Phys. Letters **68**, 455 (1979)
330a. Low, M. J. D., Lacroix, M.: Infrared Phys. **22**, 139 (1982)
331. Rockley, M. G., Devlin, J. P.: Appl. Apectrosc. **34**, 407 (1980)
332. Kanstad, S. O., Nordal, P. E.: Appl. Surface Sci. **6**, 372 (1980)
333. Ibach, H., Hopster, H., Sexton, B.: Appl. Surface Sci. **1**, 1 (1977)
334. Dubois, L. H., Hansma, P. K., Somorjai, G. A.: Appl. Surface Sci. **6**, 173 (1980)
335. Weinberg, H.: Ann. Rev. Phys. Chem. **29**, 115 (1978)
336. Walmsley, D. G.: Inelastic Electron Tunnelling Spectroscopy. In: Vibrational Spectroscopy of Adsorbates. Willis, R. F. (ed.). Berlin, Heidelberg, New York: Springer 1980, pp. 67 to 90
337. Lambe, J., Jaklevic, R. C.: Phys. Rev. **165**, 821 (1968)
338. Jaklevic, R. C., Gaerttner, M. R.: Appl. Surface Sci. **1**, 479 (1978)

339. Walmsley, D. G., Nelson, W. J., Brown, N. M. D., Floyd, R. B.: Appl. Surface Sci. **5**, 107 (1980)
340. Mazur, U., Hipps, K. W.: J. Phys. Chem. **83**, 2773 (1979)
341. Hipps, K. W., Mazur, U.: J. Phys. Chem. **84**, 3162 (1980)
342. Evans, H. E., Weinberg, W. H.: J. Chem. Phys. **71**, 4789 (1979)
343. Hall, J. T., McBride, D. E.: J. Chem. Phys. **74**, 4164 (1981)
344. Brown, N. M. D., Floyd, R. B., Nelson, W. J., Walmsley, D. G.: J. Chem. Soc. Faraday I **76**, 2335 (1980)
345. Mazur, U., Hipps, K. W.: J. Phys. Chem. **85**, 2244 (1981)
346. Evans, H. E., Weinberg, W. H.: J. Amer. Chem. Soc. **102**, 2548 (1980)
347. Evans, H. E., Weinberg, W. H.: J. Amer. Chem. Soc. **102**, 2554 (1980)
347a. Weinberg, W. H.: Inelastic Electron Tunneling Spectroscopy of Supported Homogeneous Cluster Compounds. In: Vibrational Spectra and Structure. During, J. R. (ed.). Amsterdam: Elsevier Publ. Comp., Vol. 11, in press.
348. Howard, J., Waddington, T. C.: Appl. Surface Sci. **2**, 102 (1978)
349. Hall, P. G., Wright, C. J.: Neutron Scattering from Adsorbed Molecules, Surfaces, and Intercalates. In: Chemical Physics of Solids and their Interfaces, Vol. 7. London: The Chemical Society 1978, pp. 89–117
350. Wright, C. J.: Neutron Scattering Studies. In: Vibrational Spectroscopy of Adsorbates. Willis, R. F. (ed.). Berlin, Heidelberg, New York: Springer 1980, pp. 111–124
351. Renouprez, A. J., Fouilloux, P., Coudurier, G., Tochetti, D., Stockmeyer, R.: J. Chem. Soc. Faraday I **73**, 1 (1977)
352. Hall, P. G., Pidduck, A., Wright, C. J.: J. Colloid Interface Sci. **79**, 339 (1981)
353. Boutin, H., Prask, H., Iyengar, R. D.: Advan. Colloid Interface Sci. **2**, 1 (1968)
354. Tempelhoff, K., Feldmann, K.: Z. Phys. Chem. (Leipzig) **256**, 369 (1975)
355. Leftin, H. P., Hobson, M. C.: Advan. Catal. **14**, 115 (1963)
356. Porter, R. P., Hall, W. K.: J. Catal. **5**, 366 (1966)
357. Hierl, G., Krauss, H. L.: React. Kinet. Catal. Letters **3**, 47 (1975)
358. Oelkrug, D., Radjaipour, M.: Z. Phys. Chem. (Frankfurt) **123**, 163 (1980)
359. Kessler, R. W., Wilkinson, F.: J. Chem. Soc. Faraday I **77**, 309 (1981)
360. Tench, A. J., Pott, G. T.: Chem. Phys. Letters **26**, 590 (1974)
361. Zecchina, A., Lofthouse, M. G., Stone, F. S.: J. Chem. Soc. Faraday I **71**, 1476 (1975)
362. Breysse, M., Claudel, B., Faure, L., Guénin, M.: Proc. 7th Intern. Vac. Congr. & 3rd Intern. Conf. Solid Surf., Vienna, 1977, pp. 485–488
363. Coluccia, S., Deane, M., Tench, A. J.: Proc. Int. Congr. Catal., 6th, London, 1976, **1**, 171 (1977)
364. Flockhart, B. D.: Electron Spin Resonance Studies of Adsorbed Species. In: Surface and Defect Properties of Solids, Vol. 2. London: The Chemical Society 1973, pp. 69–96
365. Deviney, M. L.: Adv. Colloid Interface Sci. **2**, 237 (1969)
366. Svejda, P., Maki, A. H.: Chem. Phys. Letters, **83**, 610 (1981)
366a. Duncan, T. M., Dybowski, C.: Surface Sci. Reports **1**, 157 (1981)
367. van Vleck, J. H.: Phys. Rev. **74**, 1168 (1948)
368. Haeberlen, U.: High Resolution NMR in Solids: Selective Averaging. New York: Academic Press 1976
369. Mehring, M.: High Resolution NMR Spectroscopy in Solids. Berlin, Heidelberg, New York: Springer-Verlag 1976
370. Deiniger, D., Michel, D.: Wiss. Z. Karl Marx Univ. Leipzig, Math.-Naturwiss. Reihe **22**, 551 (1973)
371. Michel, D.: Z. Phys. Chem. (Leipzig) **252**, 263 (1973)
372. Gay, I. D., Liang, S. H.: J. Catal. **44**, 306 (1976)
373. Kriz, J. F., Gay, I. D.: J. Phys. Chem. **80**, 2951 (1976)
374. Michel, D., Meiler, W., Pfeiffer, H., Rauscher, H. J.: J. Mol. Catal. **5**, 263 (1979)
375. Bernstein, T., Michel, D., Pfeifer, H., Fink, P.: J. Chem. Soc. Faraday Trans. I **77**, 797 (1981)
376. Ali, I. T., Gay, I. D.: J. Phys. Chem. **85**, 1251 (1981)
377. Pines, A., Gibby, M. G., Waugh, J. S.: Chem. Phys. Letters **15**, 373 (1972)

378. Andrew, E. R.: Nuclear Magnetic Resonance-Rapidly Rotating Solids. In: Magnetic Resonance. Coogan, C. K., Ham, N. S., Stuart, S. N., Pilbrow, J. R., Wilson, G. V. H. (eds.). New York: Plenum 1970, pp. 163–176
379. Taylor, P. C., Baugher, J. F., Kriz, H. M.: Chem. Rev. **75**, 203 (1975)
380. Steijskal, E. O., Schaefer, J., Henis, J. M. S., Tripodi, M. K.: J. Chem. Phys. **61**, 2351 (1974)
381. O'Reilly, D. E., Leftin, H. P., Hall, W. K.: J. Chem. Phys. **29**, 970 (1958)
382. Hall, W. K., Leftin, H. P., Cheselske, F. J., O'Reilly, D. E.: J. Catal. **2**, 506 (1963)
383. Bermudez, V. M.: J. Phys. Chem. **74**, 4160 (1970)
384. Kvilividze, V. I.: Dokl. Akad. Nauk SSSR **157**, 158 (1964)
385. Kvilividze, V. I.: Dokl. Akad. Nauk SSSR **157**, 673 (1964)
386. Kvilividze, V. I., Ievskaya, N. M., Egorova, T. S., Kiselev, V. F., Sokolov, N. D.: Kinetics and Catalysis **3**, 73 (1962)
387. Egorov, M. M., Kvilividze, V. I., Kiselev, V. F., Krassilnikov, K. G.: Kolloid Z. **212**, 126 (1966)
388. Webster, R. K., Jones, T. L., Anderson, P. J.: Proc. Brit. Ceram. Soc. **5**, 153 (1965)
389. Freude, D., Müller, D., Schmiedel, H.: Surface Sci. **25**, 289 (1971)
390. Enriquez, M. A., Domérieux-Morin, C., Fraissard, J.: Appl. Surface Sci. **5**, 180 (1980)
391. Cirillo, A. C., Dollish, F. R., Hall, W. K.: J. Catal. **62**, 379 (1980)
392. Schreiber, L. B., Vaughan, R. W.: J. Catal. **40**, 226 (1975)
393. Stübner, B., Knözinger, H., Conard, J., Fripiat, J. J.: J. Phys. Chem. **82**, 1811 (1978)
394. Kiviat, F. E., Petrakis, L.: J. Phys. Chem. **77**, 1232 (1973)
395. Pearson, R. M.: J. Catal. **46**, 279 (1977)
396. Knözinger, H.: J. Catal. **53**, 171 (1978)
397. Gay, I. D., Liang, S.: J. Catal. **44**, 306 (1976)
397a. Dawson, W. H., Kaiser, S. W., Ellis, P. D., Inners, R. R.: J. Phys. Chem. **86**, 867 (1982)
398. Liang, S. H. C., Gay, I. D.: J. Catal. **66**, 294 (1980)
399. Freude, D., Pfeifer, H., Ploss, W., Staudte, B.: J. Mol. Catal. **12**, 1 (1981)
400. Mestdagh, M. M., Stone, W. E. E., Fripiat, J. J.: J. Phys. Chem. **76**, 1220 (1973) and J. Chem. Soc. Faraday Trans. I **72**, 154 (1976)
401. Freude, D., Oehme, W., Schmiedel, H., Staudte, B.: J. Catal. **32**, 137 (1974)
402. Rauscher, H.-J., Michel, D., Pfeifer, H.: J. Mol. Catal. **12**, 159 (1981)
403. Gay, I. D.: J. Phys. Chem. **84**, 3230 (1980)
404. Duncan, T. M., Vaughan, R. W.: J. Catal. **67**, 49 (1981)
405. Slotfeldt-Ellingsen, D., Resing, H. A.: J. Phys. Chem. **84**, 2204 (1980)
406. Resing, H. A., Slotfeldt-Ellingsen, D., Garroway, A. N., Pinnavaia, T. J., Unger, K.: ^{13}C Chemical Shifts in Adsorption Systems: Molecular Motions, Molecular Orientations, Qualitative and Quantitative Analysis. In: Magnetic Resonance in Colloid and Interface Science. Fraissard, J. P., Resing, H. A. (eds.). Dordrecht, Boston: Reidel Publ. Comp. 1980, pp. 239–258
407. Chang, J. J., Pines, A., Fripiat, J. J., Resing, H. A.: Surface Sci. **47**, 661 (1975)
408. Maciel, G. E., Sindorf, D. W.: J. Chromatogr. **205**, 438 (1981)
409. Sindorf, D. W., Maciel, G. E.: J. Amer. Chem. Soc. **103**, 4263 (1981)
410. Linder, M., Hohener, A., Ernst, R. R.: J. Magn. Res. **35**, 379 (1979)
411. Roberts, M. W.: Advan. Catal. **29**, 55 (1980)
412. Seah, M. P., Dench, W. A.: Surf. Interf. Analysis **1**, 2 (1979)
413. Seah, M. P.: Surf. Interf. Analysis **2**, 222 (1980)
414. Cimino, A., deAngelis, B. A.: J. Catal. **36**, 11 (1975)
415. Gajardo, P., Declerck-Grimee, R. I., Delvaux, G., Olodo, P., Zabala, J. M., Canesson, P., Grange, P., Delmon, B.: J. Less Common Met. **54**, 311 (1977)
416. Declerck-Grimee, R. I., Canesson, P., Friedman, R. M., Fripiat, J. J.: J. Phys. Chem. **82**, 885 (1978)
417. Patterson, T. A., Carver, J. C., Leyden, D. E., Hercules, D. M.: J. Phys. Chem. **80**, 1700 (1976)
418. Gajardo, P., Grange, P., Delmon, B.: J. Catal. **63**, 201 (1980)

419. Okamoto, Y., Tomioka, H., Katoh, Y., Imanaka, T., Teranishi, S.: J. Phys. Chem. **84**, 1833 (1980)
420. Dufresne, P., Payen, E., Grimblot, J., Bonnelle, J. P.: J. Phys. Chem. **85**, 2344 (1981)
421. Abart, J., Delgado, E., Ertl, G., Jeziorowski, H., Thiele, N., Wang, X. Zh., Taglauer, E.: Appl. Catal., **2**, 155 (1982)
422. Ng, K. T., Hercules, D. M.: J. Phys. Chem. **80**, 2094 (1976)
423. Iwasawa, Y., Ichinose, H., Ogasawara, S.: J. Chem. Soc. Faraday I **77**, 1763 (1981)
424. Yermakov, Yu. I., Kuznetsov, B. N., Startsev, A. N., Zhdan, P. A., Shepelin, A. P., Zaikovskii, V. I., Plyasova, L. M., Burmistrov, V. A.: J. Mol. Catal. **11**, 205 (1981)
425. Tereczki, B., Kurth, R., Boehm, H. P.: Extended Abstracts Carbon 80, International Carbon Conference, Baden-Baden, 1980, p. 218
426. Carley, A. F., Moroney, L., Roberts, M. W.: Faraday Symposia Chem. Soc. **15**, 39 (1980)
426a. Knox, J. H.: ibid. **15**, 165 (1980)
427. Kondo, S., Muroya, M.: Bull. Chem. Soc. Japan **43**, 2657 (1970)
428. Krauss, H. L., Naumann, D.: Z. anorg. allg. Chem. **430**, 23 (1977)
429. Czanderna, A. W., Vaccum Microbalance Tech. **6**, 129 (1967)
430. Dawson, P. T., Walker, P. C.: Desorption Methods. In: Experimental Methods in Catalytic Research. Anderson, R. B., Dawson, P. T. (eds.). New York, San Francisco, London: Academic Press, 1976. Vol. 3, pp. 211–279
431. Ibok, E. E., Ollis, D. F.: J. Catal. **66**, 391 (1980)
432. Malet, P., Munuera, G.: Abstracts, Colloque International du C.N.R.S. sur l'Adsorption aux Interfaces gaz/solide et liquide/solide. Aix-en-Provence (France) 1981
433. Munuera, G., Stone, F. S.: Disc. Faraday Soc. **52**, 205 (1971)
434. Knözinger, H., Schubart, W.: unpublished results
435. Stakebake, J. L.: J. Phys. Chem. **77**, 581 (1973)
436. Lee, J. A., Newnham, C. E., Tye, F. L.: J. Colloid Interface Sci. **42**, 372 (1973)
437. Morishige, K., Kittaka, S., Morimoto, T.: Bull. Chem. Soc. Japan **53**, 2128 (1980)
438. Morishige, K., Kittaka, S., Iwasaki, S., Morimoto, T.: J. Phys. Chem. **85**, 570 (1981)
439. Robertson, S. D., McNicol, B. D., de Baas, J. H., Kloet, S. C., Jenkins, J. W.: J. Catal. **37**, 424 (1975)
440. Brenner, A., Hucul, D. A., Hardwick, S. J.: Inorg. Chem. **18**, 1478 (1979)
441. Villani, S.: Isotope Separation. American Nuclear Society 1976, p. 350
442. Shatenshtein, A. I., Kasakowa, S. S.: Isotopenanalyse des Wassers, Berlin: Verlag d. Wissenschaften 1960
443. Haldeman, R. G., Emmett, P. H.: J. Amer. Chem. Soc. **78**, 2917 (1956)
444. Arnett, E. M., Duggleby, P. McC.: Anal. Chem. **35**, 1420 (1962)
445. Boehm, H. P., Herrmann, M.: Z. anorg. allg. Chem. **352**, 156 (1967)
446. Boehm, H. P., Sappok, R.: Z. anorg. allg. Chem. **365**, 152 (1969)
447. Zettler, A., Günther, G., Boehm, H. P.: unpublished results
448. Lee, J. K., Weller, S. W.: Anal. Chem. **30**, 1057 (1958)
449. Cheselske, F. J., Hall, W. K., Wallace, W. E.: J. Phys. Chem. **63**, 505 (1959)
450. Larson, J. G., Hall, W. K.: J. Phys. Chem. **69**, 3080 (1965)
451. Kittelmann, U., Unger, K.: Progr. Colloid Polymer Sci. **67**, 19 (1980)
452. Rivin, D.: Rubber Chem. Technol. **36**, 729 (1963)
453. Fritz, G.: Z. Naturf. **10b**, 423 (1955)
454. Deuel, H., Huber, G.: Helv. Chim. Acta **34**, 1697 (1951)
455. Fripiat, J. J., Uytterhoeven, J.: J. Phys. Chem. **66**, 800 (1962)
456. Wistuba, H.: Doctoral Thesis, University of Heidelberg, 1967, p. 26; and Janssen, K.: Doctoral Thesis, University of München, 1975, p. 14
457. Unger, K., Gallei, E.: Kolloid Z. & Z. Polymere **237**, 358 (1970)
458. Sato, M., Kanbayashi, T., Kobayashi, N., Shima, Y.: J. Catal. **7**, 342 (1967)
459. Peglar, R. J., Hambleton, F. H., Hockey, J. A.: J. Catal. **20**, 309 (1971)
460. Yates, D. J. C., Dembinski, G. W., Kroll, W. R., Elliot, J. J.: J. Phys. Chem. **73**, 911 (1969)
461. Hanke, W.: Z. anorg. allg. Chem. **395**, 191 (1973)
462. Nondek, L.: Reaction Kinet. Catal. Letters **2**, 283 (1975)

463. Boonstra, A. H., Mutsaers, C. A. H. A.: J. Phys. Chem. **79**, 1694 (1975)
464. Kijenski, J., Hombek, R., Malinowski, S.: J. Catal. **50**, 186 (1977)
465. Goldstein, M. S.: Measurement of Acidity of Surfaces. In: Experimental Methods in Catalytic Research. Anderson, R. B. (ed.). New York, London: Academic Press 1968, pp. 361–401
466. Benesi, H. A., Winquist, B. H. C.: Advan. Catal. **27**, 97 (1978)
467. Trambouze, Y., de Mourgues, L., Perrin, M.: Compt. Rend. Acad. Sci. **234**, 1770 (1952)
468. Holm, V. C. F., Bailey, G. C., Clark, A.: J. Phys. Chem. **63**, 129 (1959)
469. Donnet, J. B., Hueber, F., Reitzer, C., Oddoux, J., Riess, G.: Bull. Soc. Chim. France **1962**, 1727
470. Webb, A. N.: Ind. Eng. Chem. **49**, 261 (1957)
471. Mills, G. A., Boedeker, E. R., Oblad, A. G.: J. Amer. Chem. Soc. **72**, 1554 (1950)
472. Clark, A., Holm, V. C. F., Blackburn, D. M.: J. Catal. **1**, 244 (1962); and Clark, A., Holm, V. C. F.: J. Catal. **2**, 21 (1963)
473. Hildebrandt, R. A., Skala, H.: J. Catal. **12**, 61 (1968)
474. Weil-Malherbe, H., Weiss, J.: J. Chem. Soc. (London) **1948**, 2164
475. Walling, C.: J. Amer. Chem. Soc. **72**, 1164 (1950)
476. Tamele, M. W.: Disc. Faraday Soc. **8**, 270 (1950)
477. Benesi, H. A.: J. Phys. Chem. **61**, 970 (1957)
478. Hammett, L. P., Deyrup, A. J.: J. Amer. Chem. Soc. **54**, 2721 (1932); and Hammett, L. P.: Chem. Rev. **16**, 67 (1935)
479. Mizuno, K., Take, J., Yoneda, Y.: Bull. Chem. Soc. Japan **49**, 634 (1976)
480. Bertolacini, R. J.: Anal. Chem. **35**, 599 (1963)
481. Drushel, H. V., Sommers, A. L.: Anal. Chem. **38**, 1723 (1966)
482. Frenkel, M.: Anal. Chem. **47**, 598 (1975)
483. Hirschler, A. E.: J. Catal. **2**, 428 (1963)
484. Deno, N. C., Berkheimer, H. E., Evans, W. L., Petersen, H. J.: J. Amer. Chem. Soc. **81**, 2344 (1959)
485. Take, J., Nomizo, Y., Yoneda, Y.: Bull. Chem. Soc. Japan **46**, 3568 (1973)
486. Take, J., Kawai, H., Yoneda, Y.: J. Catal. **36**, 356 (1975)
487. Deeba, M., Hall, W. K.: J. Catal. **60**, 417 (1979)
488. Taniguchi, H., Masuda, T., Tsutsumi, K., Takahashi, H.: Bull. Chem. Soc. Japan **51**, 1970 (1978)
489. Tanabe, K., Yamaguchi, T.: J. Res. Inst. Catal., Hokkaido Univ. **11**, 179 (1964)
490. Take, J., Kikuchi, N., Yoneda, Y.: J. Catal. **21**, 164 (1971)
491. Yamanaka, T., Tanabe, K.: J. Phys. Chem. **79**, 2409 (1975)
492. Boehm, H. P., Diehl, E., Heck, W.: Proc. 2nd Intern. Conf. Carbon and Graphite (1965), London, Soc. Chem. Ind. 1966, p. 369
493. Hofmann, U., Ohlerich, G.: Angew. Chem. **62**, 16 (1950)
494. Vieböck, F., Schwappach, A.: Ber. dtsch. chem. Ges. **63**, 2818 (1930); Vieböck, F., Brecher, C.: Ber. dtsch. chem. Ges. **63**, 3207 (1930)
495. Boehm, H. P.: Angew. Chem. **78**, 617 (1966); Angew. Chem. Intern. Ed. **5**, 533 (1966)
496. Brunauer, S., Deming, L. S., Deming, W. E., Teller, E.: J. Amer. Chem. Soc. **62**, 1723 (1940)
497. Brunauer, S., Emmett, P. H., Teller, E.: J. Amer. Chem. Soc. **60**, 309 (1938)
498. Sappok, R., Boehm, H. P.: Carbon **6**, 573 (1968)
499. Sappok, R.: Doctoral Thesis, University of Heidelberg, 1966
500. Förster, H. J.: Doctoral Thesis, University of Heidelberg, 1967.
501. Wistuba, H.: Doctoral Thesis, University of Heidelberg, 1967
502. Huggins, C. M.: J. Phys. Chem. **65**, 1881 (1961)
503. Krauss, H. L., Westphal, U.: Z. Naturforsch. **33b**, 1278 (1978)
504. Moeseler, R., Horvath, B., Lindenau, D., Horvath, E. G., Krauss, H. L.: Z. Naturforsch. **31b**, 892 (1976)
505. Hogan, J. P.: J. Polymer Sci. A-1 **8**, 2637 (1970)
506. Morys, P., Gerritzen, R., Krauss, H. L.: Z. Naturforsch. **31b**, 774 (1976)
507. Krauss, H. L., Rebenstorf, B., Westphal, U.: Z. anorg. allg. Chem. **414**, 97 (1975)

508. Hierl, G., Krauss, H. L.: Z. anorg. allg. Chem. **401**, 263 (1973)
509. Krauss, H. L., Naumann, D.: Z. anorg. allg. Chem. **446**, 23 (1978)
510. Krauss, H. L., Weisser, B.: Z. anorg. allg. Chem. **412**, 82 (1975)
511. Horvath, B., Strutz, J., Möseler, R., Horvath, E. G.: Z. anorg. allg. Chem. **449**, 5 (1979)
512. Horvath, B., Geyer, J., Krauss, H. L.: Z. anorg. allg. Chem. **426**, 141 (1976)
513. Roozeboom, F., Fransen, T., Mars, P., Gellings, P. J.: Z. anorg. allg. Chem. **449**, 25 (1979)
514. Barber, M., Evans, E. L., Thomas, J. M.: Chem. Phys. Letters **18**, 423 (1973)
515. Kruyt, H. R., de Kadt, G. S.: Kolloid Z. **47**, 44 (1929)
516. Kinoshita, K., Bett, J.: Carbon **11**, 237 (1973)
517. Voll, M., Boehm, H. P.: Carbon **8**, 741 (1970)
518. Boehm, H. P., Voll, M.: Carbon **8**, 227 (1970)
519. Voll, M., Boehm, H. P.: Carbon **9**, 481 (1971)
520. Garten, V. A., Weiss, D. E., Willis, J. B.: Aust. J. Chem. **10**, 295 (1957)
521. Hallum, J. V., Drushel, H. V.: J. Phys. Chem. **62**, 110 (1958); Drushel, H. V., Hallum, J. V.: J. Phys. Chem. **62**, 1502 (1958)
522. Epstein, B. D., Dalle-Molle, E., Mattson, J. S.: Carbon **9**, 609 (1971)
523. Kinoshita, K., Bett, J. A. S.: Carbon **11**, 403 (1973)
524. Randin, J. P., Yeager, E.: J. Electroanal. Chem. **58**, 313 (1975)
525. Evans, J. F., Kuwana, T.: Anal. Chem. **49**, 1632 (1977)
526. Studebaker, M. L., Huffman, E. W. D., Wolfe, A. C., Nabors, L. G.: Ind. Eng. Chem. **48**, 162 (1956)
527. Matsumara, Y., Takahashi, H.: Carbon **17**, 109 (1979)
528. Zarifyanz, Yu. A., Kiselev, V. F., Lezhnev, N. N., Nikitina, O. V.: Carbon **5**, 127 (1967)
529. Nikitina, O. V., Kiselev, V. F., Lejnev, N. N.:: Carbon **8**, 402 (1970)
530. Scharrer, E., Boehm, H. P.: unpublished
531. Sappok, R., Boehm, H. P.: Carbon **6**, 283 (1968)
532. Rüdorff, W.: Advan. Inorg. Radiochem. **1**, 223 (1959); and Lagow, R. J., Badachhape, R. B., Wood, J. L., Margrave, J. L.: J. Chem. Soc. Dalton **1974**, 1268
532a. Kita, Y., Watanabe, N., Fujci, Y.: J. Amer. Chem. Soc. **101**, 3832 (1978)
533. Cadman, P., Scott, J. D., Thomas, J. M.: Carbon **15**, 75 (1977)
534. Cadman, P., Scott, J. D., Thomas, J. M.: J. Chem. Soc. Chem. Comm. **1975**, 654
535. Wibaut, J. P., LaBastide, G.: Rec. Trav. Chim. Pays-Bas **43**, 731 (1924)
536. Boehm, H. P., Tereczki, B., Schanz, K.: Proc. Colloque Internat. sur l'Adsorption aux Interfaces gaz/solide et liquide/solide, Aix-en-Provence, 1981, Elsevier, Amsterdam, 1982, p. 395
537. Puri, B. P., Hazra, R. S.: Carbon **9**, 123 (1971)
538. López-González, J. de D., Moreno-Castilla, C., Guerrero, A., Rodriguez-Reinoso, F.: Extended Abstracts, 15th. Bienn. Conference on Carbon, Philadelphia (1981), p. 250
539. Young, G. I.: J. Colloid Sci. **13**, 67 (1958); and de Boer, J. H., Hermans, M. E. A., Vleeskens, J. M.: Proc. Kon. Nederl. Akad. Wetensch., Ser. B. **60**, 45 (1957)
540. Boehm, H. P.: Discuss. Faraday Soc. **52**, 264 (1971)
541. Parks, G. A.: Chem. Rev. **65**, 177 (1965)
542. Clauss, A., Plass, R., Boehm, H. P., Hofmann, U.: Z. anorg. allg. Chem. **291**, 205 (1957)
543. Papirer, E., Donnet, J. B., Nguyen, V. T.: Bull. Soc. Chim. France **1971**, 1187; and Papirer, E., Guyon, E., Perol, N.: Carbon **16**, 133 (1978)
544. Matsumura, Y., Hagiwara, S., Takahashi, H.: Carbon **14**, 247 (1976)
545. Donnet, J. B., Papirer, E.: Rev. Gén. Caoutchouc Plast. **42**, 389 (1965)
546. Wershaw, R. L., Pinckney, D. J.: Science **199**, 906 (1978)
547. Boehm, H. P., Diehl, E.: Z. Elektrochem. **66**, 642 (1962)
548. Jester, C. P., Rocklin, R. D., Murray, R. W.: J. Electrochem. Soc. **127**, 1979 (1980)
549. Watkins, B. F., Behling, J. R., Kariv, E., Miller, L. L.: J. Amer. Chem. Soc. **97**, 3549 (1975)
550. Koval, C. A., Anson, F. C.: Anal. Chem. **50**, 223 (1978)
551. Bastick, M., Dupupet, G., Morlot, M., Perrot, J. M., Weber, J.: Carbon **15**, 25 (1977)
552. Elliott, C. M., Murray, R. W.: Anal. Chem. **48**, 1247 (1976)

553. Papirer, E., Nguyen, V. T., Donnet, J. B.: Carbon **16**, 141 (1978)
554. Papirer, E., Guyon, E.: Carbon **16**, 127 (1978)
555. Mazur, S., Matusinovic, T., Camann, K.: J. Amer. Chem. Soc. **99**, 3888 (1977)
556. Oyama, N., Brown, A. P., Anson, F. C.: J. Electroanal. Chem. **87**, 435 (1978)
557. Firth, B. E., Miller, L. L., Mitani, M., Rogers, Th., Lennox, J., Murray, R. W.: J. Amer. Chem. Soc. **98**, 8271 (1976)
558. Lennox, J. C., Murray, R. W.: J. Amer. Chem. Soc. **100**, 3710 (1978)
559. Zwier, T. A., Burke, M. F.: Anal. Chem. **53**, 812 (1981)
560. Sandler, S. R., Karo, W.: Organic Functional Group Preparation, New York, London: Academic Press 1971, Vol. 2, pp. 205–222
561. Harttig, H., Hüttinger, K. J.: Chem. Ing.-Techn. **50**, 880 (1978)
562. Harttig, H., Hüttinger, K. J.: J. Colloid Interface Sci. **78**, 295 (1980)
563. Oyama, N., Anson, F. C.: J. Amer. Chem. Soc. **101**, 1634 (1979); and J. Electroanal. Chem. **88**, 289 (1978)
564. Willman, K. W., Rocklin, R. D., Nowak, R., Kuo, K.-N., Schultz, F. A., Murray, R. W.: J. Amer. Chem. Soc. **102**, 7629 (1980)
565. Snell, K. D., Keenan, A. G.: Chem. Soc. Revs. **8**, 259 (1979)
566. Murray, R. W.: Acc. Chem. Res. **13**, 135 (1980)
567. Yacynych, A. M., Kuwana, T.: Anal. Chem. **50**, 640 (1978)
568. Lin, A. W. C., Yeh, P., Yacynych, A. M., Kuwana, T.: J. Electroanal. Chem. **84**, 411 (1977)
569. Oyama, N., Yap, K. B., Anson, F. C.: Electroanal. Chem. **100**, 233 (1979)
570. Donnet, J. B., Henrich, G.: Bull. Soc. Chim. France **1960**, 1609
571. Donnet, J. B., Henrich, G., Riess, G.: Rev. Gén. Caoutchouc **38**, 1803 (1961)
572. Donnet, J. B., Papirer, E., Vidal, A.: Grafting of Macromolecules onto Carbon Blacks. In: Chemistry and Physics of Carbon. Walker, P. L., Thrower, P. A. (eds.). New York, Basel: Dekker 1975. Vol. 12, pp. 171–207
573. Donnet, J. B., Peter, G., Riess, G.: J. Polymer. Sci. Part C **22**, 645 (1967)
574. Donnet, J. B., Vidal, A., Riess, G.: J. Chim. Phys. **68**, 1642 (1971); and Donnet, J. B., Vidal, A., Riess, G., Geldreich, L.: Rev. Gén. Caoutchouc, Plast. **47**, 1289 (1970)
575. Donnet, J. B., Riess, G., Majowski, G.: European Polym. J. **7**, 1065 (1971)
576. Tsubokawa, N., Takeda, N., Kanamaru, A.: Carbon **18**, 378 (1980)
577. Iler, R. K.: The Chemistry of Silica, New York, Chichester, Brisbane, Toronto: J. Wiley & Sons 1979
578. Janssen, K.: Doctoral thesis, Univ. Munich (1975)
579. Wartmann, J., Deuel, H.: Chimia (Aarau) **12**, 82 (1958)
580. Boehm, H. P., Schneider, M.: Z. anorg. allg. Chem. **301**, 326 (1959)
581. Gmelin, L.: Handbuch der anorganischen Chemie, 8th Edn., Schwefel, Syst.-No. 9B, Verlag Chemie, Weinheim (1963), p. 1796
582. Fery, N., Laible, R., Hamann, K.: Angew. Makromol. Chem. **34**, 81 (1973)
583. Duele, H., Wartmann, J., Hutschneker, K., Schobinger, U., Güdel, C.: Helv. Chim. Acta **42**, 1160 (1959)
584. McDaniel, M. P.: J. Phys. Chem. **85**, 532 (1981)
585. Hair, M. L., Hertl, W.: J. Phys. Chem. **77**, 2070 (1973)
586. Locke, D. C., Schmermund, J. T., Banner, B.: Anal. Chem. **44**, 90 (1972)
587. Saunders, D. H., Barford, R. A., Magidman, D., Olszewski, L. T., Rothbart, H. L.: Anal. Chem. **46**, 834 (1974)
588. McDaniel, M. P.: J. Phys. Chem. **85**, 537 (1981)
589. Morterra, C., Low, M. J. D.: J. Phys. Chem. **73**, 321 and 327 (1969)
590. Iler, R. K.: U.S. Pat. 2,657,149 (DuPont), 1953
591. Ballard, C. C., Broge, E. C., Iler, R. K., St. John, D. S., McWhorter, J. R.: J. Phys. Chem. **65**, 20 (1961)
592. Broge, E. C.: U.S. Pat. 2,736,668 (DuPont), 1956
593. Nicolaon, C. A., Teichner, S. J.: Bull. Soc. Chim. France **1968**, 4343
594. Stöber, W., Bauer, G., Thomas, K.: Liebigs Ann. Chem. **604**, 104 (1957)
595. Ref. 577, p. 692
596. Lange, K. R.: Chem. and Ind. **1969**, 1273

597. Neimark, I. E., Chertov, U. M., Sheinfain, R. Yu., Kruglikova, N. S.: Dokl. Akad. Nauk SSSR **132**, 1356 (1960)
598. Siffert, J., Delmon, B.: Bull. Soc. Chim. France **1970**, 2833
599. Halász, I., Sebastian, I.: Angew. Chem. **81**, 464 (1969); Angew. Chem. Intern. Ed. **8**, 453 (1969)
600. Low, M. J. D., Rhodes, Y. E., Orphanos, P. D.: J. Catal. **40**, 236 (1975)
601. Utsugi, H., Horikoshi, H., Matsuzawa, T.: J. Colloid Interface Sci. **50**, 154 (1975)
602. Kessaissia, Z., Papirer, E.: J. Chim. Phys. **75**, 709 (1978)
603. Kessaissia, Z., Papirer, E., Donnet, J. B.: J. Colloid Interface Sci. **82**, 526 (1981)
604. Brust, O. E., Sebastian, I., Halász, I.: J. Chromatogr. **83**, 15 (1973)
605. Kohlschütter, H.-W., Best, P., Wirzing, G.: Z. anorg. allg. Chem. **285**, 236 (1956)
606. Stöber, W.: Kolloid-Z. **149**, 39 (1956)
607. Fink, P., Camara, B., Wolleschensky, E., Kellner, E., Welz, P.: Z. Chem. **10**, 474 (1970)
608. Hambleton, F. H., Hockey, J. A.: Trans. Faraday Soc. **62**, 1694 (1968)
609. Peglar, R. J., Hambleton, F. H., Hockey, J. A.: J. Catal. **20**, 309 (1971)
610. Koltsov, S. I., Aleskovskii, V. B.: Zhur. Fiz. Khim. **42**, 1210 (1968)
611. Kunawicz, J., Jones, P., Hockey, J. A.: Trans. Faraday Soc. **67**, 848 (1971)
612. Peri, J. B., Hensley, A. L., Jr.: J. Phys. Chem. **72**, 2926 (1968)
613. Berg, K., Unger, K.: Kolloid-Z., Z. Polym. **246**, 682 (1971)
614. Hertl, W.: J. Phys. Chem. **72**, 1248 (1968)
615. Koltsov, S. I., Kuznetsova, G. N., Aleskovskii, V. B.: Izv. Akad. Nauk SSSR, Neorg. Mat. **3**, 894 (1967)
616. Weiss, Al., Weiss, Ar.: Z. anorg. allg. Chem. **276**, 95 (1954)
617. Hertl, W.: J. Phys. Chem. **72**, 3993 (1968)
618. Hertl, W., Hair, M. L.: J. Phys. Chem. **75**, 2181 (1971)
619. Yates, D. J. C., Dembinski, G. W., Kroll, W. R., Elliott, J. J.: J. Phys. Chem. **73**, 911 (1969)
620. Boehm, H. P., Schneider, M., Arendt, F.: Z. anorg. allg. Chem. **320**, 43 (1963)
621. Nondek, L., Vyskočil, V.: J. Chromatogr. **206**, 581 (1981)
622. Wartmann, J., Deuel, H.: Helv. Chim. Acta **42**, 1166 (1959)
623. Stark, F. O., Johannson, O. K., Vogel, G. E., Chaffee, R. G., Lacefield, R. M.: J. Phys. Chem. **72**, 2750 (1968)
624. Buzek, F., Rathousky, J.: J. Collod Interface Sci. **79**, 47 (1981)
625. Unger, K., Berg, K., Gallei, E.: Kolloid-Z., Z. Polym. **234**, 1108 (1969)
626. Boucher, L. J., Oswald, A. A., Murrell, L. L.: Prepr. Div. Pet. Chem., Am. Chem. Soc. **19**, 162 (1974)
627. Engelhardt, H., Mathes, D.: J. Chromatogr. **142**, 311 (1977)
628. Majors, R. E.: J. Chromatogr. Sci. **15**, 334 (1977)
629. Unger, K.: Merck-Kontakte **1979** (2), 32 (E. Merck A. G., Darmstadt, Germany)
630. Chuiko, A. A., Pavlik, G. E., Neimark, I. E.: Dopovidi Akad. Nauk Ukr. RSR **1963**, 230; C.A. **59**, 12836 (1963)
631. Abel, E. W., Polard, F. H., Uden, P. C., Nickless, G.: J. Chromatogr. **22**, 23 (1966)
632. Burwell, R. L., Leal, O.: J. Chem. Soc., Chem. Comm. **1974**, 342
633. Knox, J. H., Pryde, A.: J. Chromatogr. **112**, 171 (1975)
634. Becker, N., Unger, K.: Fresenius Z. Anal. Chem. **304**, 374 (1980)
635. Parr, W., Grohmann, K.: Tetrahedron Lett. **1971**, 2633
636. Parr, W., Grohmann, K.: Angew. Chem. **84**, 266 (1972); Angew. Chem., Intern. Ed. **11**, 314 (1972)
637. Novotný, M., Bektesh, S. L., Denson, K. B., Grohmann, K., Parr, W.: Anal. Chem. **45**, 971 (1973)
638. Engelhardt, H., Kromidas, S.: Naturwiss. **67**, 353 (1980)
639. Foucault, A., Caude, M., Oliveros, L.: J. Chromatogr. **185**, 345 (1979)
640. Wrighton, M. S., Austin, R. G., Bocarsly, A. B., Bolts, J. M., Haas, O., Legg. K. D., Nadjo, L., Palazzotto, M. C.: J. Amer. Chem. Soc. **100**, 1602 (1978)
641. Schumann, H., Pachaly, B.: Angew. Chem. **93**, 1092 (1981); Angew. Chem. Intern. Ed. **20**, 1043 (1981)
642. Haller, I.: J. Amer. Chem. Soc. **100**, 8050 (1978)

643. Fulcher, C., Crowell, M. A., Bayliss, R., Holland, K. B., Jezorek, J. R.: Anal. Chim. Acta **129**, 29 (1981)
644. Stöber, W.: Kolloid-Z., Z. Polym. **149**, 39 (1956)
645. Kiselev, A. V., Korolev, A. Ya., Petrova, R. S., Shcherbakova, K. D.: Kolloidn. Zh. **22**, 671 (1960)
646. Unger, K., Berg, K., Gallei, E., Erdel, G.: Fortschrittsber. Kolloide Polym. **55**, 34 (1971)
647. Roumeliotis, P., Unger, K.: J. Chromatogr. **149**, 211 (1978)
648. Knözinger, H., Rumpf, E.: Inorg. Chim. Acta **30**, 51 (1978)
649. Castrillo, T., Knözinger, H., Wolf, M.: Inorg. Chim. Acta **45**, L235 (1980)
650. Horner, L., Ziegler, H.: Liebigs Ann. Chem. **1976**, 628
651. Burggraf, L. W., Kendall, D. S., Leyden, D. E., Pern, F. J.: Anal. Chim. Acta **129**, 19 (1981)
652. Leyden, D. E., Luttrell, G. H.: Anal. Chem. **47**, 1612 (1975)
653. Kochloefl, K., Liebelt, W., Knözinger, H.: J. Chem. Soc., Chem. Comm. **1977**, 510
654. Neimark, I. E., Chertov, V. M.: Dokl. Akad. Nauk SSR **138**, 877 (1961)
655. Unger, K., Berg, K.: Z. Naturforsch. **24b**, 454 (1969)
656. Weigand, N., Sebastian, I., Halász, I.: J. Chromatogr. **102**, 325 (1974)
657. Horner, L., Roth, W., Ziegler, H.: Naturwissensch. **63**, 39 (1976)
658. Hara, S., Dobashi, A.: J. Chromatogr. **186**, 543 (1979)
659. Kimura, K., Nakajima, M., Shono, T.: Anal. Lett. **13**, 741 (1980)
660. Shields, G. D., Boucher, L. J.: J. Inorg. Nucl. Chem. **40**, 1341 (1978)
661. Novotný, M., Bektesh, S. L., Grohmann, K.: J. Chromatogr. **83**, 25 (1973)
662. Allum, K. G., Hancock, R. D., Howell, I. V., McKenzie, S., Pitkethly, R. C., Robinson, P. J.: J. Organomet. Chem. **87**, 203 (1975)
663. Boehm, H. P.: Chem.-Ing.-Techn. **46**, 716 (1974)
664. Hill, J. M.: J. Chromatogr. **76**, 455 (1973)
665. Sugawara, K. F., Weetall, H. H., Schucker, G. D.: Anal. Chem. **46**, 489 (1974)
666. Bailey, D. C., Langer, S. H.: Chem. Revs. **81**, 109 (1981)
667. Jackson, S. D., Wells, P. B., Whyman, R., Worthington, P.: Metal Clusters and Cluster Catalysis. In: Catalysis, Vol. 4. Kemball, C., Dowden, D. A. (eds.). London: The Royal Society of Chemistry 1981, pp. 75–99
668. Horner, L., Schumacher, F.: Liebigs Ann. Chem. **1976**, 633
669. Smith, A. K., Basset, J. M., Maitlis, P. M.: J. Mol. Catal. **2**, 223 (1977)
670. Studer, D. W., Schrader, G. L.: J. Mol. Catal. **9**, 169 (1980)
671. Castrillo, T., Knözinger, H., Lieto, J., Wolf, M.: Inorg. Chim. Acta **44**, L 239 (1980)
672. Knözinger, H., Thornton, E. W., Wolf, M.: J. Chem. Soc., Faraday Trans. I **75**, 1888 (1979)
673. Thornton, E. W., Knözinger, H., Tesche, B., Rafalko, J. J., Gates, B. C.: J. Catal. **62**, 117 (1980)
674. Bilhou, J. L., Bilhou-Bougnol, V., Graydon, W. F., Basset, J. M., Smith, A. K.: J. Mol. Catal. **8**, 411 (1980)
675. Brown, S. C., Evans, J.: J. Chem. Soc., Chem. Comm. **1978**, 1063
676. Pierantozzi, R., McQuade, K. J., Gates, B. C., Wolf, M., Knözinger, H., Ruhmann, W.: J. Amer. Chem. Soc. **101**, 5436 (1979)
677. Castrillo, T., Knözinger, H., Wolf, M., Tesche, B.: J. Mol. Catal. **11**, 151 (1981)
678. Castrillo, T., Knözinger, H., Wolf, M.: Inorg. Chim. Acta **45**, L 235 (1980)
679. Wild, F. R. W. P., Gubitosa, G., Brintzinger, H. H.: J. Organomet. Chem. **148**, 73 (1978)
680. Howe, R. F., Davidson, D. E., Whan, D. A.: J. Chem. Soc., Faraday Trans. I **68**, 2266 (1972)
681. Howe, R. F.: Inorg. Chem. **15**, 486 (1976)
682. Adams, D. M., Gardner, I. R., Parkyns, N. D.: J. Catal. **45**, 145 (1976)
683. Anderson, J. R., Elmes, P. S., Howe, R. F., Mainwaring, D. E.: J. Catal. **50**, 508 (1977)
684. Robertson, J., Webb, G.: Proc. Roy, Soc., Ser. A **341**, 383 (1974)
685. Kuznetsov, V. L., Bell, A. T., Yermakov, Yu. I.: J. Catal. **65**, 374 (1980)
686. Smith, G. C., Chojnacki, T. P., Dasgupta, S. R., Iwatate, K., Watters, K. L.: Inorg. Chem. **14**, 1419 (1975)

687. Conrad, H., Ertl, G., Knözinger, H., Küppers, J., Latta, E. E.: Chem. Phys. Lett. **42**, 115 (1976)
688. Theolier, A., Smith, A. K., Leconte, M., Basset, J. M., Zanderighi, G. M., Psaro, R., Ugo, R.: J. Organomet. Chem. **191**, 415 (1980)
689. Howe, R. F.: J. Catal. **50**, 196 (1977)
690. Watson, P. L., Schrader, G. L.: J. Mol. Catal. **9**, 129 (1980)
691. Deeba, M., Gates, B. C.: J. Catal. **67**, 303 (1981)
692. Psaro, R., Ugo, R., Zanderighi, G. M., Besson, B., Smith, A. K., Basset, J. M.: J. Organomet. Chem. **213**, 215 (1981)
693. Smith, A. K., Besson, B., Basset, J. M., Psaro, R., Fusi, A., Ugo, R.: J. Organomet. Chem. **192**, C31 (1980)
694. Commereuc, D., Chauvin, Y., Hugues, F., Basset, J. M., Olivier, D.: J. Chem. Soc., Chem. Comm. **1980**, 154
695. Ward, M. D., Harris, T. V., Schwartz, J.: J. Chem. Soc., Chem. Comm. **1980**, 357
696. Ward, M. D., Schwartz, J.: J. Amer. Chem. Soc. **103**, 5253 (1981)
697. Ward, M. D., Schwartz, J. J. Mol. Catal. **11**, 397 (1981)
698. Karol, F. J., Wu, Ch., Reichle, W. T., Maraschin, N. J.: J. Catal. **60**, 68 (1979)
699. Young, R. P.: Can. J. Chem. **47**, 2237 (1969)
700. Jewett, D., Lawless, J.: Naturwiss. **68**, 570 (1981)
701. Lange, K. R.: Chem. and Ind. **1968**, 441
702. Bridger, K., Vincent, B.: Europ. Polym. J. **16**, 1017 (1980)
703. Boehm, H. P., Schneider, M., Wistuba, H.: Angew. Chem. **77**, 622 (1965); Angew. Chem. Intern. Ed. **4**, 600 (1965)
704. Schrader, R., Wissing, R., Kubsch, H.: Z. anorg. allg. Chem. **365**, 191 (1969)
705. Strelko, V. V., Kanibolotskii, V. A., Vysotskii, Z. Z.: Zh. Fiz. Khim. **42**, 1219 (1968)
706. Hastings, C. R., Aue, W. A., Augl, J. M.: J. Chromatogr. **53**, 487 (1970)
707. Dietz, E., Fery, N., Hamann, K.: Angew. Makromol. Chem. **35**, 115 (1974)
708. Papirer, E., Nguyen, V. T.: Angew. Makromol. Chem. **28**, 31 (1973)
709. Bridger, K., Fairhurst, D., Vincent, B.: J. Colloid Interface Sci. **68**, 190 (1979)
710. Yamashita, S., Kohjiya, S., Kita, A., Shimizu, A.: J. Appl. Polym. Sci. **17**, 2935 (1973)
711. Weetall, H. H.: Nature (London) **223**, 959 (1969)
712. Weetall, H. H.: Anal. Chem. **46**, 602 A (1974)
713. Suckling, C. J.: Chem. Soc. Revs. **6**, 215 (1977)
714. Jones, P., Hockey, J. A.: Trans. Faraday Soc. **67**, 2669 (1971)
715. Parfitt, G. D., Ramsbotham, J. R., Rochester, C. H.: Trans. Faraday Soc. **67**, 841 (1971)
716. Parfitt, G. D., Ramsbotham, J. R., Rochester, C. H.: Trans. Faraday Soc. **67**, 3100 (1971)
717. Vidal, A., Papirer, E., Donnet, J.-B.: J. Chim. Phys. **71**, 445 (1974)
718. Greenler, R. G.: J. Chem. Phys. **37**, 2094 (1962)
719. Knözinger, H., Bühl, H., Ress, E.: J. Catal. **12**, 121 (1968)
720. Jackson, P., Parfitt, G. D.: J. Chem. Soc., Faraday Trans. I **68**, 1443 (1972)
721. Peri, J. B.: J. Phys. Chem. **70**, 1482 (1966)
722. Primet, M., Basset, J., Mathieu, M.-V., Prettre, M.: J. Phys. Chem. **74**, 2868 (1970)
723. Basset, J., Mathieu, M.-V., Prettre, M.: Rev. Chim. Min. **5**, 870 (1968)
724. Basset, J., Mathieu, M.-V., Prettre, M.: J. Chim. Phys. **66**, 611 (1969)
725. Moss, J. H., Parfitt, G. D., Wright, A.: Colloid Polym. Sci. **256**, 1121 (1978)
726. Harrison, P. G., Thornton, E. W.: J. Chem. Soc. Faraday I **74**, 2604 (1978)
727. Solymosi, F., Rasko, J.: J. Catal. **65**, 235 (1980)
728. Liefländer, M., Stöber, W.: Z. Naturforsch. **15b**, 411 (1960)
729. Stöber, W., Liefländer, M., Bohn, E.: Topochemische Grenzflächenreaktionen an Oxydoberflächen. In: Beitr. Silikose-Forschung, Grundfragen der Silikoseforschung. Hauptverwaltung der Bergbau-Berufsgenossenschaften (eds.). Bochum, 1960, Vol. 4, p. 111
730. Isirikyan, A. A., Kiselev, A. V., Ushakova, E. V.: Kolloidn. Zh. **25**, 125 (1963)
731. Thornton, E. W., Harrison, P. G.: J. Chem. Soc. Faraday I **71**, 2468 (1975)
732. Evans, J. V., Whateley, T. L.: Trans. Faraday Soc. **63**, 2769 (1967)
733. Parkyns, N. D.: J. Chem. Soc., Ser. A **1969**, 410
734. Primet, M., Pichat, P., Mathieu, M.-V.: J. Phys. Chem. **75**, 1221 (1971)
735. Jackson, P., Parfitt, G. D.: J. Chem. Soc. Faraday I **68**, 896 (1972)

736. Boehm, H. P.: Kolloid Z., Z. Polymere **227**, 17 (1968)
737. Kiselev, A. V., Uvarov, A. V.: Surface Sci. **6**, 399 (1967)
738. Fink, P.: Z. Chem. **7**, 284 (1967)
739. Deo, A. V., Chuang, T. T., Dalla Lana, I. G.: J. Phys. Chem. **75**, 234 (1971)
740. Fry, H. S., Schulze, E. L.: J. Amer. Chem. Soc. **48**, 958 (1926)
741. Steyrer, R., Boehm, H. P.: unpublished
742. Ahmadi, A., Blouri, B.: Bull. Soc. Chim. France **1969**, 171
743. Seebald, H. J., Schunack, W.: Arch. Pharmaz. **305**, 406 and 785 (1972)
744. Harrison, P. G., Thornton, E. W.: J. Chem. Soc. Faraday I **72**, 2484 (1976)
745. Takezawa, A.: J. Chem. Soc. Chem. Comm. **1971**, 1451
746. Buckland, A. D., Rochester, C. H., Topham, S. A.: J. Chem. Soc. Faraday I **76**, 302 (1980)
747. Lorenzelli, V., Busca, G., Sheppard, N.: J. Catal. **66**, 28 (1980)
748. Flaig-Baumann, R., Herrmann, M., Boehm, H. P.: Z. anorg. allg. Chem. **372**, 296 (1970)
749. Fujihira, M., Ohishi, N., Osa, T.: Nature (London) **268**, 226 (1977)
750. Fujihira, M., Kubota, T., Osa, T.: J. Electroanal. Chem. **119**, 379 (1981)
750. Fujihira, M., Ohishi, N., Osa, T.: J. Electroanal. Chem. **119**, 379 (1981)
751. Fox, M. A., Nobs, F. J., Voynick, T. A.: J. Amer. Chem. Soc. **102**, 4029 (1980)
752. Ghosh, P. K., Spiro, T. G.: J. Amer. Chem. Soc. **102**, 5543 (1980)
753. Harrison, P. G., Thornton, E. W.: J. Chem. Soc. Faraday I **72**, 1317 (1976)
754. Snell, K. D., Keenan, A. G.: Chem. Soc. Revs. **8**, 259 (1979)
755. Parfitt, G. D., Ramsbotham, J., Rochester, C. H.: J. Chem. Soc. Faraday I **68**, 17 (1972)
756. Moses, P. R., Wier, L. M., Lennox, J. C., Finklea, H. O., Lenhard, J. R., Murray, R. W.: Anal. Chem. **50**, 576 (1978)
757. Finklea, H. O., Murray, R. W.: J. Phys. Chem. **83**, 353 (1979)
758. Wrighton, M. S., Palazzotto, M. C., Bocarsly, A. B., Bolts, J. M., Fischer, A. B., Nadjo, L.: J. Amer. Chem. Soc. **100**, 7264 (1978)
759. White, H. S., Murray, R. W.: Anal. Chem. **51**, 236 (1979)
760. Willmann, K. W., Greer, E., Murray, R. W.: Nouv. J. Chim. **3**, 455 (1979)
761. Moses, P. R., Wier, L., Murray, R. W.: Anal. Chem. **47**, 1882 (1975)
762. Osa, T., Fujihira, M.: Nature (London) **264**, 349 (1976)
763. Hawn, D. D., Armstrong, N. R.: J. Phys. Chem. **82**, 1288 (1978)
764. Tomkiewicz, M.: J. Electrochem. Soc. **127**, 1518 (1980)
765. Armstrong, N. R., Lin, A. W. C., Fujihira, M., Kuwana, T.: Anal. Chem. **48**, 741 (1976)
766. Moses, P. R., Murray, R. W.: J. Amer. Chem. Soc. **98**, 7435 (1976)
767. Davis, D. G., Murray, R. W.: Anal. Chem. **49**, 194 (1977)
768. Firth, B. E., Miller, L. L.: J. Amer. Chem. Soc. **98**, 8272 (1976)
769. Bauer, F., Hüttinger, K. J.: Chem. Ing. Techn. **51**, 742 (1979)
770. Hugues, F., Smith, A. K., Taarit, Y. B., Basset, J. M., Commereuc, D., Chauvin, Y.: J. Chem. Soc. Chem. Comm. **1980**, 68
771. Hieber, W., Leutert, F.: Z. anorg. allg. Chem. **204**, 145 (1932)
772. Hieber, W., Brendel, G.: Z. anorg. allg. Chem. **289**, 324 (1957)
773. Smith, A. K., Hugues, F., Theolier, A., Basset, J. M., Ugo, R., Zanderighi, G. M., Bilhou, J. L., Bilhou-Bougnol, V., Graydon, W. F.: Inorg. Chem. **18**, 3105 (1979)
774. Brenner, A., Burwell, R. L.: J. Amer. Chem. Soc. **97**, 2565 (1975)
775. Burwell, R. L., Brenner, A.: J. Mol. Catal. **1**, 77 (1975/76)
776. Bilhou, J. L., Theolier, A., Smith, A. K., Basset, J. M.: J. Mol. Catal. **3**, 245 (1977/78)
777. Bowman, R. G., Burwell, R. L.: J. Catal. **63**, 463 (1980)
778. Smith, J., Howe, R. F., Whan, D. A.: J. Catal. **34**, 191 (1974)
779. Bjorklund, R. B., Burwell, R. L.: J. Colloid Interface Sci. **70**, 383 (1979)
780. Parkyns, N. D.: Proc. Int. Congr. Catal., 3rd., Amsterdam, 1964, **2**, 914 (1965)
781. Brenner, A., Hucul, D. A.: Inorg. Chem. **18**, 2836 (1979)
782. McVicker, G. B., Vannice, M. A.: J. Catal. **63**, 25 (1980)
783. Smith, A. K., Theolier, A., Basset, J. M., Ugo, R., Commereuc, D., Chauvin, Y.: J. Amer. Chem. Soc. **100**, 2590 (1978)

784. Watters, K. L., Howe, R. F., Chojnacki, T. P., Fu, C.-M., Schneider, R. L., Wong, N. B.: J. Catal. **66**, 424 (1980)
785. Anderson, S. L. T., Watters, K. L., Howe, R. F.: J. Catal. **69**, 212 (1981)
786. Knözinger, H., Zhao, Y.: J. Catal. **71**, 337 (1981)
787. Deeba, M., Streusand, B. J., Schrader, G. L., Gates, B. C.: J. Catal. **69**, 218 (1981)
788. Deeba, M., Scott, J. P., Barth, R., Gates, B. C.: J. Catal. **71**, 373 (1981)
789. Knözinger, H., Zhao, Y., Tesche, B., Barth, R., Epstein, R., Gates, B. C., Scott, J. P.: Faraday Sisc. **72**, 53 (1982)
790. Eady, C. R., Johnson, B. F. G., Lewis, J.: J. Chem. Soc. Dalton **1976**, 1706
791. Dent, A. L., Kokes, R. J.: J. Phys. Chem. **73**, 3772 (1969)
792. Boccuzzi, F., Borello, E., Zecchina, A., Bossi, A., Camia, M.: J. Catal. **51**, 150 (1978)
793. Coluccia, S., Boccuzzi, F., Ghiotti, G., Mirra, C.: Z. phys. Chem. (Wiesbaden) **121**, 141 (1980)
794. Chang, C. C., Kokes, R. J.: J. Catal. **38**, 491 (1975)
795. Chang, C. C., Kokes, R. J.: J. Catal. **28**, 92 (1973)
796. Dent, A. L., Kokes, R. J.: J. Amer. Chem. Soc. **92**, 1092 and 6709 (1970)
797. Dent, A. L., Kokes, R. J.: J. Phys. Chem. **75**, 487 (1971)
798. Dent, A. L., Kokes, R. J.: J. Phys. Chem. **74**, 3653 (1970)
799. Chang, C. C., Kokes, R. J.: J. Amer. Chem. Soc. **92**, 7517 (1970)
800. Nguyen, T. T., Lavalley, J. C., Saussey, J., Sheppard, N.: J. Catal. **61**, 503 (1980)
801. Nollen, K., Kaden, V., Hamann, K.: Angew. Makromol. Chem. **6**, 1 (1969)
802. Vidal, A., Schultz, J., Papirer, E., Donnet, J.-B.: Compt. Rend. Acad. Sci. (Paris), Ser. C **277**, 1331 (1973)
803. Grunwald, P., Gunsser, W.: Naturwissensch. **65**, 60 (1978)
804. Weetall, H. H., Detar, C. C.: Biotechnol. Bioeng. **17**, 295 (1975)
805. Tauster, S. J., Fung, S. C., Garten, R. L.: J. Amer. Chem. Soc. **100**, 170 (1978)
806. Tauster, S. J., Fung, S. C., Baker, R. T. K., Horsley, J. A.: Science (Washington) **211**, 1121 (1981)
807. Ioffe, M. S., Kuznetsov, B. N., Ryndin, Yu. A., Yermakov, Yu. I.: Proc. 6th Congr. Catal., London, 1976, **1**, 131 (1977)
808. Lunsford, J. H.: Catal. Rev. **8**, 135 (1973)
809. Che, M., Naccache, C., Imelik, B.: J. Chim. Phys. **65**, 1301 (1968)
810. Naccache, L., Mériaudeau, P., Che, M., Tench, A. J.: Trans. Faraday Soc. **67**, 506 (1971)
811. Chon, H., Pajares, J.: J. Catal. **14**, 257 (1969)
812. Tanaka, K., Blyholder, G.: J. Phys. Chem. **76**, 3184 (1972)

Chapter 3

Kinetics of Chemical Processes on Well-defined Surfaces

Gerhard Ertl

Institut für Physikalische Chemie, Universität München, Sophienstraße 11,
D-8000 München 2, FRG

Contents

1. Introduction

The rate of product formation in a catalytic reaction depends on 'extrinsic' parameters (which can be varied for a given catalyst) such as partial pressures, temperature, flow rate etc. as well as on 'intrinsic' parameters which are defined by the nature of the catalyst itself. The steady-state rate is in general a consequence of a rather complex series of elementary steps consisting of adsorption, desorption, surface reaction and migration processes whose rate constants in turn depend on various parameters. It is therefore practically impossible to draw any firm conclusions on the reaction mechanism or the individual rate constants on the basis of kinetic measurements for the overall reaction. In order to elucidate the microscopic processes it is obviously necessary to perform studies along the following two lines:

i) Apart from the extrinsic also the intrinsic parameters should be varied in a systematic manner. That means the use of model systems with well-defined surface structure (single crystals with varying orientation and eventually defect geometry and concentrations) and chemical composition (either atomically clean or with well characterized additives) rather than of 'real' catalysts with their complex surface properties. This includes necessarily the need of ultrahigh vacuum (UHV) techniques as well as of methods for preparing and characterizing the structural and chemical properties of these surfaces.

ii) Instead of analyzing the reactant and product molecules before and after interaction with the catalyst, the properties of the surface species themselves should be investigated. The development of a great variety of surface spectroscopic techniques [1] during the last two decades offers indeed the possibility for in depth characterization of not too complicated systems. Since these methods are based for the most part on the interactions of particles (electrons, ions, neutrals) with matter, a serious disadvantage is associated with their application: In situ measurements can only be performed at pressures $\lesssim 10^{-2}$ Pa which is of course far outside the atmospheric pressure range usually applied in catalysis.

Table 1 lists the essential differences between the 'surface science approach' and the conditions of 'real catalysis' which appear at a first glance to be so marked that hardly any connection between both fields might be suspected. There exist however, various ways along which this apparent gap can be bridged:

a) Systematic studies with various single crystal planes including the introduction of well-defined defects (such as ordered step arrays [2]) can for example be compared with measurements using varying particle size, since

Table 1. Relation between surface science and catalysis

	Surface Science	'Real' catalysis
Surface composition	clean surface	undefined (promotors?)
Surface structure	single crystal	polycrystalline ('active sites'?)
Pressure	$<10^{-2}$ Pa	$>10^{5}$ Pa

in the latter case the relative portions of exposed crystal planes will systematically vary [3].

b) Analysis of the actual surface composition of a catalyst (as close as possible to its working conditions) provides the starting point for deliberately modifying the chemical nature of a model surface. Even a badly defined catalyst surface will frequently turn out to become essentially clean under reaction conditions, i.e. by removal of impurities or reduction of oxides by the molecules involved.

c) The pressure gap is often considered to offer the most serious problem. High pressure cells have been developed [4] which allow to run the reaction under atmospheric (or even higher) pressures and to analyse the surface afterwards after evacuation without exposing it to air. What is more important, however, is the fact that the kinetics will be primarily determined by the surface concentration rather than by the partial pressure. Higher coverages can, however, also be achieved by lowering the temperature instead of rising the pressure. Apart from this it would certainly be wrong to conclude that catalytic reactions will always proceed at high surface concentrations: The stationary coverage of a certain species will in general be determined by the kinetics of various parallel, consecutive or competitive reaction steps and may thus be indeed rather low even if high pressures are applied. Information on the reaction mechanism may further be obtained by studying the reverse reaction (at low pressure) since the catalyst will only accelerate the rate by which the equilibrium is established and since forward and backward reaction have necessarily proceed through the same microscopic steps.

The present chapter will start with an outline of the theoretical concepts describing the kinetics of elementary surface reactions and of the corresponding experimental verification. Then several examples will be presented for which the 'surface science approach' has yielded a more or less closed picture, whereby special attention will always be paid to corresponding relations with 'real' catalytic systems. Since the whole area is still in its infancy, this contribution has necessarily to be incomplete and will in several respects even be rudimentary. Most probably also major revisions of some of the conclusions will become necessary in the coming years. On the other hand, the detailed and unique information which can be obtained may serve to explain the strongly growing interest in this field of chemistry.

2. Kinetic Concepts

A. General

The basic types of elementary processes involved in a catalytic reaction are schematically illustrated by Figure 1. These include adsorption with or without bond breaking (dissociative and non-dissociative) and the reverse steps of desorption, as well as surface migration and finally chemical transformation, i.e. the formation of new molecules. The rate of product formation will then be determined by the individual rate constants (which are a function of temperature and coverage) and by the surface concentrations c_i as well as by the surface configurations of the individual species. The latter aspect is often neglected in simplified treatments (*i.e.* a random distribution of the adsorbed particles is assumed), but may indeed become rather important. A more refined analysis has to take into account the possible existence of intermediates between the gaseous and the chemisorbed state (so-called 'precursor' states) which may also be of vital importance for the kinetics.

The organization of this chapter will start with some basic concepts on the ground-state properties of adsorbed layers and will then describe the kinetic aspects associated with increasing complexity.

B. The Adsorbed State

The properties of chemisorbed systems have been investigated quite extensively during the past years [5]. With regards to the electronic, energetic and structural properties at least for non-dissociative chemisorption on transition metal surfaces remarkable analogies with corresponding complex compounds are observed [6]. A single adsorbed molecule will exhibit a most favourable configuration with respect to the locations of the substrate atoms within the unit cell of a (single crystal) surface as for example shown for CO/Pd(100) in Figure 2 [7]. The energy difference between that state and a CO molecule in the gas phase equals the adsorption energy (at zero coverage) which in the present case was determined to be 160 kJ mol^{-1}. If the ligand is displaced from its equilibrium position in a direction perpendicular to the surface the potential energy will vary as shown schematically in Figure 3. Information on the shape of this potential near its minimum is contained in the force constant for vibration: For the CO/Pd(100) system the frequency of the M—C vibration mode was determined to be 236 cm^{-1}

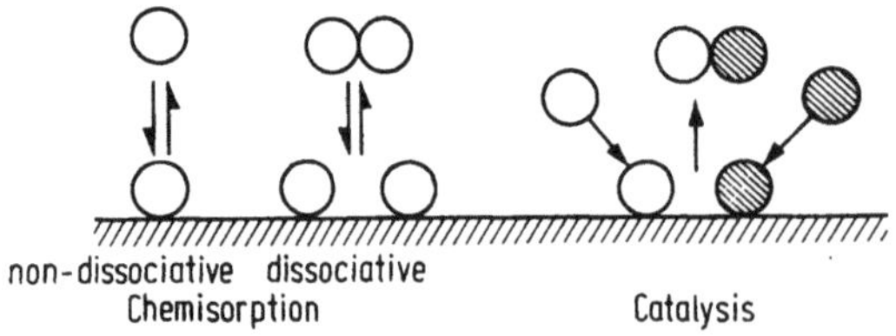

Figure 1. Schematic illustration of the basic elementary steps involved in a heterogeneously catalysed reaction

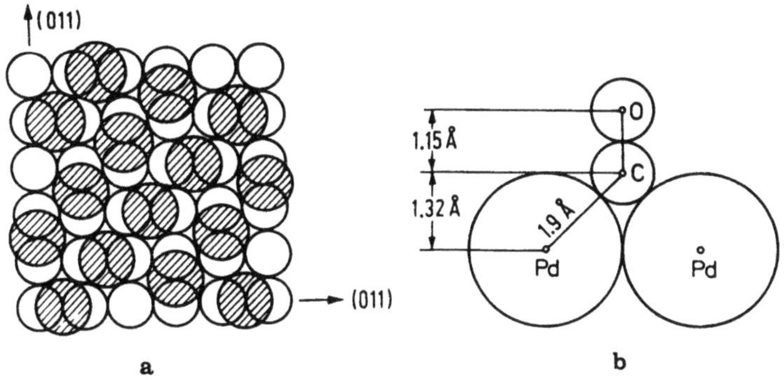

Figure 2. Structure of CO adsorbed on a Pd (100) surface at relative coverage. $\theta = 0.5$. (Reproduced with permission from ref. [7])

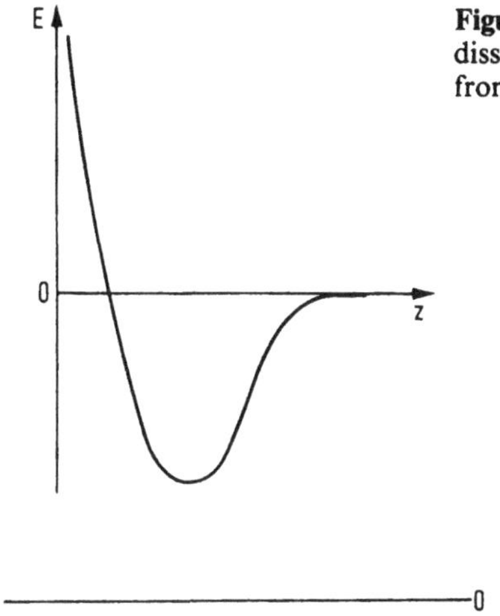

Figure 3. Schematic variation of the potential for a non-dissociatively adsorbed particle with vertical distance z from the surface

[8]. Since the surface exhibits two-dimensional periodicity the depth of the potential minimum will vary accordingly if the adsorbed particle is moved parallel to the surface. This effect is illustrated schematically in Figure 4 for one-dimensional motion along a certain crystallographic orientation. The barrier height E^* determines the activation energy for surface diffusion in that particular direction.

This activation barrier will now determine the actual state of the particle on the surface [9].

i) If $E^* > kT$, the adsorption is localized. The translational (and rotational) degrees of freedom of the adsorbed particle are mostly transformed into vibrational modes localized near the 'adsorption sites'. Surface migration

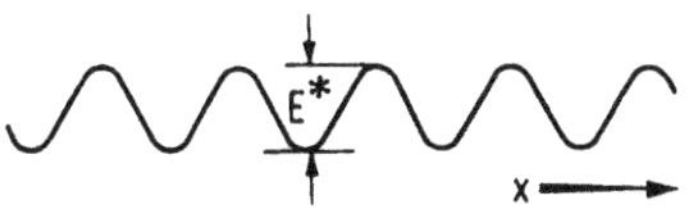

Figure 4. Variation of the potential of an adsorbed particle along a particular direction parallel to the surface

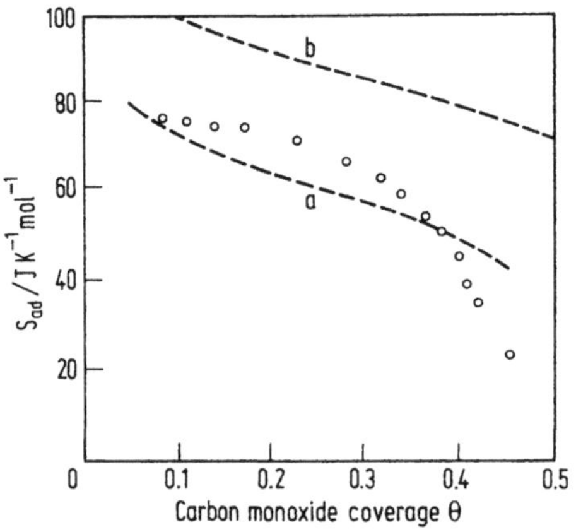

Figure 5. Adsorption entropy, S_{ad}, for CO on Pd (100) as a function of relative coverage, θ. Dark circles: Experimental data. Curve *a*: Calculated for localised adsorption. Curve *b*: Calculated for delocalised adsorption. (Reproduced with permission from ref. [7])

occurs by hopping from one potential minimum to a neighboring one and to a good approximation the adsorbed particle can be considered as occupying sites on a two-dimensional lattice.

ii) If $E^* < kT$, the adsorption is delocalized, *i.e.* the adsorbed particles exhibit a high mobility parallel to the surface and may fairly well be described by a two-dimensional gas.

The degree of localisation will reflect itself also in macroscopic properties of an adsorbed system, such as for example the adsorption entropy which can be derived from measured adsorption isotherms (*i.e.* the equilibrium adsorbed amount as a function of pressure at various temperatures) [10]. Figure 5 shows experimentally determined values for the differential entropy of CO adsorbed on Pd(100) around 450 K as a function of coverage together with theoretical curves approximately calculated for the two limiting cases of perfect localization and two-dimensional gas [7]. At low coverages the experimental data fall well between the two theoretical limits and thus justify the underlying models, but at higher coverages they drop even below

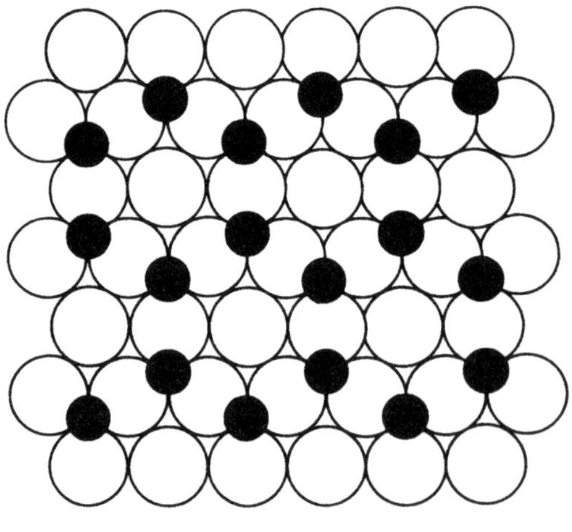

Figure 6. Structure formed by adsorbed H atoms on a Ni(111) surface at relative coverage $\theta = 0.5$ and $T \leqq 300$ K. (Reproduced with permission from ref. [11])

the curve for a lattice model with random occupation of the sites: Obviously additional effects come now into play causing a higher degree of order.

Numerous investigations with low energy electron diffraction (LEED) demonstrated that the formation of ordered adsorbed phases with unit cells differing from those of the substrate lattice is more the rule than the exception. The periodic structure formed by CO/Pd(100) at a coverage $\theta = 0.5$ is reproduced in Figure 2b. Figure 6 shows the structure formed by H atoms on a Ni(111) surface at $\theta = 0.5$ and low temperatures as another example [11]. The reason for this effect has to be sought in the operation of mutual interactions between the adsorbed particles: At low enough temperatures the free energy will be minimized by the formation of long-range order through lowering of the internal energy which overcompensates the loss in entropy.

Classification may be performed into direct (through space) or indirect (through bond) interactions [12] in analogy to the effects discussed for organic molecules [13], although pure through space interactions may be difficult to isolate. The basic underlying mechanisms comprise dipole-dipole interactions, orbital overlap, elastic interactions as well as interactions mediated through the valence electrons of the substrate which are involved in the chemisorption bond. It is of essential importance that these interactions may be repulsive as well as attractive, depending on the mutual configuration of two neighboring particles. In addition, also non-pairwise interactions may come into play [14]. Their strength is frequently only of the order of a few kcal/mole which nevertheless may give rise to several consequences which are important in the present context.

i) Structure of the Adsorbed Layer. Superposition of the two-dimensional potential produced by the substrate lattice (Figure 4) and of the sum of the interactions with other adsorbates will determine the actual energy of an adsorbed particle. If the substrate corrugation dominates, the particles will form a lattice with identical local configuration, such as the H/Ni(111) system shown in Figure 6. As a consequence of short-range repulsion usually not all *a priori* equivalent sites can be occupied. If the adparticle interaction dominates, so-called incoherent structures may be formed in

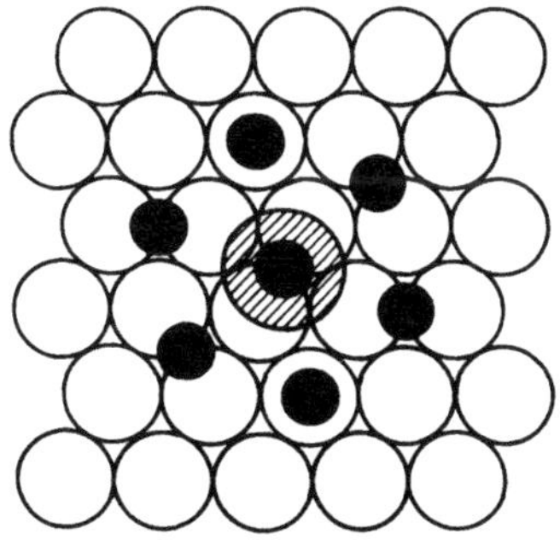

Figure 7. Structure of CO adsorbed on Ni(111) at relative coverage $\theta = 0.57$. (Reproduced with permission from ref. [15])

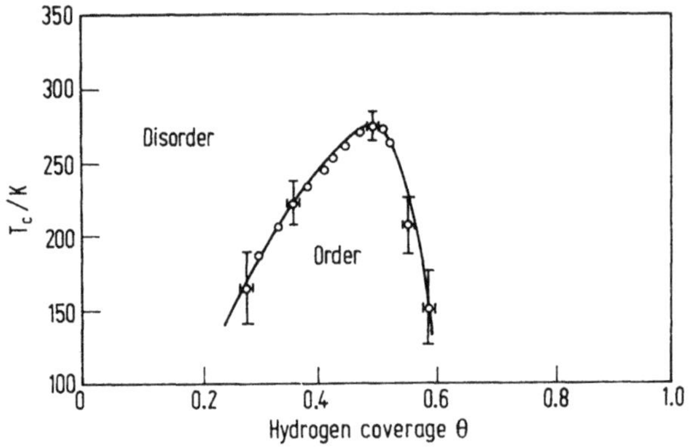

Figure 8. Phase diagram for H adsorbed on Ni(111) [11]. The data points mark the order-disorder transition temperatures, T_c, for varying relative coverages, θ, as determined by LEED.

which the registry with the substrate varies. The structure model for CO/Ni(111) at high coverage shown in Figure 7 is an example of this kind [15]. It is quite evident that these effects will cause a break-down of the simple concept of a fixed number of well-defined adsorption sites.

If the temperature is increased the interaction energies can be overcome by kT giving rise to order-disorder transitions, *i.e.* discontinuous changes of the mutual configurations of the adsorbed particles. Figure 8 shows the relatively simple phase diagram for the system H/Ni(111) [11], but these may also be much more complicated (as with three-dimensional) systems. If repulsive interactions are dominating, the adsorbed particles tend to stay apart from each other and will therefore uniformly cover the whole available surface. If, on the other hand, attraction prevails there will be a tendency for condensation: The surface then consists of islands with high local coverage and of patches which are almost bare.

ii) Thermodynamic Properties. The effects mentioned last will of course be of direct consequence for the adsorption energy. In the case of repulsive interactions the mean distance between the adsorbed particles decreases with increasing coverage and the repulsion then causes a lowering of the effective adsorption energy. If attractions dominate, the adsorption energy will at first increase until islands are formed which then continuously grow, the adsorption energy then remaining constant. Generally speaking, the operation of interactions will cause an induced heterogeneity of surfaces which are a priori energetically homogeneous. Adsorption isotherms are no longer expected to follow the Langmuir equation (which is based on a constant adsorption energy and a fixed number of adsorption sites), and necessary modifications by including interactions can indeed be found already in the earlier literature [16].

iii) Kinetics of Adsorption. The non-constant adsorption energy as well as deviations from a random distribution of the adsorbed particles will influence the variation of the sticking coefficient with coverage, as will be discussed later.

iv) Surface Diffusion. Interaction potentials will of course also influence the barriers of surface diffusion. Additional effects are to be expected if order formation comes into play.

v) Kinetics of Desorption. The rate of desorption can generally be written in Arrhenius form as

$$r_d = n_s^x \cdot \nu_d \cdot \exp(-E_d/RT) \tag{1}$$

(see section 2.E), where n_s is the density of adsorbed particles, x the reaction order for desorption, ν_d the preexponential, and E_d the activation energy for desorption.

Since E_d is directly related to the depth of the adsorption potential, any variation of the adsorption energy will of course also be reflected in this quantity. In view of transition state theory, ν_d is determined by the entropy difference between the adsorbed state and the transition state, which in turn will be affected by the degree of order on the surface, yielding ν_d to be coverage dependent.

Finally also the reaction order x will be affected. If, for example, due to attractive interactions desorption occurs preferentially from the edges of islands, x tends to become zero at higher coverages. The simple concept of $x = 1$ for unimolecular desorption and $x = 2$ for desorption determined by recombination of two surface particles will obviously only be valid for random distributions, *i.e.* if interactions are negligible.

vi) Kinetics of Surface Reactions. If two (or more) different species are present on the surface, additional interactions will come into play which render the whole situation even more complex.

C. Kinetics of Non-dissociative Adsorption

The kinetics of adsorption may generally be expressed in terms of the sticking coefficient s which is equal to the probability that a particle striking the surface becomes chemisorbed. If

$$\mathrm{d}n/\mathrm{d}t = p \cdot (2\pi\, MRT_g)^{-1/2} \tag{2}$$

is the number of particles impinging 1 cm^2 of the surface per sec, and n_s the density of adsorbed species per cm^2, then

$$r_a = \mathrm{d}n_s/\mathrm{d}t = s \cdot \mathrm{d}n/\mathrm{d}t \tag{3}$$

If the mean lifetime in the adsorbed state becomes very short this definition (as well as the experimental determination) of s becomes somewhat ambiguous. We shall then denote a particle to be adsorbed if it reaches the ground-state of the potential well (Figure 3) and stays there long enough in order to become completely thermally equilibrated with the solid.

The most accurate method for deriving the sticking coefficient (if it is not too low) consists in determining the ratio of the fluxes of incoming and reflected molecules in a molecular beam experiment [17]. More convenient (but always associated with some degree of uncertainty due to calibration problems) are methods in which the surface concentration n_s is followed as a function of the gas exposure. The latter is usually defined as $E = \int p \, dt$, with 1 L (Langmuir) = 10^{-6} Torr sec. Combination of equations (2) and (3) yields

$$\begin{aligned} \frac{dn_s}{dt} &= s(2\pi MRT_g)^{-1/2} p \\ n_s &= s(2\pi MRT_g)^{-1/2} \int p \, dt = s(2\pi MRT_g)^{-1/2} E \\ s &= (2\pi MRT_g)^{1/2} \frac{dn_s}{dE} \end{aligned} \tag{4}$$

This means that the function $s(n_s)$ can then simply be derived by differentiating the n_s vs. E curve. n_s is usually determined by Auger electron spectroscopy (AES), X-ray or UV photoelectron spectroscopy (XPS or UPS), work function changes ($\Delta\varphi$) or from the area below thermal desorption spectra (TDS).

If first the sticking coefficient on a clean surface (*i.e.* at zero coverage), s_0, is considered, it becomes obvious from Figure 3 that its value is determined by the efficiency with which the impinging particle gets rid of its translational energy perpendicular to the surface. Otherwise it will be inelastically reflected back into the gas phase. Figure 9, curve *a*, shows a polar plot the angular distribution of NO molecules coming off a clean Pt(111) surface at 300 K if a molecular beam strikes the surface under the indicated angel [18]. The specular lobe clearly indicates that these molecules have undergone only incomplete momentum exchange upon collision. The flux of reflected molecules is in this case only 12% of the primary intensity, the remainder being adsorbed, *i.e.* $s_0 = 0.88$. The release of translational energy

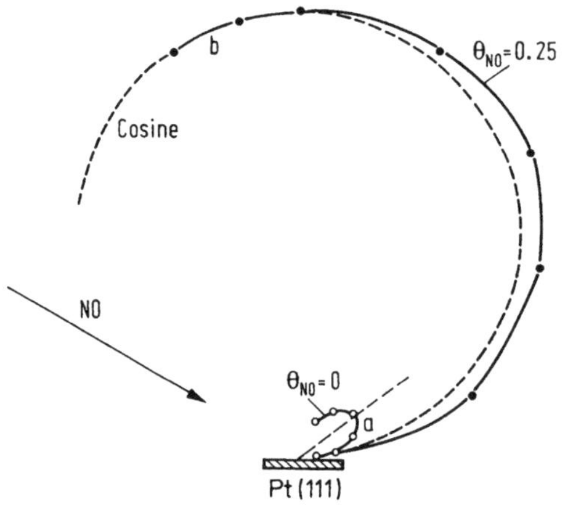

Figure 9. Polar plot of the angular distribution of NO molecules scattered from a Pt(111) surface at 300 K. **a** Clean surface ($\theta_{NO} = 0$); **b** Surface saturated with adsorbed NO ($\theta_{NO} = 0.25$). Temperature 300 K. (Reproduced with permission from ref. [18])

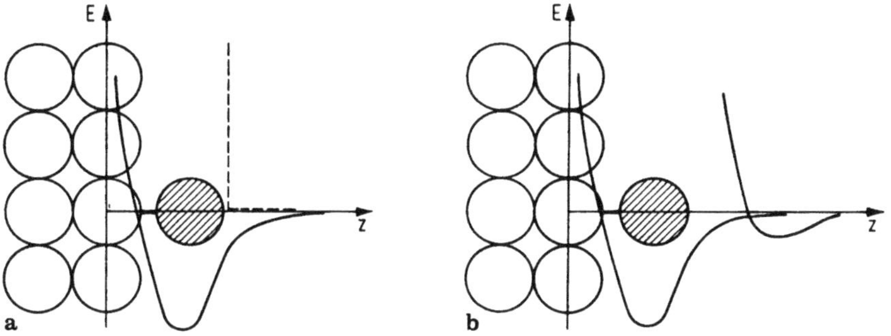

Figure 10. Potential for the interaction of a gaseous particle with an occupied adsorption site. **a** Hard-wall potential (Langmuir case). **b** Realistic (van der Waals-like) potential

may in principle occur through the following processes: Phonon excitations, electronic excitations, diffraction, transformation of translational into internal energy. While the possible role of electronic excitations in the solid is still under dispute [19], (multi-) phonon creation will certainly be in most cases the dominant mechanism. Diffraction effects (selective adsorption) come only into play under special conditions with light particles at samples with pronounced modulation of the potential parallel to the surface [20]. Transformation of translational into rotational energy upon scattering at surfaces has recently been observed experimentally ('rotational rainbow') [21], but will usually also be of minor importance.

The initial sticking coefficient for non-dissociative adsorption of molecules such as CO or NO on transition metal surfaces is mostly close to unity, but can also be rather small (for example only about 10^{-2} for N_2/Fe(111) [22]). Although so far not yet studied in great detail, it is expected to decrease with increasing surface temperature (since then the probability for pickup of additional energy upon collision increases) as well as with increasing trans-

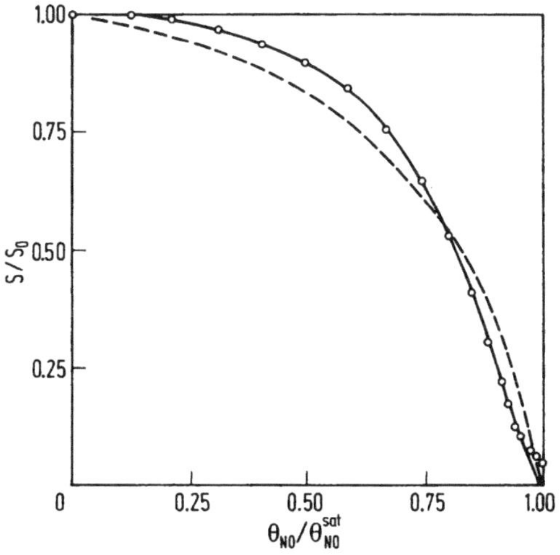

Figure 11. Variation of the relative sticking coefficient, s/s_0, with relative coverage, for NO adsorption on a Pt(111) surface at 300 K. Data, —·—·—; calculation, — — — (Kisliuk model). (Reproduced with permission from ref. [18])

lational energy in the direction normal to the surface (since then more energy has to be released). For the normal temperatures which apply in the present context these effects can, however, be widely neglected.

The simplest model for the variation of s with coverage is that underlying the derivation of the Langmuir adsorption isotherm. If a particle strikes an empty site it becomes adsorbed with probability s_0, otherwise it is reflected, yielding $s = s_0(1 - \theta)$ where $\theta = n_s/n_{s,\,max}$ is the relative coverage. This model would imply a hard-well potential for the interaction of a gaseous particle with an already occupied site as sketched in Figure 10a, which is of course quite unrealistic. Indeed quite frequently s is observed to vary non-linearly with coverage as shown in Figure 11 for the example NO/Pt(111) [18]. Such a behavior can be explained by a 'precursor' model [23, 24]. The interaction potential will in fact exhibit a shallow minimum (arising for example by van der Waals-forces) as indicated by Figure 10b, into which an incoming particle can be trapped with high probability. In this second layer the mobility will be rather high so that the trapped particle can eventually find an empty chemisorption site before it desorbs again. With increasing coverage the mean diffusion length needed to reach an empty site will continuously increase so that the sticking coefficient decreases, however less than linear. The angular distribution of NO molecules scattered at an already saturated surface (Figure 9, curve b) is almost of the cosine type, and the mean velocity of these molecules corresponds to the surface temperature: Both effects indicate effective translational energy accommodation due to trapping in the precursor state. Determination of the populations of rotational levels by means of laser-induced fluorescence demonstrated that in this case also the rotational energy of NO molecules scattered at an NO covered Pt(111) surface became completely accommodated to the surface temperature [25].

The precursor state model can be refined if the adsorbed layer is not randomly distributed but exhibits order due to the operation of lateral interactions [26]. The variation of s with coverage for the system CO/Pd(100) as shown in Figure 12 is an example which has been described by this theory [7].

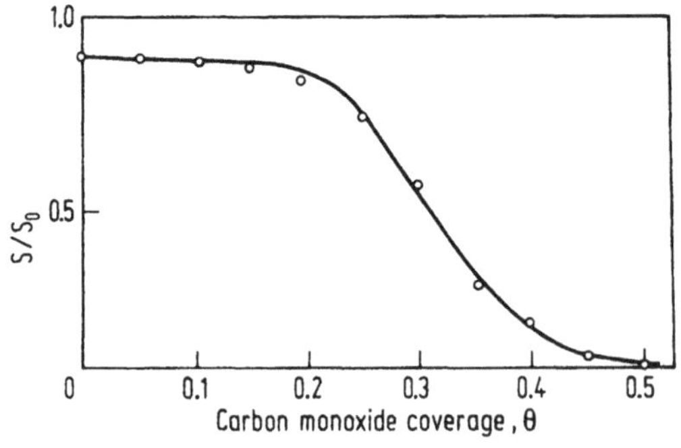

Figure 12. Variation of the relative sticking coefficient, s/s_0, with relative coverage, θ, for CO adsorption on a Pd(100) surface. (Reproduced with permission from ref. [7])

D. Kinetics of Dissociative Adsorption

The sticking coefficient now denotes the probability that a molecule striking a surface becomes dissociatively adsorbed. This process occurs quite frequently with only small or even not noticeable activation energy. The bond energy in the molecule may obviously be overcome by the formation of *two* new bonds with the surface. The one-dimensional potential energy diagram of the type first proposed by Lennard-Jones [27] as reproduced in Figure 13 is well-known and illustrates the energetic conditions. A diatomic molecule approaching a surface becomes first attracted by a relatively flat potential minimum for non-dissociative adsorption (= the molecular 'precursor' state), but can also be transformed through an adiabatic transition into the dissociated state, whereby the crossing point between the two potential curves determines the effective activation barrier. Even more instructive is a schematic two-dimensional potential diagram (Figure 14) which shows the variation of the energy as a function of the interatomic distance within the molecule as well as of the distance from the surface. Diagrams of this type are commonly used in homogeneous gas phase reaction dynamics [28]. Depending on whether the saddle point (= transition state) is located in the entrance or the exit channel, either translational or vibrational excitation is needed for transformation into the dissociated state. Dissociative adsorption may in principle proceed 'directly', *i.e.* if the incoming particle has enough energy to cross the activated state without previously being trapped in the molecular state. H_2/Cu is probably the best studied example of this kind [29]. As can be seen from Figure 15 the probability for dissociative adsorption rises steeply as soon as a critical value of the velocity normal to the surface is reached.

Another possibility is offered by the 'indirect' mechanism. The gaseous particle is first trapped in the molecular 'precursor' state from where it is thermally activated into the dissociated state. This molecular state is of course of a completely different nature than the precursor state in the preceding

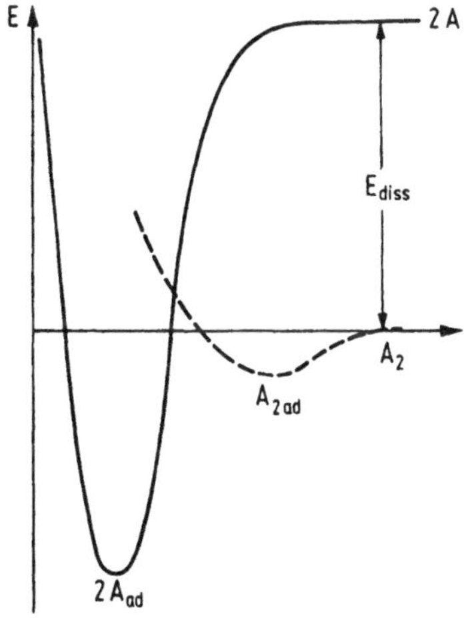

Figure 13. One-dimensional potential diagram illustrating dissociative chemisorption. (After Lennard-Jones [27])

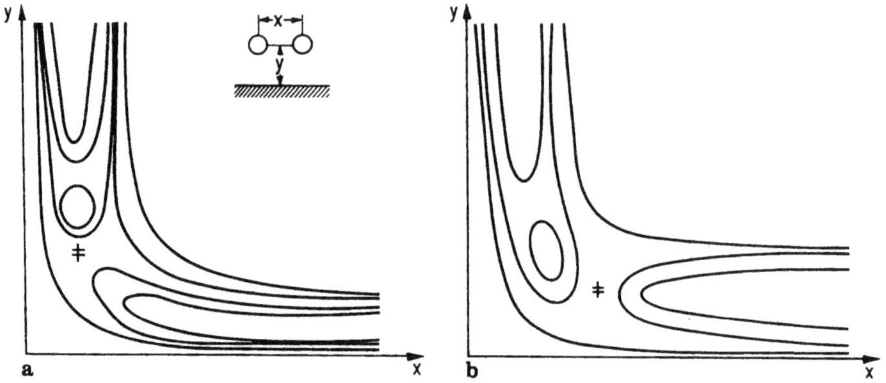

Figure 14. Two-dimensional potential diagrams illustrating dissociative chemisorption. **a** With translational energy activation, **b** with vibrational activation

section (which comprises a weakly held molecule in a second layer). Clear identification of a molecularly chemisorbed species which is able to dissociate on the surface has for example been possible with the systems O_2/Pt(111) [30] and N_2/Fe(111) [22] with adsorption energies of around 33 kJ mol^{-1} in both cases. If this mechanism prevails, the sticking coefficient will in fact be composed of elementary rate constants for trapping into the 'precursor' state (α), desorption from this state (k_{-1}), as well as transformation into the dissociated state (k_2) (see Figure 13):

$$s_0 \approx \frac{\alpha}{2} \cdot \frac{k_2}{k_{-1}} \,.$$

If the trapping coefficient α is assumed to be independent of temperature, then the variation of s_0 with temperature obviously yields the activation energy $E^* = E_2 - E_1$. E^* can of course also be negative (if $E_1 > E_2$), such as for example with N_2/Fe(111) where $E^* = -3.3$ kJ mol^{-1} was determined for the limit of zero coverage [17], whereas values of $+12$ kJ mol^{-1} and $+\,27$ kJ mol^{-1} were derived for the Fe(100) and (110) planes, respectively [31].

In general a clear distinction between the direct and indirect mechanisms will be rather difficult and presumably both mechanisms will often run parallel. One should then be cautious in interpreting apparent activation energies as derived from the dependence of the sticking coefficient on the surface temperature ($s = s' \cdot \exp(-E'/RT)$) in terms of actual energy barriers. Apart from this, the parameters s' and E' derived in this way are of course of vital importance for a quantitative description of the phenomenological reaction kinetics.

Quite in contrast to non-dissociative adsorption, in the present case the sticking coefficient frequently decreases more strongly than linear with coverage. As an example Figure 16 shows the variation of s with coverage at $T_2 = 320$ K for the system O_2/Pt(111) [32]. The Langmuir model predicts a variation of

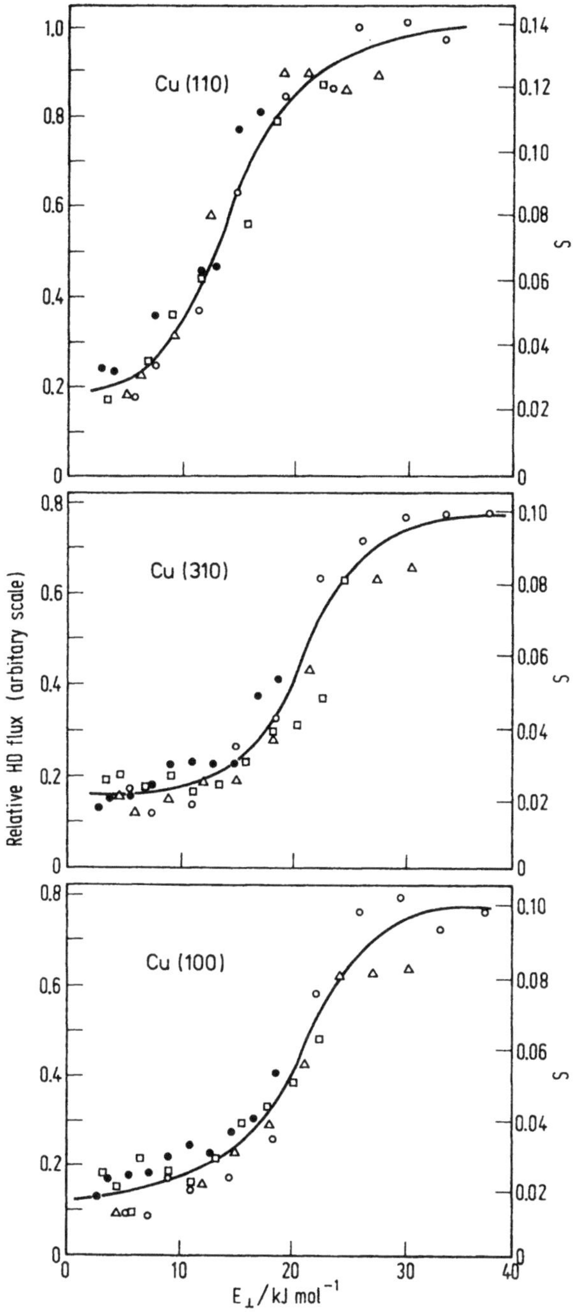

Figure 15. Results from molecular beam studies of H_2 adsorption on (100), (110) and (310) faces of Cu. The crystal face is exposed to a molecular beam of H_2 together with a highly dissociated beam of deuterium. The data show the variation of the dissociative sticking probability, *s*, of H_2, and the relative flux of desorbed HD, with variation of the normal component of the H_2 velocity, $E_{\perp}$. Angles of incidence, θ_i: ○, 25°; △, 35°; □, 45°; ●, 50°. Surface temperature 850 K. (Reproduced with permission from ref. [29])

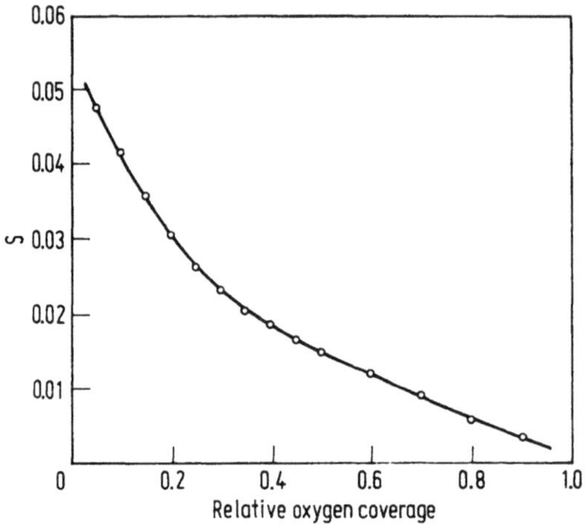

Figure 16. Variation of the dissociative sticking coefficient, s, with relative coverage, θ, for O_2 on Pt(111) at 320 K. (Reproduced with permission from ref. [32])

the kind $s = s_0(1 - \theta)^2$ if a molecule has to find two neighboring empty sites in order to be dissociated. But often the decrease is even stronger and eventually exponential. An obvious reason for such effects can be sought in the coverage dependence of the activation energy which is expected to increase parallel to the decrease of the adsorption energy (due to mutual repulsion). Such a behavior has for example been found for dissociative nitrogen adsorption on Fe(100) as can be seen from Figure 17 [31]. It can furthermore easily be shown that a linear increase of E' with coverage leads to the well-known Elovich equation which has the form of a rate law logarithmic with time [33].

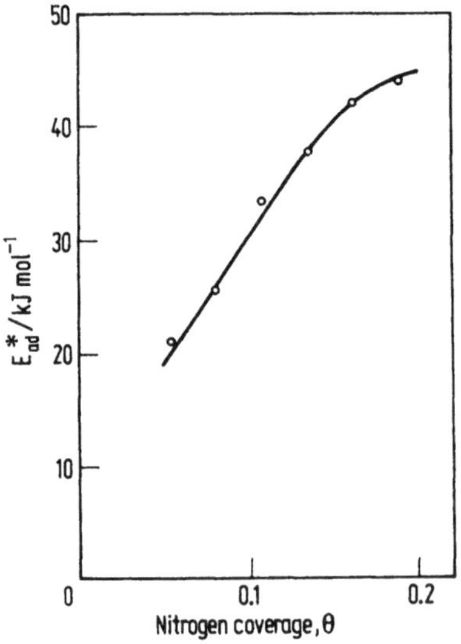

Figure 17. Variation of the activation energy for dissociative adsorption, E^*_{ad}, with relative coverage, θ, for N_2 on Fe(100). (Reproduced with permission from ref. [31])

E. Kinetics of Desorption

Thermal desorption at least of the product molecules is essential for heterogeneous catalysis and will occur if the temperature is high enough to overcome the adsorption energy and possibly existing additional activation barriers. The experimentally simplest way to study these processes consists in the 'flash desorption' technique (also called thermal desorption spectroscopy, TDS) where the temperature of the adsorbate covered sample is continuously increased and simultaneously the variation of the partial pressure is recorded [10, 34]. Simple material balance considerations [35] leads to the following expression for the change of partial pressure

$$\frac{dp}{dt} = -A\frac{dn_s}{dt}\frac{kT}{V} - \frac{Sp}{V} \tag{5}$$

(≈ rate of gas evolution — rate of gas removal by pumping) where A = surface area, n_s = density of adsorbed particles per cm^2, V = volume of the vacuum system, S = effective pumping rate. Equation (5) may be rewritten as

$$\frac{-AkT}{V}\cdot\frac{dn_s}{dt} = \frac{AkTr_d}{V} = \frac{dp}{dt} + \frac{Sp}{V} \tag{6}$$

with $r_d = -dn_s/dt$ = rate of desorption.

If the pumping rate S is high enough (which is usually fulfilled experimentally), then $\frac{Sp}{V} \gg \frac{dp}{dt}$, yielding

$$r_d \approx \frac{Sp}{AkT} \tag{7}$$

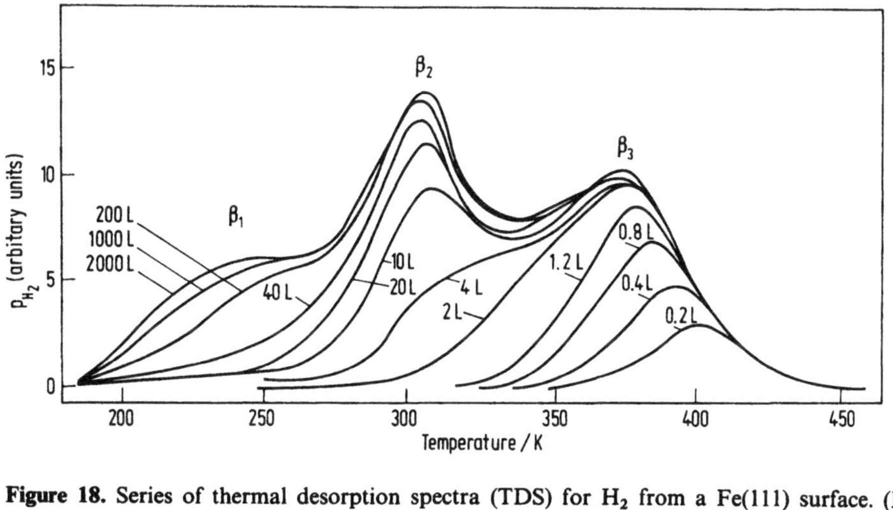

Figure 18. Series of thermal desorption spectra (TDS) for H_2 from a Fe(111) surface. (Reproduced with permission from ref. [36])

That means the rate of desorption is simply proportional to the partial pressure. Integration of this equation yields

$$n_s = \frac{S}{AkT} \int p \, dt \tag{8}$$

i.e. the area below such a thermal desorption trace is directly proportional to the surface concentration (coverage).

As an example Figure 18 shows a series of thermal desorption spectra for H_2 desorbing from a Fe(111) surface after varying exposure at low temperature [36]. The variation of the area reflects the increase of the adsorbed amount with exposure (reflecting adsorption kinetics), whereas the shapes of these curves are determined by the kinetics of desorption. Procedures for analysing these data can be found in the extensive literature existing on this subject [47–51]. Instead of following the partial pressure also a quantity directly proportional to the coverage (such as $\Delta\varphi$ [42] or AES signal intensity [43]) may applied to record the rate of desorption, which enables better elimination of spurious effects arising not from the crystal plane under investigation.

Separation of the relevant kinetic parameters, namely the reaction order as well as the preexponential and activation energy for desorption, from TDS data is, however, generally complicated and mostly based on certain assumptions. More direct information may be obtained either by isothermal measurements, where the rate of desorption is followed as a function of time at constant temperature [42, 44], or by applying modulated molecular beam techniques which provide directly the mean lifetime on the surface [45]. Finally it should be mentioned that in the case of non-activated adsorption the activation energy for desorption E_d, practically equals the isosteric adsorption energy which can for example be extracted from (equilibrium) adsorption isotherms using Clausius-Clapeyron's equation [46]

$$\mathrm{d}\ln p/\mathrm{d}(1/T))_{\theta=\mathrm{const}} = -E_{ad}/R \tag{9}$$

A non-dissociatively adsorbed species will desorb in the limit of zero coverage with first-order kinetics:

$$-\mathrm{d}n_s/\mathrm{d}t = r_d = k_d n_s \tag{10}$$

At a constant temperature this corresponds to a exponential decay of the surface concentration, $n_s = n_s^0 \, e^{-t/\tau}$, where the mean surface lifetime τ is simply related with the rate constant k_d through

$$\tau = 1/k_d \tag{11}$$

If k_d is written in Arrhenius form

$$k_d = \gamma_d \exp(-E_d/RT) \tag{12}$$

this implies

$$\tau = \tau_0 \exp(E_d/RT) \tag{13}$$

with $\tau_0 = 1/\nu_d$.

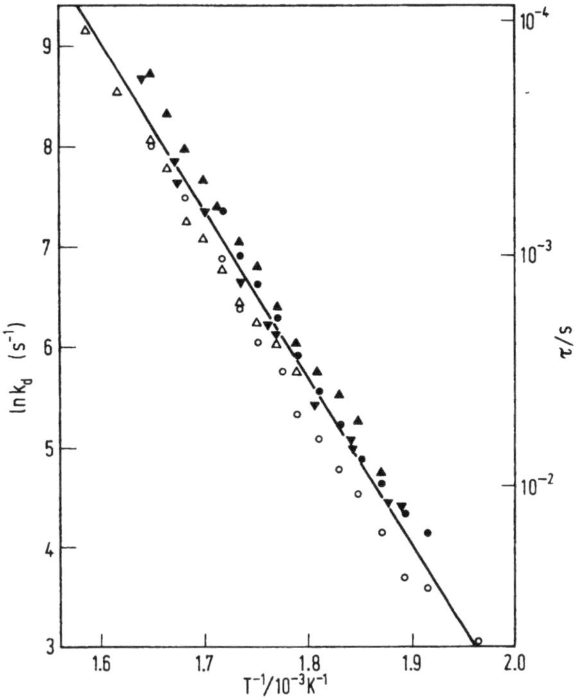

Figure 19. Modulated molecular beam data for NO on Pt(111) at various temperatures. The data show the mean surface lifetime, τ, and the corresponding desorption rate constants, k_d, at limiting conditions of zero coverage. The points were obtained at varying modulation frequencies: ●, 19 Hz; ■, 23 Hz; ▽, 45 Hz; ▲, 43 Hz; , 91 Hz. The line was calculated: for details see ref. [18]. (Reproduced with permission from ref. [18])

Equation (13) is called the Frenkel equation according to a simple theoretical treatment based on statistical mechanics (47]. In this theory τ_0 turned out to be the vibrational period of the adsorbed particle against the surface which should be of the order of 10^{-13} sec. Although a preexponential of $\nu_d = 10^{13}\ s^{-1}$ is frequently assumed for first-order desorption kinetics, such a value is often not verified experimentally. Direct determination of τ can be performed by recording the phase lag between the primary and scattered signal in a modulated molecular beam experiment [45]. As an example Figure 19 shows a plot of the logarithm of τ *vs.* $1/T$ for the system NO/Pt(111) [18]. The steady-state coverage of NO was far less than a few percent throughout these measurements which can therefore be regarded to reflect the zero-coverage limit. The occurrence of a straight line indicates indeed the validy of an Arrhenius expression over this range, the kinetic parameters being $E_d = 138.3\ kJ\ mol^{-1}$ and $\nu_d = 10^{15.5}\ s^{-1}$. Since the molecules form a stronger bond with defect sites (steps, kinks etc.) they will rapidly diffuse on the surface to these sites (even if present in very low concentration) and spend there most of their lifetime. It should therefore be noted that these numbers reflect desorption from defect sites rather than from a perfect (111) plane.

Preexponentials exceeding 10^{13} s^{-1} (as in the present case) can qualitatively easily be explained on the basis of transition state theory (TST) [48]. The rate constant can accordingly be written as

$$k_d \approx \frac{kT}{h}\frac{f^{\neq}}{f_a}\exp(-E_d/RT) \quad (14)$$

where $kT/h \approx 10^{-13}$ s^{-1} is a universal frequency factor, and $f^{\neq}$ and f_a the partition functions for the transition state and the adsorbed state, respectively. Since motions in the adsorbed state will generally be more restricted than in the transition state (which represents a molecule just in front of the surface) $f^{\neq}/f_a$ is always expected to be >1. Estimates for limiting cases have been made by Yates [38]. If the adsorbed state is completely immobile and the transition state is described by a two-dimensional gas, then for a molecule with mass 30 at 300 K $f^{\neq}/f_a \approx 10^3$ or $\nu_d \approx 10^{16}$ s^{-1}. This number can even increase to about 10^{18} s^{-1} if internal degrees of freedom and possible ordering phenomena in the adlayer (at higher coverages, see below) are taken into account [38, 42, 49]. A whole series of careful determinations of the preexponential factor reported in the literature indeed confirms these conclusions [18, 42, 49–53].

Failure of transition state theory is generally treated by introduction of a 'transmission factor' $\varkappa$

$$k = \varkappa \cdot k_{TST} \quad (15)$$

Based on considerations on the microscopic reversibility of adsorption and desorption it can be shown [54, 55] that in the present case $\varkappa$ is equal to the sticking coefficient s, *i.e.* $k_d = sk_{d,TST}$. So, values $\nu_d > 10^{13}$ s^{-1} are only to be expected for systems with high sticking coefficients (the 'intermediate coupling case') which is in fact the case with the cited examples. If, in addition, the sticking coefficient is independent of the normal component of the translational energy, the molecules are predicted to desorb with a cosine angular distribution (Knudsen law) and with a Boltzmann distribution of the translational energy which corresponds to the surface temperature [56], as indeed frequently observed.

If s is small ('weak coupling'), also the preexponential for desorption should decrease due to the inefficiency of energy transfer from the solid to the adsorbate. The system (molecular) N_2/Fe(111) is such an example of 'slow' desorption for which $s_0 \approx 10^{-2}$ and $\nu_d \approx 10^{11}$ s^{-1} were found [22]. In this case the angular distribution should be broader than cosine and the desorbing molecules should be 'colder' than the surface [55]. Experimental verification of these effects has been obtained for Ar desorbing from Pt(111) [57].

Several complicating factors may come into play at higher coverages:

i) Since in the case of nonactivated adsorption $E_d \approx E_{ad}$ any variation of the adsorption energy with coverage due to lateral interactions equally manifest itself in the activation energy for desorption. This effect gives rise to a broadening of the thermal desorption spectra and eventually to the appearance of multiple peaks similar as shown in Figure 18 which are

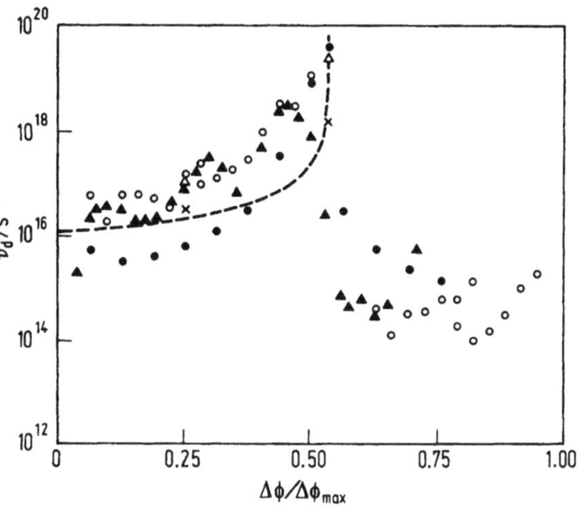

Figure 20. Variation of the pre-exponential for desorption, ν_d, with relative coverage for CO on Ru(0001). The relative coverage is indicated by the work-function change ratio $\Delta\varphi/\Delta\varphi_{max}$. The data have been collected from various experiments: open symbols from desorption experiments; closed symbols from equilibrium isosteric heat of adsorption data together with sticking coefficient data. The line (— — —) indicates the behaviour expected from a model described in ref. [42]. (Reproduced with permission from ref. [42])

sometimes assigned as different adsorption 'states' (with labeling such as α, β_1, β_2 etc.). Multiple states can arise from molecules desorbing from sites with identical local geometry but varying occupancy of neighboring sites or from molecules on different (*e.g.* linear and bridge) sites. Both effects have to be attributed to the operation of lateral interactions since a single isolated molecule with always occupy only that type of sites which is associated with the highest adsorption energy (provided the rate of desorption is lower than that for surface diffusion). Distinction between these two models can only be made by probing the local geometry of the adsorbed particles, *e.g.* by vibrational spectroscopy.

ii) If E_a varies with coverage this will also influence the partition function of the adsorbed state f_a and through equation (14) also the preexponential ν_d. Even more pronounced effects can come into play if ordering causes a substantial lowering of the configurational entropy, leading to an increase of the term $f^{\neq}/f_a$. Figure 20 shows the variation of ν_d with coverage for the system CO/Ru(0001) which has been attributed to such an effect [42, 58].

iii) If adsorption proceeds through a mobile precursor state (see section 2.C) also passage through this state during thermal desorption has to be taken into account because of microscopic reversibility. Theoretical treatments of this problem [59–61] lead to rather complicated expressions from which however some important general conclusions can be drawn. Not only the measurable preexponential and activation energies for desorption are affected, but also the apparent reaction order may increase and the denominator of the rate expression may even contain a $(1 - \theta)$ term.

iv) The apparent reaction order for desorption is not only affected by passage through a precursor state, but also if desorption takes place from an ordered phase. In fact it will become rather questionable to separate the reaction order and the preexponential from experimental data, if both quantities are dependent on coverage. Analysis can then only be rationalized by *assuming* a constant reaction order and by referring the total coverage dependence to ν_d.

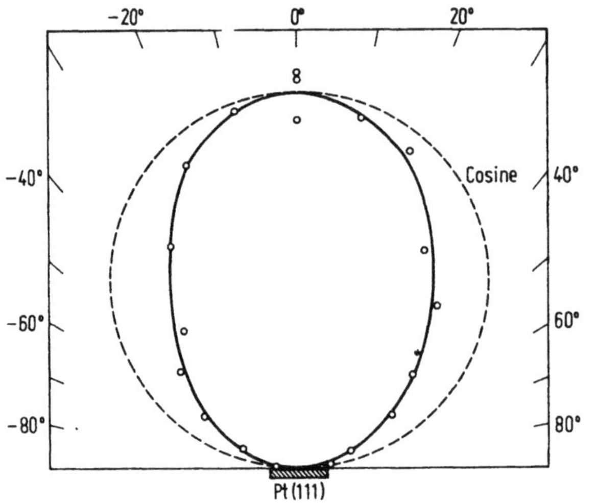

Figure 21. Polar plot of the angular distribution of O_2 molecules desorbing (through recombination of 2 O_{ad}) from a Pt(111) surface. Initial relative coverage, $\theta = 0.5$. (Reproduced with permission from ref. [32])

Associative desorption through recombination of two surface species proceeds through the same channels as discussed in section 2.D for the reverse process (dissociative adsorption) and according to the potential diagrams of Figures 13 and 14. With the indirect mechanism the molecule formed is first accommodated in this molecularly adsorbed state (intrinsic precursor) from where desorption takes place, *i.e.* $2A_{ad} \rightarrow A_{2,ad} \rightarrow A_2$. In this case the desorbing molecules will usually exhibit a cosine angular distribution and a Boltzmann velocity distribution characterized by the surface temperature. If desorption occurs directly ($2A_{ad} \rightarrow A_2$) over a barrier which is above the energy zero (cf. Figure 13) the angular distribution is expected to be sharper than cosine and the velocity distribution differs from the Boltzmann behavior [56, 62]. An example for a non-cosine angular distribution is shown in Figure 21 [32]. A nice example of both Boltzmann and 'hot' molecules leaving the surface has been investigated by Comsa *et al.* [63]. Figure 22 shows time-of-flight spectra of D_2 molecules desorbing from Pd(100) at various exit angles. Further complications arise in this case, however, since the D atoms were initially not adsorbed on the surface but were diffusing through the bulk from the backside of a membrane.

As with dissociative adsorption a clear distinction between these two limiting mechanisms will be rather difficult and is also not essential for the formulation of the rate law. This will in both cases be given by

$$r_d = k_d[A_{ad}]^2 \tag{16}$$

as long as the coverage is low enough so that the adsorbed particles are randomly distributed on the surface. If entropy differences are neglected, the preexponential for second-order desorption will be given by

$$\nu_d = \frac{kT}{hn_s} \tag{17}$$

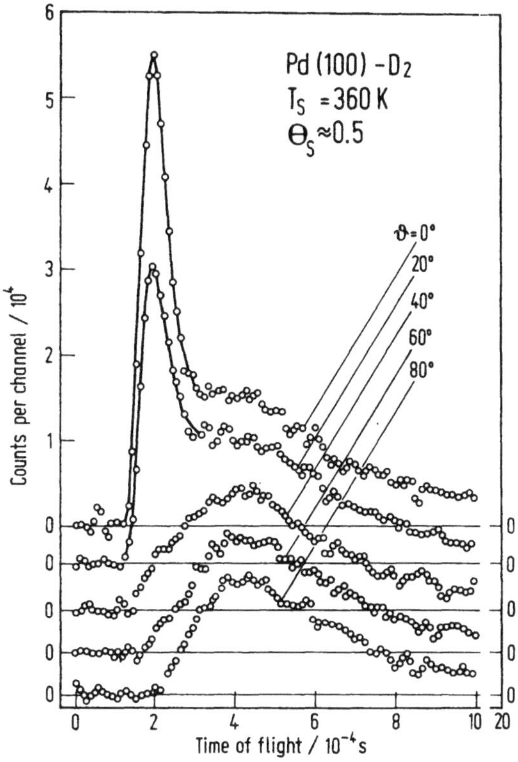

Figure 22. Time-of-flight distributions of D_2 molecules desorbing from a sulfur covered Pd(100) surface at various angles θ. Temperature 360 K. (Reproduced with permission from ref. [63])

for which typical values between 10^{-3} and 10^{-1} cm^2 s^{-1} are estimated [38]. As with first order desorption this number may increase by several orders of magnitude if effects of the partition functions of the adsorbed state and the transition state are taken into account. With increasing coverage additional effects due to lateral interactions similar as discussed above for the case of non-dissociative adsorption will come into play. The activation energy as well as the preexponential vary with coverage and the reaction order may no longer be two. The latter effect becomes particularly evident if attractive interactions exist which then cause island formation. The reaction order for desorption may then eventually tend even towards zero as for example observed with H_2/Ni(110) [64].

F. Coadsorption

During a catalytic reaction usually more than one kine of species will be present on the surface which effect will be called coadsorption. Although marked interactions between different particles may exist, these will retain their molecular identity which can experimentally be probed, for instance by ultraviolet photoelectron spectroscopy or vibrational spectroscopy. For the sake of simplicity only systems with two kinds of adspecies, A and B, will be considered in the following, but the qualitative conclusions

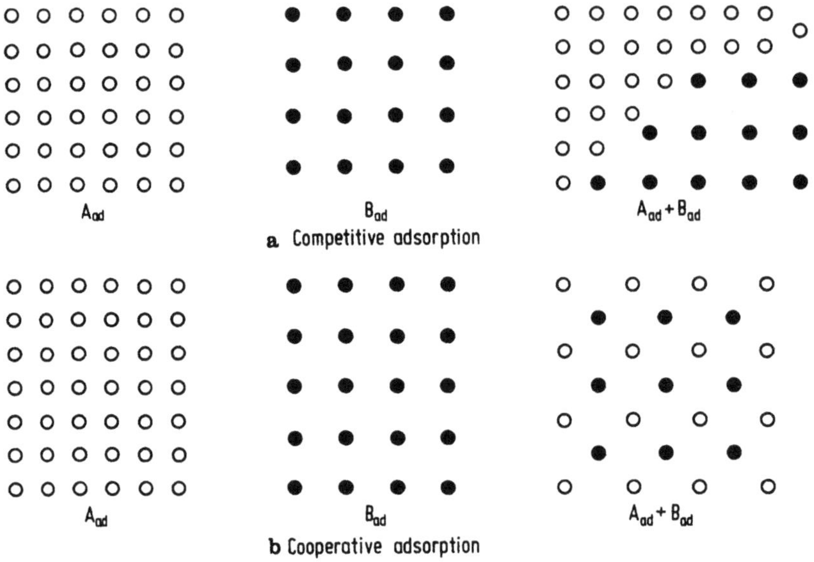

Figure 23. Schematic surface configurations of two different kinds of particles in the case of (a) competitive, (b) cooperative coadsorption. (Reproduced with permission from ref. [65])

can easily be generalized. The presence of a second species will influence both the ground-state as well as the kinetic properties.

If two kinds of particles are present on the surface the following basic interactions will come into play: A—A, B—B and A—B, with the corresponding interaction energies ε_{AA} etc. If these interactions are assumed to act independently and pairwise (which is a crude simplification) the following two limiting situations may occur which can experimentally be distinguished by LEED [65]:

a) If $\varepsilon_{AA} + \varepsilon_{BB} - 2\varepsilon_{AB} < 0$, A and B effectively repel each other so that the surface consists of two phases, namely domains consisting either of pure A or pure B as illustrated schematically by Figure 23a. The LEED pattern then consists of a superposition of diffraction spots arising from either of the adsorbed species alone. This effect is called 'competitive adsorption', since only that fraction of the total surface area which is not already covered by A can be occupied by B particles. Reaction between A and B is then restricted to the boundaries of these domains which will have obvious consequences on the reaction kinetics as discussed later.

b) If $\varepsilon_{AA} + \varepsilon_{BB} - 2\varepsilon_{AB} > 0$, there will be an effective attraction between A and B (or less repulsion than between equal particles) which then form a mixed phase as sketched in Figure 23b. In this case LEED data characteristic for a single phase are observed. This situation is called 'cooperative adsorption'. A and B are then in intimate contact, but not randomly distributed, which of course will again affect the kinetics of an eventual surface reaction.

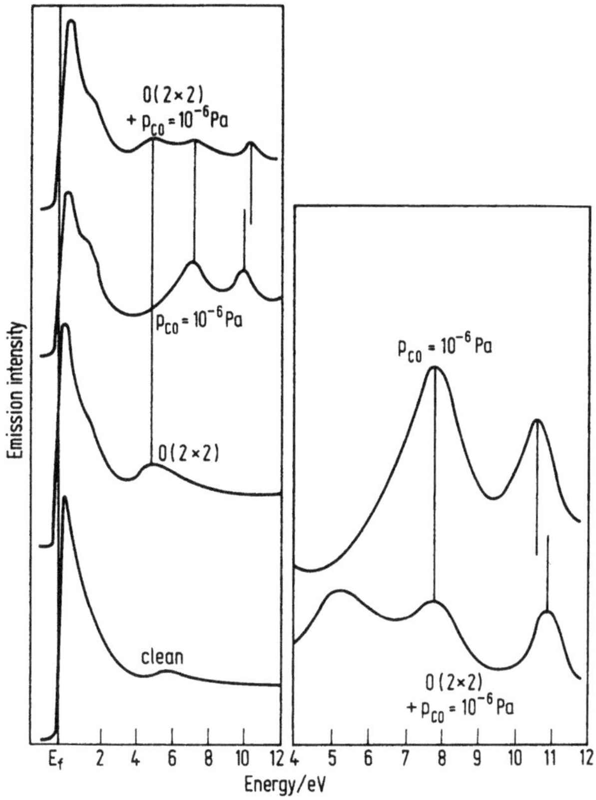

Figure 24. UPS data for O and CO coadsorbed on Ni(111). (Reproduced with permission from ref. [12b])

Realistic situations are of course much more complicated. Depending on the strengths of the interaction energies and on temperature a more or less high degree of disorder will exist and new phases may be formed if the fractional surface concentrations are changed. The system O + CO/Pd(111) offers an example where various of these structural effects were observed [66].

Apart from these effects on the surface concentrations and lateral distributions other consequences on the properties of the adlayer have to be considered:

i) The effective adsorption energy may be either lowered (as for CO in the presence of O_{ad} on Ru(0001) [67]) or increased (as for N_2 [68] or CO [69] in the presence of K on iron surfaces).

ii) An adsorbed particle is displaced onto another type of adsorption site. Experiments with NO + CO/Ru(0001) using vibrational spectroscopy nicely demonstrated such effects [70].

iii) If the surface is not perfectly uniform, but contains, for instance, defects or other low concentration irregularities, these sites may be preferentially occupied by A and are therefore no longer accessible for B. This forms the basis for the frequently made observation that rather low surface concentrations of an impurity may poison a catalyst.

iv) The electronic properties may be affected as manifested for instance in the charge distribution (to be probed by chemical shifts in XPS or by measuring dipole moments through the work function change) or in the ionisation potentials of the valence levels. An example of the latter kind is exhibited by Figure 24. On Ni(111) the CO-derived 4σ-level as analyzed by UPS is shifted by about 0.5 eV towards higher binding energies in the presence of O_{ad} [12b].

v) Variations of the surface bond are of course also reflected in the vibration frequencies of adsorbed molecules. With Ru(0001) the presence of O_{ad} shifts the frequency of the C—O stretch vibration from 1980 to 2048 m^{-1} and that for the metal-CO vibration from 445 to 387 cm^{-1} [67]. Recently a rather dramatic lowering of the C—O vibration frequency for CO adsorbed on platinum caused by the presence of potassium was observed [71] which is mentioned as another example.

The consequences on the kinetics of adsorption and desorption are quite obvious. Due to the occupation of adsorption sites by a second species usually the sticking coefficient will be lowered. This may lead to the effect that under steady-state conditions not the most strongly held species is most abundant on the surface. In catalytic oxidation of CO over the platinum metals, dissociative oxygen adsorption is associated with a higher adsorption energy than CO adsorption [72]. However, preadsorbed oxygen atoms affect the sticking coefficient for CO only slightly, whereas in turn CO_{ad} inhibits quite effectively dissociation of O_2. The consequences on the overall reaction kinetics will be discussed in section 3.A.

The presence of a second species may, however, also increase the sticking coefficient and acts thereby as a promoter. The influence of potassium on the kinetics of dissociative nitrogen adsorption on iron is an example of this kind [73] which will be outlined in more detail in section 3.C.

Changes of the effective adsorption energy due to the operation of lateral interactions (either repulsive or attractive) enter of course directly into the activation energy for desorption and therefore into the rate of desorption. Numerous examples are known where addition of a second species 'displaces' adsorbed particles from the surface. The mechanism of these displacement

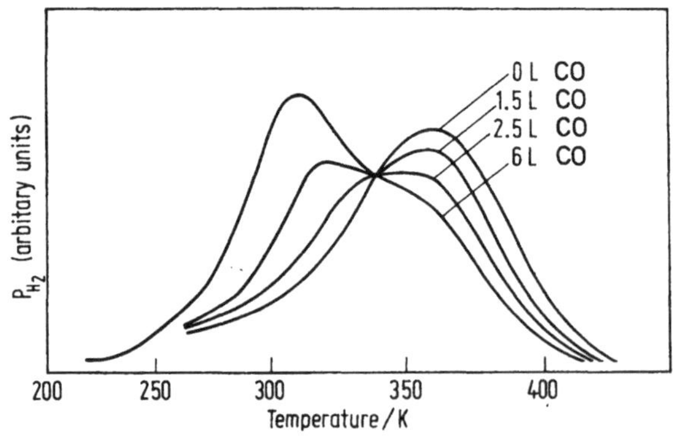

Figure 25. Thermal desorption spectra for H_2 from Ni(111) with various predoses of CO. (Reproduced with permission from ref. [75])

processes is most likely rather similar to that operating with SN2 reactions in coordination chemistry [74]. Attachment of a second ligand lowers the bond strength (through repulsive interactions) to an extent that desorption takes place. If the temperature is low enough this lowering of the activation energy can be probed by thermal desorption spectroscopy. As an example Figure 25 shows corresponding TDS data for hydrogen + CO coadsorbed on Ni (111) where CO would 'displace' H_2 if the sample is held at room temperature [75].

G. Surface Reactions

Although chemical transformations on surfaces consist frequently of a complex series of elementary steps, the latter can be classified into a small number of basic types.

a) $A \rightarrow B + C$ involves the breaking of at least one chemical bond and is from this point of view completely analogous to dissociative chemisorption for which the relevant features have already been discussed in preceding sections. In summary this reaction may proceed either directly via $A \rightarrow B_{ad} + C_{ad}$, or indirectly through $A \rightarrow A_{ad} \rightarrow B_{ad} + C_{ad}$, whereby clear distinction between these two mechanisms may be rather difficult. If it is assumed that the steady-state concentration of A_{ad} is small and constant then also the overall kinetics of both mechanisms will be analogous. If A_{ad} itself is only formed as an intermediate in a surface reaction (such as NH_{ad} during NH_3 decomposition on Fe(110) [76]) its identification will present the key for elucidating the reaction mechanism.

If one of the products appears subsequently in the gas phase it needs not necessarily be intermediately adsorbed, that is, the elementary process might be of the kind

$$A \rightarrow B + C_{ad}$$

(this would exhibit some resemblance to the reverse Eley-Rideal mechanism, see below), although so far no clear example for this mechanism is known.

b) Associative reaction of the type $A + B \rightarrow C$ has much in common with desorption via recombination of two adsorbed particles, but now also additional aspects have to be taken into consideration. Two basic mechanisms have been discussed over many years in the literature:

i) With the Langmuir-Hinshelwood (LH) mechanism [77] it is assumed that both species are in the chemisorbed state from which they react by passage through a potential hyperface similar to that shown in Figure 14:

$$A_{ad} + B_{ad} \rightarrow C\,.$$

The product molecule is then either intermediately adsorbed or probably (if the surface reaction is exothermic) leaves the surface with excess energy, that is, not in thermal equilibrium with the surface. Catalytic oxidation of CO over platinum is a rather well understood example of this type which will be discussed in detail in section 3.A.

ii) The Eley-Rideal mechanism [77] assumes that only one reactant is in the chemisorbed state whereas the other interacts either through direct collision from the gas phase or from a weakly held (probably physisorbed) state exhibiting a rather short lifetime on the surface: $A + B_{ad} \rightarrow C$. Although this mechanism is often postulated in order to interprete kinetic rate laws, so far no example is known where it has been unequivocally established. It is quite evident that kinetic measurements alone will never be able to allow a clear decision of this kind.

iii) Some sort of intermediate mechanism ('precursor' mechanism) has recently been proposed [78] whereafter one of the reactants is not fully accomodated at the surface but reacts from a state which is vibrationally excited with respect to the motion normal to the surface. Although this mechanism was used in order to explain experimental data on the catalytic oxidation of hydrogen on platinum [79] it is felt that further clarification is needed before any firm conclusions can be drawn.

Formulation of rate equations for reactions of the kind $A + B \rightarrow C$ can be started with expressions of the type

$$r = k\theta_A\theta_B \text{ (LH)} \tag{18}$$

or

$$r = k'\theta_B p_A \quad \text{(ER or precursor mechanism)} \tag{19}$$

provided that the adsorbed particles are randomly distributed over the surface (= no lateral interactions) and that no interference with adsorbed product molecules occurs. In view of the discussions during the preceding sections already these assumptions will probably only rarely be fully justified. As soon as the surface is energetically heterogeneous (either *a priori* or induced) and if the adlayer is ordered, the rate constants will not only be a function of temperature but also of the surface concentrations and the reaction orders may deviate from those based on this simple model.

The situation becomes still far worse if an attempt is made to correlate the coverages with the respective partial pressure. If adsorption/desorption steps are fast compared to the rate of chemical transformation, these can be assumed to be in equilibrium and the θ_i's can be related with the p_i's through adsorption isotherms. The latter should in principle be carefully determined experimentally since now all the effects of coadsorption come into play, apart from the fact that even for a single component system the adsorption isotherm is generally complex due to the various effects discussed in sections 2.B to 2.E. If finally a 'real' surface is used which is *a priori* energetically heterogeneous then there is apparently no hope that a proper description can be reached on the basis of simple expressions derived from Langmuir-type isotherms. Nevertheless this approach is frequently made because of its simplicity and surprisingly it yields sometimes even a reasonable modeling of the actual situation, probably because several effects are able to compensate each other [80].

Using the crude (and by no means justified) assumptions the fractional coverages θ_A and θ_B in equations (18) and (19) are related with the respective partial pressures through

$$\theta_A = \frac{b_A p_A}{1 + b_A p_A + b_B p_B} \tag{20}$$

$$\theta_B = \frac{b_B p_B}{1 + b_A p_A + b_B p_B} \tag{21}$$

This extension of the regular Langmuir adsorption isotherm is based on a model whereby an adsorption site may be occupied by either A or B (= competitive adsorption) and that the product molecule C is so weakly adsorbed that its steady-state surface concentration is negligible.

The rate equation for the LH reaction (equation (18)) then writes

$$r_{LH} = k\theta_A\theta_B = \frac{b_A b_B p_A p_B}{(1 + b_A p_A + b_B p_B)^2} \tag{22}$$

For constant T and p_B the rate will pass through a maximum upon variation of p_A as can easily be checked.

If

$$b_A p_A \ll b_B p_B + 1 ,$$

r_{LH} will increase linearly with p_A, whereas the other limiting case is given if

$$b_A p_A \gg 1 + b_B p_B , \qquad \text{then}$$

$$r_{LH} \approx bp/b_A p_A ,$$

that is, A inhibits the reaction. The maximum rate is obviously reached if both reactants exhibit equal surface concentrations, that is,

$$\text{if} \quad \theta_A = \theta_B .$$

The kinetics of the ER reaction (equation (19)) is, on the other hand, given by

$$r_{ER} = k' \frac{b_B p_A p_B}{1 + b_B p_B} \tag{23}$$

that is, it will always be of first order with respect to p_A whereas the order with respect to p_B changes from one to zero.

The derivation of rate equations on equal footing but with different basic assumptions (*e.g.* including strong adsorption of the product or assuming only stationary coverages instead of rapid adsorption/desorption equilibria) is straightforward and will not be further outlined in this context. Numerous examples for this approach can be found in the literature [80]. Probably the most successful and famous case is the so-called Temkin equation [81] for describing the kinetics of ammonia synthesis where the unrealistic

Langmuir adsorption isotherm was replaced by an isotherm equation based on the variation of the adsorption energy with coverage.

The reaction rate needs not necessarily to be a unique function of the partial pressures, but it may happen that two (or even more) reaction branches exist which then eventually may give rise to kinetic oscillations. These phenomena have already been extensively treated theoretically [82–85] and an experimental example will be briefly described in section 3.A.

3. Model Studies on Simple Catalytic Reactions

Numerous examples of studies on the interactions between gases and well-defined surfaces which were performed along the outlined routes can be found in the literature. In this section a few simple reactions catalysed by metal surfaces will be discussed in order to illustrate the kind of information which is now accessible. Particular emphasis will be given to the relevance for 'real' catalysis.

A. Oxidation of Carbon Monoxide

The reaction $2\,CO + O_2 \rightarrow 2\,CO_2$ is readily catalyzed by the platinum group metals and has extensively been studied with single crystal surfaces. Its mechanism is basically rather simple and it therefore represents at present the best understood catalytic reaction. Nevertheless several features need still further clarification and its kinetics is rather complex and can by no means be described by simple and generally valid rate laws. A more extensive review can be found in two recent articles [72, 86].

1. Adsorption of Carbon Monoxide

CO couples to metal surfaces through the carbon atom, similar as with carbonyl compounds. On low index planes the molecular axis is perpendicular to the surface (such as *e.g.* with Pd(100), see Figure 2 [7]), while with more open planes the ligand may be tilted in order to enable maximum orbital overlap. The system CO/Pd(210) offers an example of this kind [87]. Quite recently the first experimental evidence for CO lying flat on a Cu(311) surface was obtained [88], but this configuration will certainly be rather exceptional and will not be of more general importance in the present context. Although the CO molecule may dissociate (particularly if adsorbed on 3*d* metals) this aspect is also only of minor significance, since oxidation proceeds certainly via non-dissociated CO. The analogy with carbonyl compounds is also reflected in the frequency of the CO stretch vibration which is generally lowered upon chemisorption (due to backdonation of metallic *d*-electrons in the antibonding $2\pi^*$ level) and which can furthermore be used to identify the local geometry of the adsorption site, that is, to determine whether terminal, bridge or three-fold coordination takes place. So far no predictions can be made which type of coordination will be preferred, and also simultaneous occupation of different types of sites may occur [89]. The bond strength on different sites varies usually by less than a few kJ mol^{-1} which indicates that for CO the surfaces are energetically rather smooth and enable rapid surface diffusion.

At higher coverages a general tendency for the formation of densely packed layers is observed (see *e.g.* Figure 7), and the saturation density is essentially determined by the van der Waals diameter of the CO molecule and is usually in the range of about 1×10^{15} cm^{-2}, irrespective of the surface structure.

The initial adsorption energy ranges between 120 and 170 kJ mol^{-1} and is usually somewhat higher with the more open crystal planes. It decreases at higher coverages due to the operation of repulsive interactions. A rare example of attractive interactions is offered by the system CO/Ru(0001) [42]. The initial sticking coefficient is always high (>0.8). The variation with coverage is usually determined by a precursor state as can be seen in Figure 12 for the example CO/Pd(100). Since adsorption is not activated, the activation energy for desorption is equal to the adsorption energy. Typical values for the preexponential for desorption are in the range of 10^{15} s^{-1}.

2. Adsorption of Carbon Dioxide

CO_2 is so weakly adsorbed (estimated adsorption energy ~ 20 kJ mol^{-1}) that its lifetime on the surface above room temperature is immeasurably short. In the formulation of the reaction mechanism and kinetics therefore no distinction between $CO_{2,\,ad}$ and CO_2 will be necessary. Figure 26 shows the angular distribution of CO_2 scattered at a Pt(111) surface [90] which exhibits a pronounced specular lobe apart from a cosine (= trapping/desorption) part. This indicates only limited energy exchange between a CO_2 molecule and the surface which will be of some relevance for the reaction dynamics to be discussed.

3. Interaction with Oxygen

Interaction of the platinum metals with oxygen is much more complex and far from being completely understood. The essential features will be outlined for Pt(111) [30, 32], but similar effects occur with other surfaces. In brief, three chemically different states exist. A molecular state (desorbing around 170 K), an atomic state chemisorbed on top of the surface atoms

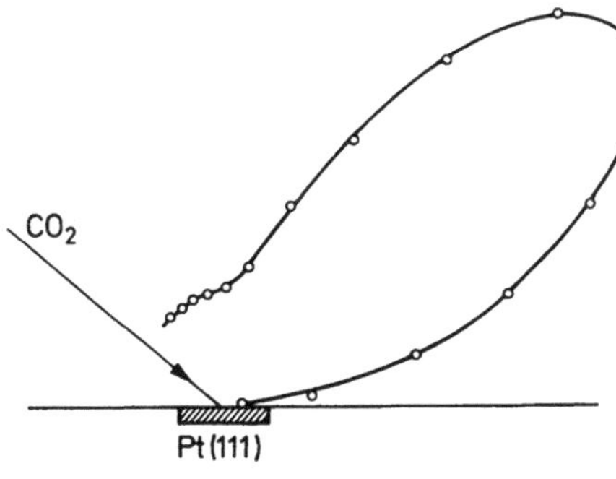

Figure 26. Polar diagram of the angular distribution for CO_2 molecules scattered at a Pt(111) surface at 300 K. (Reproduced with permission from ref. [90])

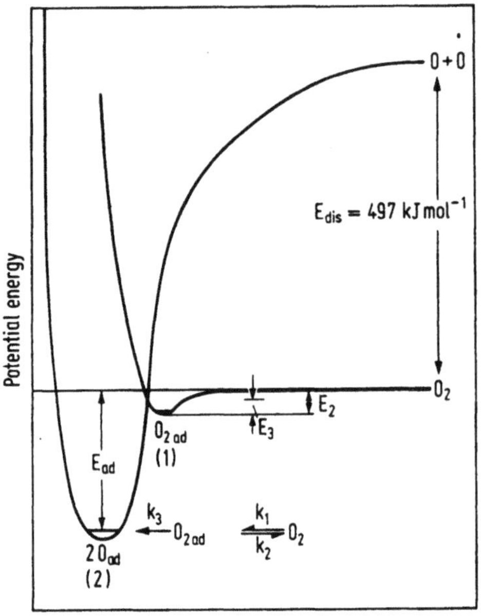

Figure 27. Lennard-Jones potential diagram for adsorption of oxygen. (Reproduced with permission from ref. [32])

(desorbing between 500 and 1000 K), and another (subsurface) atomic species (desorbing above 1250 K).

The molecular state, $O_{2,\,ad}$, is presumably lying flat on the surface and has clearly been identified by vibrational spectroscopy [91]. Its adsorption energy is around 30 kJ mol^{-1}. While the trapping probability in this state is near unity around 100 K [30], it drops to a constant value of around 0.25 for surface temperatures between 300 and 800 K (where the catalytic reaction takes place) [32]. Since this species is only weakly adsorbed (and will in addition rapidly dissociate at elevated temperatures) its steady-state surface concentration will always be negligibly small under the conditions to be considered in this context. Dissociative adsorption proceeds most probably in parallel, through both the direct and indirect paths (see section 2.D). The Lennard-Jones type potential diagram including the relevant energies is reproduced in Figure 27 [32]. The effective activation energy for dissociative adsorption is initially slightly negative, giving rise to a decrease of the (atomic) sticking coefficient s_0 from 0.06 at 300 K to 0.025 at 600 K. Since the adsorption energy E_{ad} decreases continuously from 210 to 175 kJ mol^{-1} with increasing coverage (due to mutual repulsion between the adsorbed O atoms) also the crossing point between the two potential curves in Figure 27 increases continuously giving rise to a positive activation energy at higher coverages. It is believed that this is also the main reason why at constant temperature the sticking coefficient decreases strongly with coverage as can be seen from Figure 16. The probability for dissociative adsorption increases markedly upon the introduction of surface defects such as steps [92]. s_0 differs also markedly between various metals and reaches for example values around 0.3 for Ir(111) [93] and 0.4 for

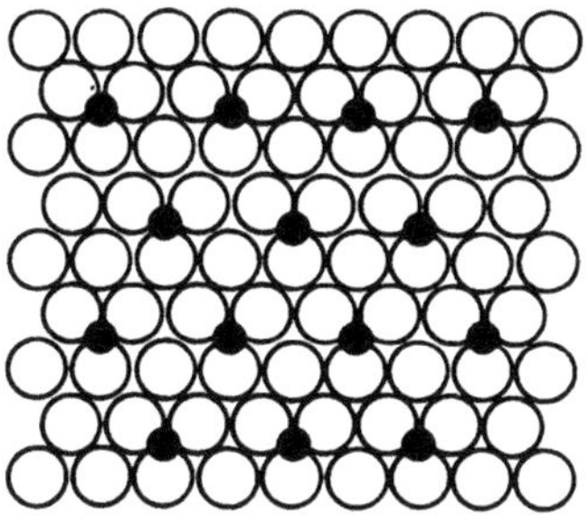

Figure 28. Structure model for O_{ad} on Pt(111) at $\theta = 0.25$

Pd(111) [94]. The preexponential for associative desorption was determined to be around 10^{-2} $cm^2\,s^{-1}$, which is a 'normal' value in the light of transition state theory.

Saturation with O_{ad} upon exposure of Pt(111) to O_2 is reached at $\theta = 0.25$ and is characterized by a 2×2-LEED pattern for which a structural model is reproduced in Figure 28. Analogous structures were observed with the most densely packed planes of other platinum metals. With Ru(0001) the occupation of threefold coordinated sites was concluded from the observed frequency of 516 cm^{-1} for the metal-oxygen vibration [95]. For O/Ir(111) an interlayer spacing of 1.30 Å and an Ir—O bond length of 2.04 Å was determined from a LEED intensity analysis [96]. The limitation of the saturation coverage at $\theta = 0.25$ with an O—O distance of 5.5 Å (which leaves enough space for additional CO uptake, see below) is obviously not determined by strong repulsive interactions between the oxygen atoms, but rather by the need of an O_2 molecule for a larger 'ensemble' of neighboring surface atoms in order to dissociate. A higher local O-coverage of 0.33 within islands could be reached by coadsorption with CO on Pd(111) which 'compresses' the O-adlayer [66] and recent experiments with NO_2/Pt(111) showed that in this case a considerably higher surface concentration of O_{ad} can be built up [97].

4. Coadsorption

Since CO_2 formation proceeds via the LH mechanism (see below) both reactants, CO_{ad} and O_{ad}, have to be simultaneously on the surface and govern by their concentrations and mutual configurations the reaction rate. The situation is rather complex.

Since CO_{ad} tends to form a densely packed layer and dissociative oxygen adsorption needs, on the other hand, a fairly large ensemble of neighboring unoccupied surface atoms, adsorbed CO inhibits oxygen adsorption. As soon as the CO coverage exceeds a critical value the sticking coefficient for oxygen drops steeply and catalytic reaction practically ceases. This is the reason why under steady-state conditions the temperature has always to be high enough ($\gtrsim$420 K) in order to establish a low enough CO coverage through the adsorption-desorption equilibrium of this species.

O_{ad} forms, on the other hand, a rather open structure so that additional uptake of CO is easily possible. Transient measurements with Pt(111) revealed that the sticking coefficient for CO on an oxygen-saturated surface is not markedly lowered [98]. Qualitatively similar observations have been made for Pd, Ru, Ir, Rh as well as other Pt surfaces [86]. If both O_{ad} and CO_{ad} are present on the surface their mutual configuration will depend on their respective concentrations (and temperature) as will be outlined by means of the observations made with Pd(111) [66]:

- At very low O_{ad} and CO_{ad} concentrations both species will be randomly distributed over the surface;
- If CO is adsorbed first to a coverage $\theta_{CO} = {}^1/_3$ (where a $\sqrt{3} \times \sqrt{3}$ R30° surface structure is completed) oxygen adsorption is inhibited. For $\theta_{CO} < < {}^1/_3$ additional oxygen is adsorbed and both species form separate domains (competitive adsorption). Catalytic reaction is then restricted to the boundaries of these domains for which effect experimental evidence has indeed been obtained [99].
- If oxygen is adsorbed first up to saturation ($\theta_o = 0.25$) additional uptake of CO leads first to a compression of the oxygen adlayer into domains with a higher local coverage ($\theta_{o,\,loc} = 0.33$) and eventually to a mixed phase consisting of both species (cooperative adsorption). Due to the operation of repulsive interactions the effective bond strength to the surface is appreciably lowered. This in turn causes a decrease of the activation energy for product formation so that reaction occurs already far below room temperature.

The repulsion between O_{ad} and CO_{ad} (if in intimate contact on the surface) manifests itself also with other properties. The ionisation energies of the CO derived valence levels are increased, and the CO induced work function change is negative while CO alone causes the work function to increase [66] (except with Pt(111) [100]). With Ru(0001) the frequency of the C—O stretch vibration was observed to be shifted from 1980 to 2048 cm^{-1} and that of the metal-CO vibration from 445 to 387 cm^{-1} while the adsorption energy of CO was simultaneously lowered by several kJ mol^{-1} [67]. All these observations can be explained by the following model (Figure 29): O_{ad} is a strongly electronegative ligand which withdraws electronic charge from the metal. Electron transfer from the metal to the antibonding $2\pi^*$ level of CO ('backdonation' effect) is, however, also responsible for part of the adsorption energy of CO. Thereby the C—O stretching frequency is lowered and the work function increases. If both O_{ad} and CO_{ad} come into intimate contact there will be a net flow of electronic charge from CO to O. As a consequence the CO molecule will be more positive (= increase of the ionisation potentials and decrease of the work function), less back-donation causes an increase of the C—O stretch frequency and a lowering of the strength of the adsorption bond (= decrease of the adsorption energy as well as of the metal—

Figure 29. Indirect interaction between neighboring O_{ad} and CO_{ad} particles, giving rise to a net flow of electronic charge from CO to O

CO vibration frequency). This effect is of general importance in so far as it explains qualitatively the existence of an activation barrier for CO_2 formation since in that case both species approaching each other have first to overcome their effective mutual repulsion before the new chemical bond may be formed.

5. *Surface Reactions*

The possible steps leading to CO_2 formation can now be formulated as follows

$$\left.\begin{array}{lr} CO + * \underset{k_{-1}}{\overset{k_1}{\rightleftharpoons}} CO_{ad} & \text{(i)} \\ O_2 + * \xrightarrow{k_2} 2O_{ad} & \text{(ii)} \\ CO_{ad} + O_{ad} \xrightarrow{k_3} CO_2 & \text{(iii)} \\ CO + O_{ad} \longrightarrow CO_2 & \text{(iv)} \\ 2CO_{ad} + O_2 \longrightarrow CO_2 & \text{(v)} \end{array}\right\} \quad (23)$$

* denotes schematically a free adsorption site (which consists, however, of a different ensemble of surface atoms). Desorption of O_{ad} starts only above 700 K and is therefore neglected. The possible effect of less reactive 'surface oxide' species is also not included in this scheme. The ER reaction (v) can experimentally easily be ruled out: If a CO saturated surface is exposed to O_2, only negligible CO_2 formation is observed. Obviously the oxygen molecule has first to dissociate, the M—O bond being weaker ($\sim$370 kJ mol^{-1}) and therefore more reactive than the O—O bond (500 kJ mol^{-1}). Clear distinction between reaction steps (iii) (LH) and (iv) (ER) is obviously more difficult since exposure of an O_{ad} layer to CO will lead in any case to adsorption of this species. Time-resolved measurements by means of the modulated molecular beam technique were in this case able to provide a unequivocal answer. Since CO_2 will practically instantaneously appear in the gas phase after formation (due to its low heat of adsorption) there should also be a immeasurably short time lag between impingment of CO on an O_{ad} layer and detection of CO_2 if the reaction proceeds via the ER mechanism. In the case of the LH reaction, on the other hand, the CO molecule is first chemisorbed and spends probably a measurable time on the surface before product formation occurs. Figure 30 shows the results of an experiment [98] in which a O_{ad}-covered Pt(111) surface was suddenly exposed to a modulated CO beam. The phase lag between the primary CO beam and the CO_2 coming off the surface served for determining the mean surface residence time, τ, of CO. τ was found (under these conditions) to be of the order of 10^{-3} s which can only be reconciled with a LH mechanism, so that indeed the steps (i)–(iii) comprise the actual mechanism of this reaction. Quite analogous experiments were performed with Pd(111) and led to the

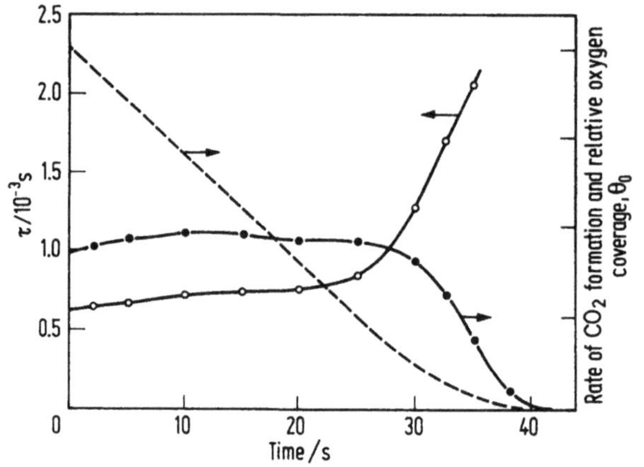

Figure 30. Modulated molecular beam results for CO_2 formation on Pt(111) [98]. At $t = 0$ a O-covered surface was exposed to a modulated CO beam. The figure shows the relative oxygen coverage, θ_0(– – – –), the relative rate of CO_2 formation, ●, as well as the relaxation time (mean surface lifetime), τ (—○—). Temperature 442 K, effective $p_{CO} = 1.5 \times 10^{-5}$ Pa, modulation frequency 100 Hz, angle of beam incidence 55°

same result [101]. Isotopic tracer studies with Pt(111) at lower temperatures also ruled out the ER mechanism [102] and since the overall kinetic behavior is quite similar for all platinum metals it can safely be concluded that this reaction will always proceed via the LH mechanism. For low O_{ad} and CO_{ad} concentrations the activation energy for the step $O_{ad} + CO_{ad} \rightarrow CO_2$ was determined to be $E^*_{LH} = 100$ kJ mol^{-1} for both Pd(111) [101] and Pt(111) [98]. The preexponentials are of the order of 10^{-2} and 10^{-1} cm^2 s^{-1}, respectively, which is of the 'normal' order of magnitude in view of the transition state theory. Since with increasing coverage repulsive interactions lower the effective adsorption energies it becomes plausible why also E^*_{LH} is decreasing. Values ranging between 20 and 60 kJ mol^{-1} have been reported for various systems investigated under these conditions [103]. Parallel to this decrease of E^*_{LH} also the preexponential drops. This is a phenomenon called 'compensation effect' which is quite generally observed with kinetic studies and for which only plausible arguments but no quantitative explanation can yet be offered.

The fact that the kinetic parameters vary with coverage (due to lateral interactions) even with a single crystal surface is one of the reasons why simple Langmuir-type formulations of the reaction kinetics provide no appropriate description of the actual situation.

Figure 31a illustrates the progress of the reaction (at low surface concentrations) by means of a none-dimensional potential energy diagram. It is interesting to notice that a large fraction of the reaction enthalpy ($\Delta H = 284$ kJ mol^{-1}) is already liberated in the chemisorption steps of the reactants, so that the surface reaction itself is only weakly exothermic. The progress of the reaction becomes probably even clearer by the schematic two-dimensional potential diagram of Figure 31b, where a contour map of the potential energy as a function of the O ... CO distance as well as of the distance of the center of gravity of the CO + O complex from the surface is plotted. As outlined in the preceding section, the activation barrier will be dominated by repulsive interaction between CO_{ad} and O_{ad}. The activated complex has most probably a bent configuration whereby the

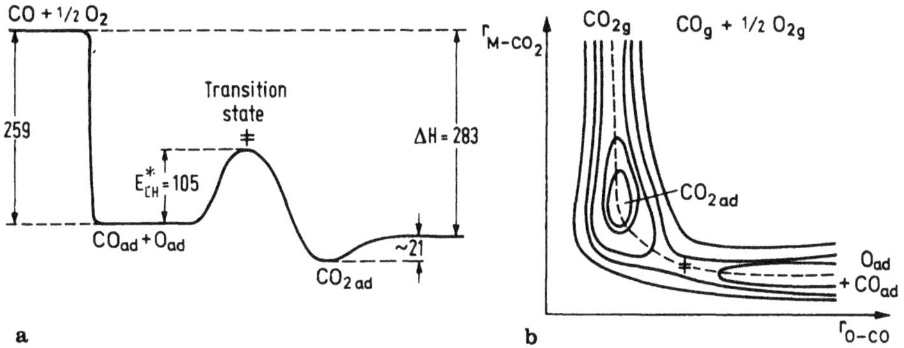

Figure 31. Potential diagrams illustrating the progress of the reaction $\frac{1}{2} O_2 + CO \rightarrow CO_2$ on Pt(111) at low coverages. **a** one-dimensional, **b** two-dimensional. Energies in kJ mol^{-1}

bonds to the surface are already weakened. One may wonder if after passing the transition state the CO_2 molecule formed is briefly trapped in the potential well for $CO_{2,\,ad}$ and becomes thermally accommodated with the surface. Experiments with Pt surfaces showed that this is indeed not the case: the angular distribution of the evolved CO_2 molecules is sharper than cosine [98, 104, 105] which indicates that they carry excess translational energy [56]. Figure 32 shows polar plots of CO_2 angular distributions at various surface temperatures which reflect the sensitive influence of phonon excitations of the substrate on energy transfer. Time-of-flight measurements [105] revealed that CO_2 coming off along the direction of the surface normal have an excess translational energy of about 30 kJ mol^{-1}, and quite recent experiments using infrared emission revealed that the product molecules are also vibrationally considerably hotter than the surface [106, 107]. All these results indicate that CO_2 is energetically only weakly coupling to a Pt surface.

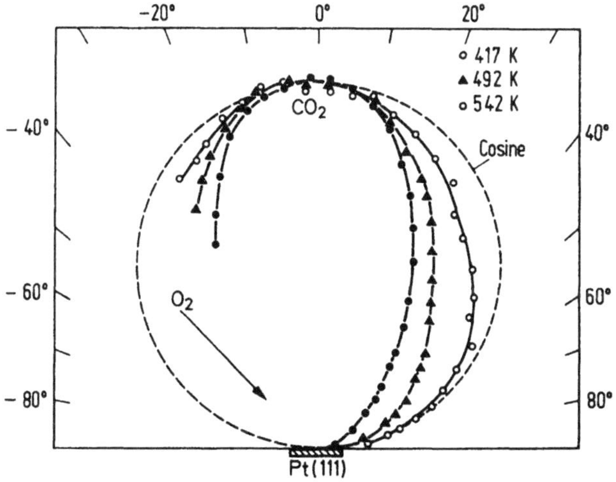

Figure 32. Angular distributions of CO_2 molecules formed at a Pt(111) surface at various temperatures with $p_{CO} = 5 \times 10^{-6}$ Pa and $p_{O_2} = 7 \times 10^{-5}$ Pa [98]

Scattering of CO_2 at Pt reveals indeed that the trapping probability is small and that a large fraction undergoes direct inelastic scattering [90, 108] (see Figure 26).

It is interesting to notice that with Pd(111) a cosine angular distribution of CO_2 formation was observed which indicates complete translational accommodation in this case [101] although the overall energetics is almost identical to that derived for Pt(111).

6. Reaction Kinetics

Steady-state reaction kinetics will of course be determined by all three processes, (i)–(iii), which are involved in the reaction mechanism, and either of these (chemisorption of oxygen, adsorption or desorption of CO, as well as the LH reaction) may become rate-limiting. Since all rate constants depend — even with a uniform single crystal surface — on the coverage, it appears to be rather impossible to formulate a generally valid rate equation $r = f(T, p_{CO}, p_{O_2})$. An elaborate attempt of this kind for Ir(110) was only able to reproduce the qualitative features [109].

As a typical example Figure 33 shows the stationary rate of CO_2 formation on Ir(111) in a CO + O_2 mixture of about 9×10^{-6} Pa total pressure as a function of temperature, together with the variations of the surface concentrations of O_{ad} and CO_{ad} [110]. A maximum in the reaction rate is quite typical for all systems of this kind and is reached at a

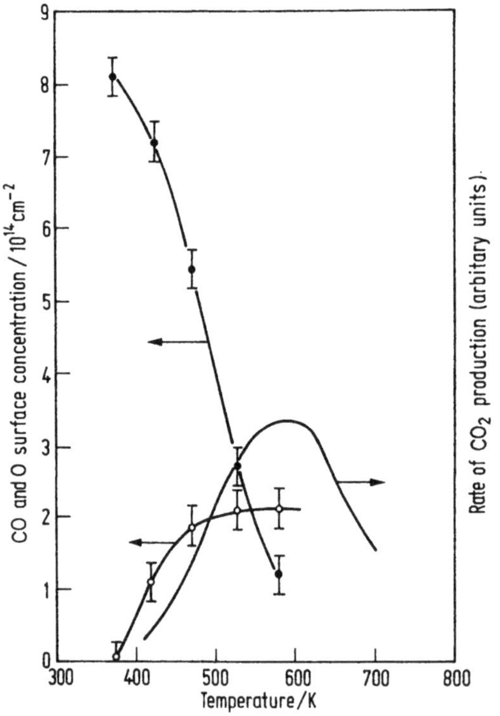

Figure 33. Steady-state rate of CO_2 formation on Ir(111) as a function of temperature, together with the surface concentrations of O_{ad} and CO_{ad}. $p_{O_2} = 3p_{CO} = 2.39 \times 10^{-5}$ Pa. ○; O: ●; CO. (Reproduced with permission from ref. [110])

temperature where θ_O and θ_{CO} are about equal (which is also to be expected in view of the simplest formulation of a LH reaction). But even formulation of the rate equation in terms of $r = k\theta_O\theta_{CO}$ will only be valid as long as both reactants are randomly distributed over the surface and breaks down as soon as island formation etc. comes into play.

A rough classification of the various reaction regimes follows a study with Pd(111) [101] but can equally be applied to other surfaces of the platinum group metals:

a) Low surface concentrations of O_{ad} and CO_{ad}.

The surface temperature has to be high enough (*i.e.* above the maximum of the *r(T)*-curve, see fig. 33) in order to ascertain that the $CO \rightleftharpoons CO_{ad}$ equilibrium is shifted to the left hand side and θ_{CO} is small. In addition $p_{O_2} < p_{CO}$ in order to prevent the build-up of an appreciable O_{ad} concentration (which is only removed by reaction, but not by desorption). Under these conditions oxygen adsorption becomes rate-limiting, *i.e.* $r \approx k_2 p_{O_2}$. Since k_2 (= sticking coefficient for oxygen) is only weakly dependent on temperature the apparent activation energy should be small. Both conclusions (as well as independence of p_{CO}) are in agreement with experimental observation under properly chosen conditions.

b) Low CO coverage, higher O_{ad} concentrations.

The temperature has still to be high enough in order to keep θ_{CO} (through desorption) small, but now $p_{O_2} > p_{CO}$ so that a higher O_{ad} concentration is build up. Since this latter species tends to form islands it becomes plausible why under these conditions the rate is frequently found to be independent of θ_o (this becomes also evident from an inspection of Figure 30), *i.e.* $r \approx k'\theta_{CO}$. Since $\theta_{CO} \propto p_{CO}$ in the low CO coverage regime, the rate becomes proportional to the CO pressure, and the apparent activation energy is approximately given by

$$E^*_{app} \approx E_{ad,CO} - E^*_{LH},$$

which is of the order of about 40 kJ mol^{-1}.

c) Higher CO concentrations

As long as *r(T)* is below its maximum an appreciable fraction of the surface is covered by CO which in turn inhibits adsorption of oxygen. As a consequence the rate is now given to a first approximation by $r \approx k'' p_{O_2}/p_{CO}$, where the apparent activation energy is mainly determined by the adsorption energy of CO. This will be the situation usually governing the kinetics at higher total pressure since of course the *r(T)* maximum (where θ_{CO} drops) will be shifted to even higher temperatures with increasing CO pressure. Rate laws of this type have in fact been derived in several studies under more 'real' conditions [111].

Another interesting aspect of the kinetics of this reaction concerns the possible occurrence of hysteresis effects and of temporary oscillations in the rate of CO_2 formation. There exists already a large amount of experimental data (mostly with Pt at higher pressures) as well as theoretical models [82–85, 112] whose full discussion would go far beyond the scope of this chapter.

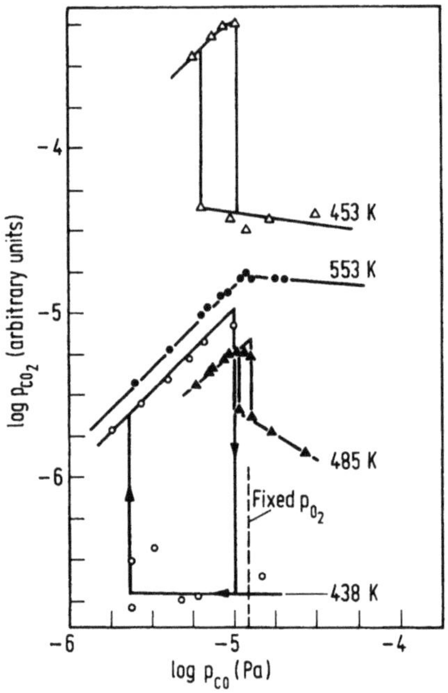

Figure 34. Rate of CO_2 formation on a Pt-wire as a function of CO pressure at fixed p_{O_2} and temperatures. Hysteresis effects, depending on whether p_{CO} is increased or lowered. (Reproduced with permission from ref. [114])

A typical example for hysteresis effects [113, 114] is shown in Figure 34. The rate of CO_2 formation as a function of CO pressure at constant p_{O_2} and T increases continuously until is suddenly drops to a low value which is also maintained if the CO pressure is lowered again until it switches back to the original values. There exists obviously a pressure region where the rate is either high or low depending on the procedure, and these metastable states can be maintained over periods of hours. By using the reaction mechanism as presented above with simplified but reasonable kinetic equations, it is possible [85, 115] to model this effect. Figure 35 shows the rate as a function of p_{CO} calculated for special conditions on the basis of this model which indeed exhibits the qualitative features of the described effect [85].

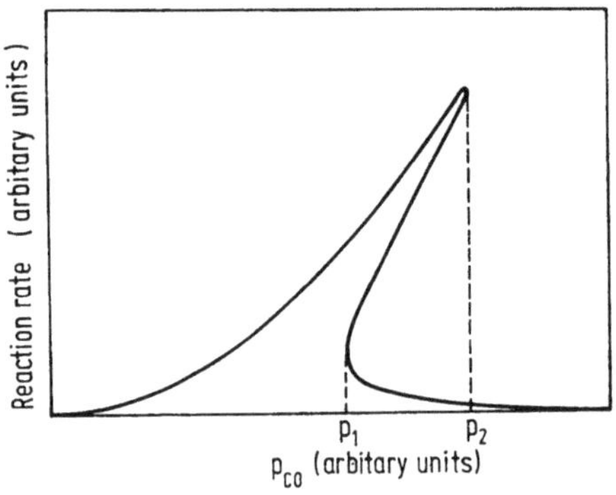

Figure 35. Illustration of modelling the hysteresis effect in CO oxidation by a suitable kinetic model The figure illustrates the dependence of the calculated reaction rate on p_{CO}. Temperature 548 K, platinum catalyst. (Reproduced with permission from ref. [85])

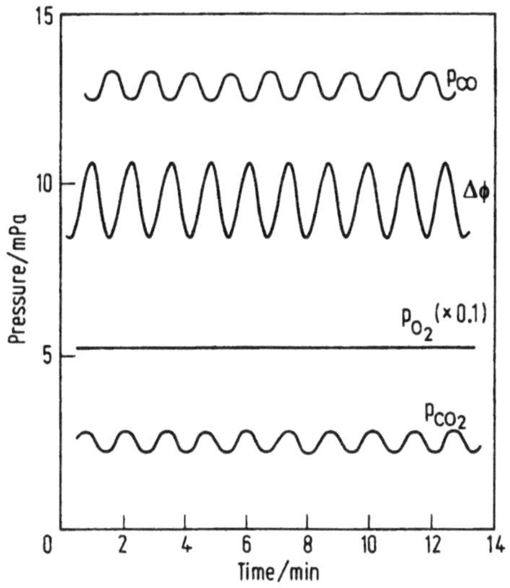

Figure 36. Temporary oscillations in the rate of CO_2 formation over a polycrystalline, clean Pt wire. $T = 502$ K. (Reproduced with permission from ref. [117])

At higher pressures within the multiple state region temporary oscillations instead of hysteresis effects are observed. While previous studies reported on these phenomena only for pressures $\gtrsim 100$ Pa, similar effects were recently also observed at pressures of about 10^{-2} Pa for the oxidation of CO by NO [116] or O_2 [117], — in the latter case even with Pt single crystal surfaces. As an example Figure 36 shows the variation with time of the rate of CO_2 formation, the CO pressure, and the work function change (monitoring the state of the surface) being recorded with a Pt wire exposed to $1 \cdot 33 \times 10^{-2}$ Pa CO and $5 \cdot 3 \times 10^{-2}$ Pa O_2 at 502 K in a flow system [117]. While the model proposed in ref. [85, 115] is able to explain the occurrence of multiple kinetic states there is still a need for a driving force which causes periodic switching from one state to the other, for which several possibilities have been proposed. Transport processes and temperature variations can clearly be ruled out for the examples discussed here. The formation of less reactive subsurface oxide species [85] appears to be more probable. Measurements with a Pt(100) single crystal surface suggested still another possibility, namely the adsorbate-induced $5 \times 20 \rightarrow 1 \times 1$ phase transformation of the surface structure [117]. A similar phase transformation occurs also with Ir(100) where quite recently again oscillations were observed [118]. The same paper reports, however, also on oscillations with Pd surfaces which undergo no phase transitions. It appears thus at present as if not a single mechanism but several possibilities exist for triggering these interesting kinetic phenomena.

7. Catalysis under 'Real' Conditions

Correlation of the data obtained for clean single crystals at low pressure with 'real' conditions is in this case relatively easy and straightforward. Although the surface composition of practical catalysts is unknown, the common impurities sulfur and carbon are reacted off by oxidation during the reaction

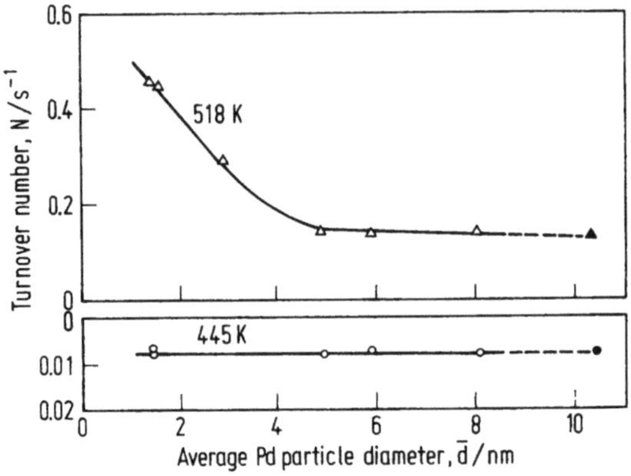

Figure 37. Turnover rates for CO_2 formation over palladium as a function of particle size at $p_{O_2} = 1.3 \times 10^{-4}$ Pa, $p_{CO} = 1.2 \times 10^{-4}$ Pa. △, ○; Pd/{1012} α-Al_2O_3: ●, ▲; (111) Pd. (Reproduced with permission from ref. [120])

[119], so that there is strong evidence that the reaction proceeds on an essentially clean surface even without UHV conditions. Measurements with various Pd single crystal surfaces revealed only very little differences in activity [119], indicating that the reaction is structure-insensitive. Studies with Pt(111) surfaces containing various step concentrations [92] revealed that defect sites exhibit a higher sticking coefficient for oxygen but a lower reaction probability for CO_2 formation so that both effects essentially compensate each other. The 'pressure gap' again offers no serious problem. Since under conditions of practical interest the reaction rate is approximately proportional to p_{O_2}/p_{CO}, it should be determined by the *ratio* rather than by the *sum* of the partial pressures.

In this context a recent paper by Ladas *et al.* [120] is most remarkable and confirms the stated conclusions. In this work small Pd crystallites of defined average diameters were evaporated onto a α-Al_2O_3 single crystal support and characterized by transmission electron mixroscopy. The turnover number for CO_2 formation (*i.e.* the number of CO_2 molecules produced per surface Pd atom per second), N, was then determined at various temperatures as a function of particle size, $\bar{d}$. Results are reproduced in Figure 37. N is constant for $\bar{d} \geqq 40$ Å and is also equal to the numbers derived from studies with a Pd(111) single crystal surface [101]. The increase of N with decreasing particle size for T = 518 K was attributed to a geometric effect whereafter small crystallites are (at the same pressure) exposed to a higher flux of gas

Table 2. Turnover rates N at 450 K for CO_2 formation at high and low pressures [120]

Catalyst	5% Pd/SiO_2	Pd/{1012} α-Al_2O_3, 8 nm particles
p_{O_2}/Pa	6.5×10^2	3.6×10^{-5}
p_{CO}/Pa	1.3×10^2	7.2×10^{-5}
N/s^{-1}	0.03	0.012

molecules. In addition the turnover rate was also compared with results obtained with a 5 % Pd/SiO_2 catalyst near atmospheric pressure [121]. As can be seen from Table 2, N varies by less than a factor of 3 (which is still within the limits of error) although the total pressure differed by 7 orders of magnitude.

B. Oxidation of Hydrogen

The gas phase reaction $2\,H_2 + O_2 \rightarrow 2\,H_2O$ is also readily catalyzed by the platinum group metals. Although it is of minor practical importance (electrochemical aspects relevant for fuel cells are not considered in this context) it represents one of the earliest examples for the scientific study of the phenomenon of catalysis. There exists now general agreement on the overall reaction mechanism and on the surface species involved, although yet much less detailed information than for CO oxidation is available [122].

1. Adsorption of Hydrogen

Hydrogen adsorbs dissociatively on all platinum group metals without any appreciable activation energy. Since this question is of relevance for the discussion of the reaction mechanism, it has to be emphasized that so far no evidence for the existence of a molecularly adsorbed species above 100 K (where reaction starts) has been obtained. This species — if existing at all — is expected to have a very low adsorption energy and would therefore play no role in the reaction with oxygen.

The initial sticking coefficient is typically of the order of about 0.1, but may also reach higher values (*e.g.* 0.5 for Pd(100) [123]). The adsorption energy ranges typically between 60 and 100 kJ mol^{-1} corresponding to strengths of the M—H bond around 250 to 270 kJ mol^{-1}. Reports on a fairly low adsorption energy of H_2 on Pt(111) ($\sim$45 kJ mol^{-1} [124]) could not be confirmed in more recent investigations which suggest a value around 80 kJ mol^{-1} [125]. Maxima of the desorption rates in TDS experiments are typically observed between 300 and 500 K, depending on the kind of surface as well as on coverage. The preexponential factor for desorption is usually that typical for second order kinetics, *e.g.* $10^{-2}\,cm^2\,s^{-1}$ for H_2/Pd(100) [123]. The adsorption energy varies roughly by about $\pm 10\%$ between different crystal planes of the same metal and is also somewhat higher at defect (*e.g.* step) sites [126, 127]. The maximum coverage on the low-index planes is around unity, that is, one H-atom per surface atom. In a few cases ordered overlayer structures are observed by LEED. On Pd(100) a $c2 \times 2$ structure forms around $\theta = 0.5$ which disorders already below room temperature. The interaction energies between neighboring H atoms were estimated on the basis of the phase diagram to be of the order of 2 kJ mol^{-1} [123]. On Pt(111) the H atoms are located in threefold coordinated sites as determined by vibrational spectroscopy [128], quite in analogy to the structure derived for H/Ni(111) from a LEED intensity analysis [11].

From field emission experiments Lewis and Gomer [129] estimated the activation energy for surface diffusion of hydrogen on platinum to be

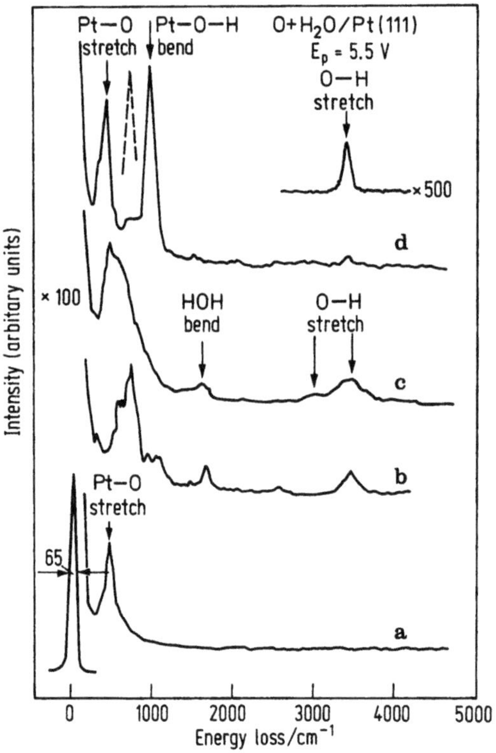

Figure 38. High resolution electron energy loss spectra from Pt(111) showing vibrational excitations of various surface species. (Reproduced with permission from ref. [136])

about 20 kJ mol^{-1}. More recently for H/W(110) activation energies for surface diffusion around 22 kJ mol^{-1} were determined for T $\geqq$ 130 K, whereas at even lower temperatures tunneling dominates [130]. This appears to be a typical number, although the strong observed variations with coverage certainly need further clarification. The adsorbed hydrogen atoms are thus even at 100 K sufficiently mobile. Bulk diffusion and surface segregation of hydrogen atom forms a complicating factor with Pd [126, 131], but will not be further discussed in the present context.

2. Adsorption of Oxygen

The properties of oxygen interacting with platinum groups metals have already briefly been outlined in section 3.A.3.

3. Adsorption of Water

The interaction with H_2O has been studied in some detail with the Pt(111) surface so that the discussion will be restricted to this system. Quite similar features are, however, observed with other surfaces [132]. UPS [133] as well as HRELS data [134] at 100 K clearly indicate molecular adsorption in the submonolayer range, whereby hydrogen bonding between the H_2O molecules comes additionally into play. The LEED pattern exhibits the formation of an ordered $\sqrt{3}\times\sqrt{3}$ R 30° structure [133, 135]. The existence of hydrogen bonds (being equivalent to attractive interactions) manifests itself also in the thermal desorption spectra. The peak temperature increases slightly with increasing coverage suggesting a reaction order smaller than unity. The adsorption energy was estimated to be about 60 kJ mol^{-1} [133] which is only slightly higher than the heat of sublimation.

It is interesting to notice that the desorption temperature increases if an oxygen covered surface was exposed to water. This effect is ascribed to the fact that H_2O then dissociates below 190 K and recombines again near 210 K [133].

4. Adsorbed Hydroxyl

Evidence for low temperature dissociation of water on an oxygen precovered Pt(111) surface was obtained by HRELS and UPS and led also to the identification of OH_{ad} [136]. Figure 38 reproduces a series of vibrational spectra exhibiting the pertinent features. Curve *a* represents the single band arising from the Pt—O stretch vibration of adsorbed oxygen. Curves *b* and *c* were obtained after adsorption of H_2O at 100 K on a clean and oxygen-covered surface, respectively, and are characteristic for non-dissociatively adsorbed water. Curve *c* converts into curve *d* after heating to 155 K. The H—O—H bond vibration is now absent and the O—H stretch vibration has shifted and narrowed. A new band at 1015 cm^{-1} was assigned to the bending vibration of OH_{ad}. It was concluded that this species is formed in appreciable concentrations through a reaction of the type

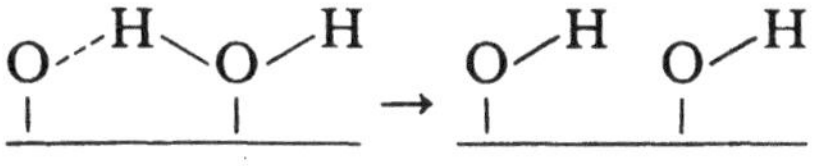

for which additional support was obtained from the UPS data. Since adsorbed oxygen atoms are needed for this process it is not surprising that no detectable quantities of OH_{ad} could be observed with a clean Pt(111) surface.

Although so far restricted to an oxygen covered surface (which obviously has a stabilizing effect), the spectroscopic identification of OH_{ad} sheds important light on the mechanism of the water formation reaction. Evidence for this intermediate also on essentially oxygen free platinum surfaces is obtained from the work by Lin *et al.* [137, 138] where desorption of OH was detected by means of laser induced fluorescence (LIF). If a platinum foil was exposed to a $H_2 + O_2$ mixture this species appeared above ~800 K with an activation energy of 130 kJ mol^{-1} which was attributed to the process $OH_{ad} \rightarrow OH$. At given O_2 pressure and temperature the production of OH passed through a maximum with increasing H_2 pressure which becomes plausible in view of the competing step $OH_{ad} + H_{ad} \rightarrow H_2O$ [137]. Both the rotational and vibrational temperatures of the desorbing OH radicals were found to be equal to the surface temperature [138], suggesting sufficiently strong coupling to the surface and the absence of an activation barrier for the reverse process, *i.e.* for $OH \rightarrow OH_{ad}$. Although these experiments were not performed under UHV conditions it is nevertheless believed that the results are representative for the behavior of a clean Pt surface and that the measured activation energy is indeed equal to the strength of the Pt—OH bond. (If its value of ~130 kJ mol^{-1} appears to be unusually small in view of the ~250 kJ mol^{-1} for the Pt—H bond it has to be remembered that OH makes generally much weaker bonds than H: 213 kJ mol^{-1} for HO—OH *vs.* 430 kJ mol^{-1} for H—H, 192 kJ mol^{-1} for HO—OC$(CH_3)_3$ *vs.* 440 kJ mol^{-1} for H—OC$(CH_3)_3$, 250 kJ mol^{-1} for HO—Cl *vs.* 430 kJ mol^{-1} for H—Cl etc. [139]).

5. *Surface Interactions and Reaction Mechanism*

Recent low-temperature studies with Pt(111) by Fisher *et al.* [140] provided the most detailed insight into the actual reaction mechanism. A oxygen-precovered surface ($\theta_O = 0.25$) was exposed to H_2 at 100 K. UPS (Figure 39) and XPS indicated no H_2O formation at this temperature. However H_2O_{ad} was detectable after 10 min. heating to 120 K, which obviously represents the low-temperature limit for a noticeable reaction rate and which is also below the desorption temperature of water (~180 K). The activation energy for the formation of H_2O_{ad} was estimated to be about 35 kJ mol^{-1} (at

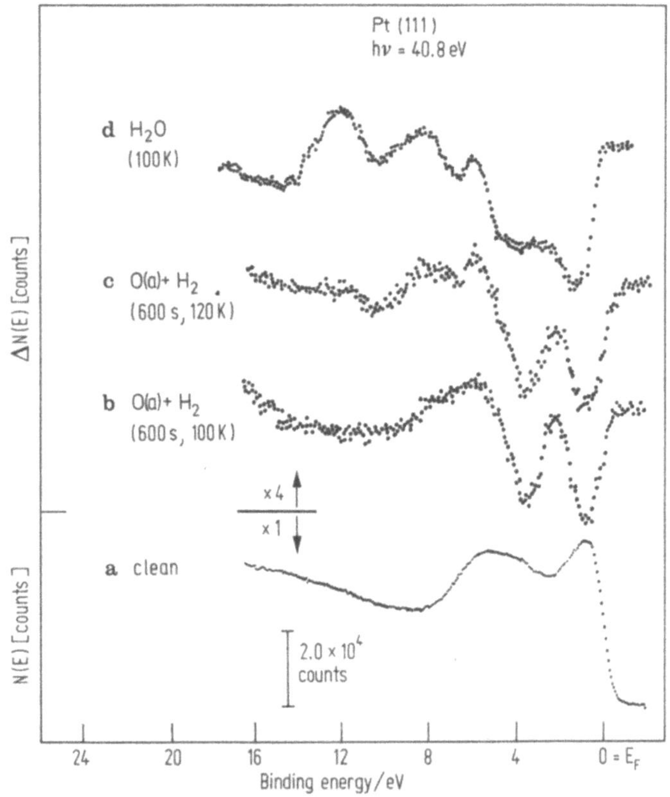

Figure 39. UPS data from Pt(111) reflecting the formation of adsorbed H_2O. (Reproduced with permission from ref. [140])

high oxygen coverage). No intermediate formation of OH_{ad} could be detected, suggesting that the step $OH_{ad} + H_{ad} \rightarrow H_2O_{ad}$ is much faster than the step $O_{ad} + H_{ad} \rightarrow OH_{ad}$. The results were quite similar in the presence or absence of an H_2 atmosphere. This indicates that the reaction proceeds indeed through thermally accommodated H_{ad} rather than through any 'precursor'-like species such as $H_{2\,ad}$ or 'hot' H atoms [79]. A mechanism involving molecular hydrogen can also be ruled out on the basis of the observation of isotopic scrambling (if a $H_2 + D_2$ mixture is used) [141] and by the fact that no experimental evidence for the existence of a H_{2ad} species has so far been found (see sect. 3.B.1). The reaction probability for a hydrogen molecule striking an oxygen-covered Pt surface was found to be of the order of 0.5 [141–143] which is higher than the sticking coefficient of hydrogen on a clean Pt surface. Obviously the sticking coefficient is increased by the presence of an oxygen adlayer, but this point needs further clarification. The influence of an H_{ad} layer on the adsorption of oxygen is another point which has to be studied in more detail, but there is no indication for strong inhibition (as in the case of CO preadsorption).

The reaction mechanism has most probably to be formulated as follows [140, 141] (which is also in agreement reached with Pd(111) from modulated molecular beam experiments [131])

$$\begin{array}{lll} H_2 & \rightarrow 2\,H_{ad} & \text{(i)} \\ O_2 & \rightarrow 2\,O_{ad} & \text{(ii)} \\ O_{ad} + H_{ad} & \rightarrow OH_{ad} & \text{(iii)} \\ OH_{ad} + H_{ad} & \rightarrow H_2O_{ad} & \text{(iv)} \\ H_2O_{ad} & \rightarrow H_2O & \text{(v)} \end{array} \qquad (24)$$

Step (iii) is slower than step (iv), but has a lower activation energy ($35\ kJ\ mol^{-1}$) than step (v) ($60\ kJ\ mol^{-1}$).

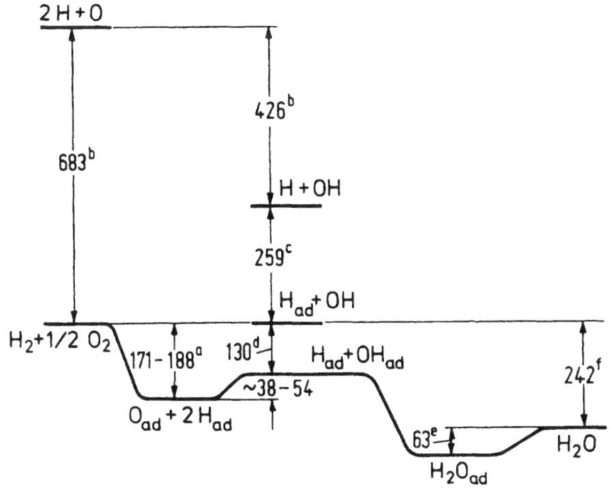

Figure 40. Schematic potential energy diagram illustrating the progress of catalytic hydrogen oxidation on a Pt(111) surface. The designations a–f refer to the processes so identified in the text. Energies in $kJ\ mol^{-1}$

The progress of the reaction is illustrated by the potential energy diagram of Figure 40. The energy differences are based on the following data:

a: The heat of adsorption for H_2 was assumed to be 80 kJ mol^{-1} (see 3.B.1), that for $O_2 \rightarrow 2\,O_{ad}$ ranges between 210 and 175 kJ mol^{-1} with increasing coverage [32], thus yielding 170 to 190 kJ mol^{-1} as the energy associated with the reaction $H_2 + \frac{1}{2} O_2 \rightarrow 2\,H_{ad} + O_{ad}$.

b: The energy for the reaction $H_2 + \frac{1}{2} O_2 \rightarrow 2\,H + O$ is given by the dissociation energies of H_2 and O_2, the dissociation energy of OH is 426 kJ mol^{-1} [139].

c: The strength of the M—H bond (*i.e.* the energy for the reaction Pt + H → Pt — H) is determined from the adsorption energy of H_2 (80 kJ mol^{-1}): $E_{M-H} = \frac{1}{2}(E_{diss,H_2} + E_{ad,H_2})$.

d: The energy for the process $OH_{ad} \rightarrow OH$ (130 kJ mol^{-1}) is taken from Lin's LIF data [137] (see section 3.B.4). The energy difference between $O_{ad} + 2\,H_{ad}$ and $OH_{ad} + H_{ad}$ thus results to be of the order 38 to 55 kJ mol^{-1} which is very close to the observed activation energy for the surface reaction (where $O_{ad} + H_{ad} \rightarrow OH_{ad}$ is rate-limiting) on Pt(111) (~35 kJ mol^{-1} [140]) as well as Pd(111) (~30 kJ mol^{-1} [131]).

e: The adsorption energy for H_2O is about 60 kJ mol^{-1} (see section 3.B.3).

f: Reaction enthalpy for the overall reaction, see ref. [139].

This potential diagram is consistent with all the experimental findings without assuming additional activation barriers. It differs from that proposed by Fisher *et al.* [140] in so far as these authors assumed a considerably higher activation energy for OH desorption (~200 kJ mol^{-1}) for which, however, no real justification can be given. The potential diagram shown in Figure 40 looks qualitatively quite similar to that for CO oxidation (Figure 31 a), however with two important differences: i) The state with the highest energy on the surface is the transition state for the CO oxidation (presumably existing only for a time corresponding to a period of vibration), but represents the long-living $H_{ad} + OH_{ad}$ configuration in the water reaction; ii) The adsorption energy for the product molecule is larger for H_2O than for CO_2. Both effects suggest more efficient energy accommodation for the water reaction than for CO oxidation which is supported by experimental observation. While CO_2 comes off a Pt(111) surface with a sharper-than-cosine angular distribution and excess translational energy (see section 3.A.5), molecular beam experiments showed that with the water formation reaction just the opposite effect occurs [144]; the H_2O molecules are evolved in a broader-than-cosine angular distribution and their mean translational energy is lower than corresponding to the surface temperature. This interesting dynamic effect not only requires thermal accommodation of H_2O

prior to its release into the gas phase, but also indicates strong coupling of this molecule to the phonons of the solid [55, 56]. Interestingly with Pd(111) H_2O formation exhibits a cosine angular distribution [131] as was also observed for CO_2 evolution [101], indicating different coupling of both molecules to this substrate.

6. Reaction Kinetics

The reaction kinetics in low pressure studies are complicated by the fact that the reactants will usually not be randomly distributed over the surface but tend to form islands. TDS studies may therefore not only reveal a peak associated with the step $H_2O_{ad} \rightarrow H_2O$ (what would be expected on the basis of the energetics of the reaction), but two additional reaction-limited maxima at higher temperatures [141, 145]. A model was proposed which limits the reaction to a zone near the perimeters of oxygen islands and whose elaborate mathematical evaluation yields results which are in qualitative agreement with the experimental observations [141].

Since there exists obviously no pronounced inhibition of the adsorption of one of the reactants by the presence of the other, the steady-state turnover numbers are expected to strongly increase with total pressure (quite in contrast to the CO oxidation reaction) and heat transfer may become a serious problem.

Hanson and Boudart [146] studied the steady-state kinetics over a Pt/SiO_2 catalyst at atmospheric pressure either in excess oxygen or hydrogen. In the former case the rate was found to be proportional to the H_2 pressure and independent of p_{O_2}, just what one would expect for reaction at an oxidised surface where hydrogen adsorption is rate-limiting. In reducing atmosphere the opposite dependence on the partial pressures was observed. Now the surface is saturated with adsorbed hydrogen and the rate is limited by oxygen adsorption. Measurements with varying particle sizes revealed that the reaction is structure-insensitive in excess oxygen (which is presumably due to oxide formation), while the specific rate varied by about one order of magnitude in excess hydrogen, which effect has to be ascribed to the dependence of the oxygen sticking coefficient. Systematic studies are certainly still needed in order to obtain a closed picture of the steady-state kinetics of this interesting and 'classical' catalytic reaction.

C. Synthesis of Ammonia

In contrast to the two preceding examples the application of our approach to the catalytic synthesis of ammonia from the elements (Haber process) presents two serious problems:

i) Appreciable yields at the necessary temperatures can — for thermodynamic reasons — only be obtained at pressures much higher than applicable for surface spectroscopic studies:

ii) Actual catalysis is not performed with a clean one-component metal surface but with a promoted iron catalyst so that the choice of a proper model system is by no means evident *a priori*. This section will concentrate on studies modelling the actually used iron catalysts. First the nature of the catalyst surface will be elucidated. Then it will be demonstrated how the pressure gap can be bridged, so that finally low-pressure studies with clean single crystal surfaces can provide insight into the actual reaction mechanism and the various parameters influencing the activity.

Numerous papers on various aspects of ammonia synthesis at 'real' conditions have been published in the past, among which only reference to two more recent contributions [147, 148] is made. Two recent review articles [149, 150] cover the results of low-pressure studies with various metal surfaces which are pertinent to this topic.

1. The Nature of the Catalyst Surface

Industrial (unreduced) iron catalysts consist of Fe_3O_4 and typically about 1% Al_2O_3 and 1% K_2O to which sometimes small amounts of other oxides such as CaO, SiO_2 or MgO are added [151]. Al_2O_3 can enter the magnetite lattice up to a concentration of a few percent [152]. Electron microprobe studies showed that K (and C) are concentrated between the magnetite domains [153, 154]. The addition of Al_2O_3 causes a pronounced increase of the surface area of the catalyst while K_2O has no such effect [155]. It is therefore commonly accepted that Al_2O_3 is a 'structural' promoter which prevents sintering, while potassium acts as an 'electronic' promoter entering directly the surface chemistry. The model studies will therefore only concentrate on the role of K.

From the results of adsorption and isotopic exchange studies Emmett *et al.* [155–157] concluded that the promoters cover a large fraction of the iron surface, even if their nominal bulk concentration is only of the order of 1%. These conclusions are confirmed by the application of surface spectroscopic techniques. Table 3 lists typical compositions of the surface regions of an unreduced industrial catalyst (BASF S 6–10) as determined by XPS in comparison with its bulk composition [158]. Reduction of the catalyst caused increases of the surface concentrations of Fe (by up to a factor of 3) and Ca, while Al and K remained essentially unchanged and the oxygen concentration decreased appreciably. It is remarkable to notice that the sulfur content of the surface (which is a strong catalyst poison) was always very low, if noticeable at all, although iron has a great affinity for this abundant element. It seems as if one of the effects of the promoters consists in the prevention of sulfur deposition.

Even more detailed insight into the surface composition was obtained by the use of scanning Auger electron spectroscopy [159]. A finely focussed

Table 3. Surface and bulk compositions (in at. %) of an unreduced industrial ammonia synthesis catalyst (BASF S6-10) [158]

Element	Al	Na	K	Fe	O	Ca	Si	Cl
Surface concentration (%)	6–10	1	6–20	1–6	50–70	1–5	2	0.5–3.5
Bulk concentration (%)	2	—	0.35	40.5	53.2	1.7	0.25	—

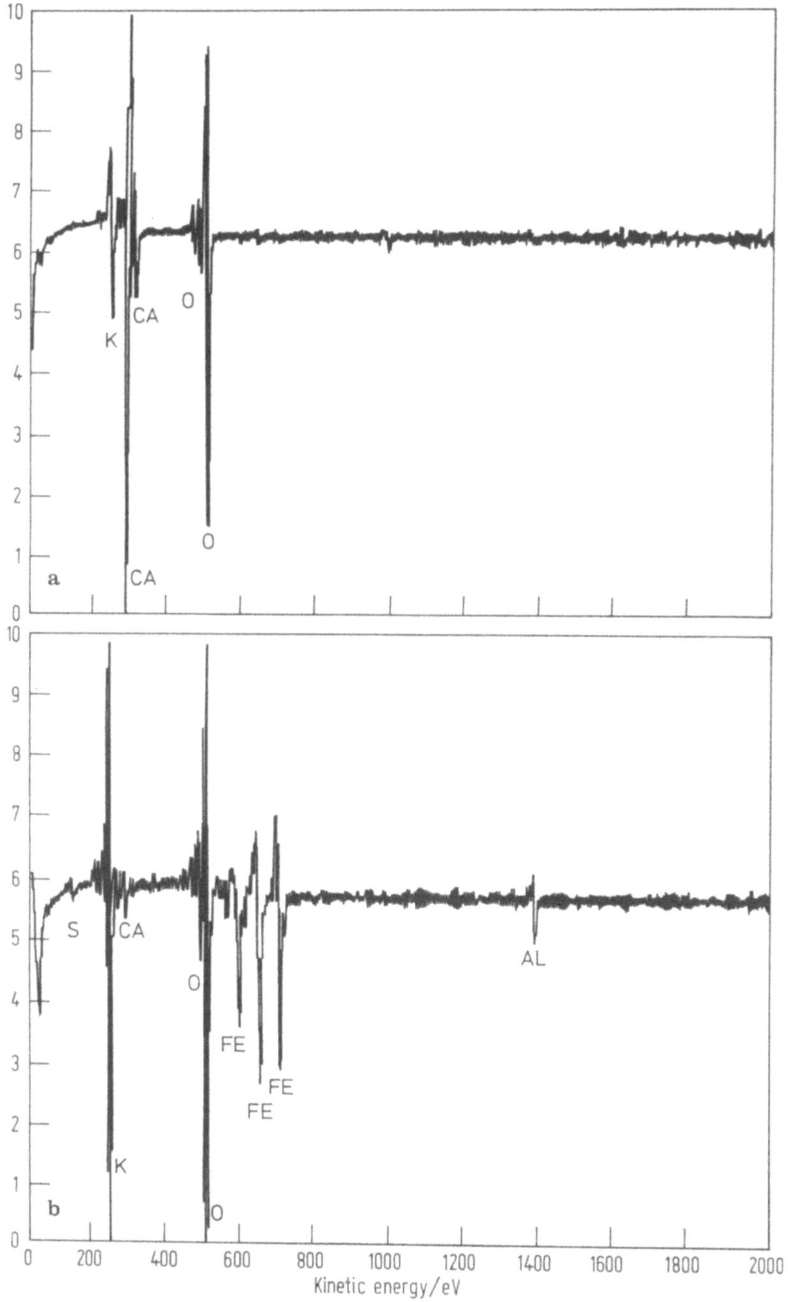

Figure 41. Auger electron spectra from an industrial ammonia synthesis catalyst. (Reproduced with permission from ref. [159])

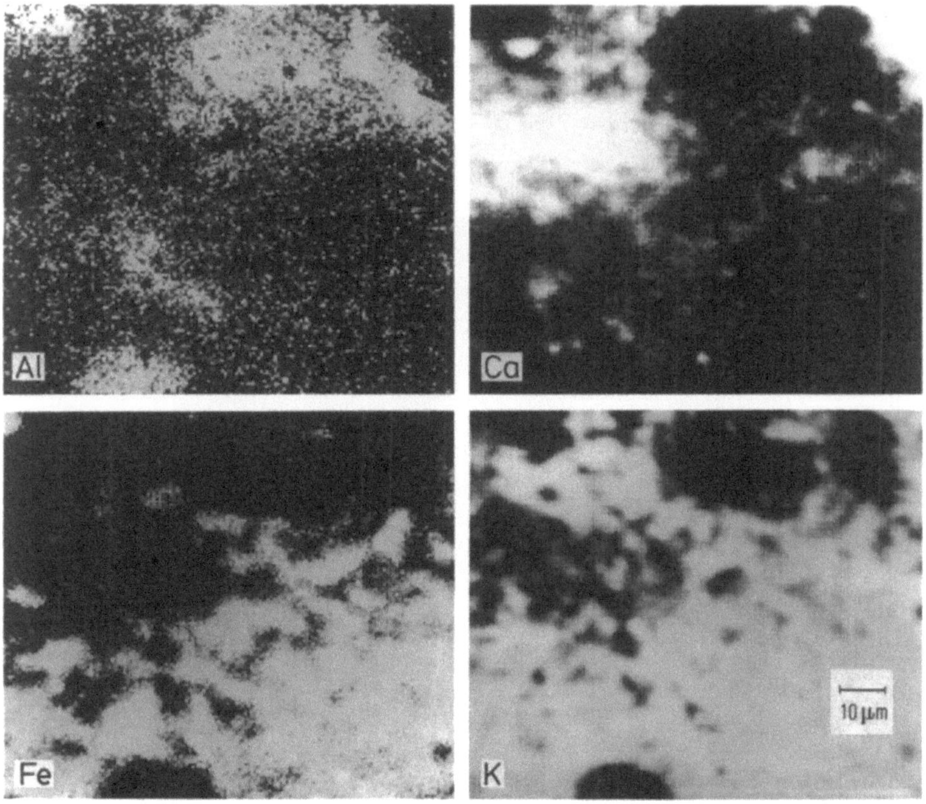

Figure 42. 'Auger maps' from an industrial ammonia synthesis catalyst illustrating the lateral surface distribution of Al, Ca, Fe and K. (Reproduced with permission from ref. [159])

electron beam analyzes only a small spot (diameter 0.2 µm) on the surface. A typical spectrum from a K- and Ca-rich surface region is reproduced in Figure 41a, together with an overall spectrum, b (where the beam was defocussed). It is quite obvious that there exist pronounced lateral variations of the surface composition. If the electron analyser is tuned to a specific energy characteristic for a certain element and the primary electron beam is scanned across the surface, the variation of the intensity (as made visible on a fluorescent screen) reflects directly the lateral distribution of this element. Typical 'Auger maps' for four elements are reproduced in Figure 42. These data indicate that Al_2O_3 and CaO tend to segregate while the Fe surface is more or less uniformly covered by a K(+O) layer.

The chemical nature of the Fe atoms under working conditions can again be derived from XPS measurements [158]. Figure 43 shows the variation of the shape of the Fe $2p_{3/2}$ level at various stages of reduction. The broad maximum centered at 711.3 eV and characteristic for $FeAl_2O_4 + Fe_2O_3$ is continuously decreasing in intensity while simultaneously a narrower peak at 706.8 eV is growing. The latter is characteristic for metallic α-Fe

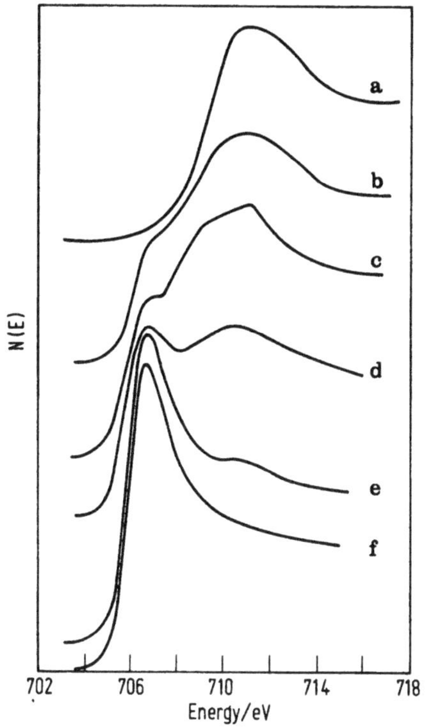

Figure 43. Variation of the Fe $2p_{3/2}$ core level spectrum of an ammonia synthesis catalyst with degree of reduction increasing in the sequence $a \rightarrow f$. (Reproduced with permission from ref. [158])

which is the only iron species left after complete reduction. Observations with the O1*s* level led further to the conclusion that upon reduction aluminium is transformed from its spinel structure in $FeAl_2O_4$ into Al_2O_3.

The presence of a K + O overlayer does obviously not affect the metallic nature of the Fe substrate. Clean Fe single crystal surfaces therefore represent proper models for studying the elementary steps of this reaction. The role of K + O will then be discussed at the end of this section.

2. High Pressure Studies

A high pressure (20 atm.) study on the kinetics of ammonia synthesis on well-defined Fe single crystal surfaces was recently performed by Spencer *et al.* [160] who found that at 773 K the activity varied in the order (111) $>$ $>$ (100) $>$ (110) by more than two orders of magnitude, thus indicating that the reaction is highly structure sensitive. That the (111) plane exhibits the highest activity was indeed already suggested earlier from more indirect evidence [161–163]. If the results with Fe (111) of ref. [160] are transformed into the reaction probability of a N_2 molecule striking the surface, a value of the order of 10^{-7} results which will be of significance during the further discussion. Surface analysis after evacuation revealed a small, but noticeable concentration of atomic nitrogen (which might, however have been affected by surface segregation processes [31]).

More detailed measurements of the latter type revealed valuable insight into the actual reaction mechanism as well as the nature of the rate-limiting step [163]. In order to draw any firm conclusions from such experiments (*i.e.* in which the reaction is run under steady-state conditions near atmospheric pressure and subsequently the reaction chamber is evacuated prior to surface analysis while the sample is kept at elevated temperature) of course information on the thermal stability of possible surface species *in vacuo* is needed. As will be outlined in more detail in the following sections, atomic nitrogen (N_{ad}) recombines and desorbs only above 700 K while all other possible species disappear below 500 K. If the reaction is run at around 600 K surface analysis will afterwards therefore provide the actual N_{ad} concentration under *working conditions*. N_{ad} is formed through

$$N_2 \rightleftharpoons N_{2\,ad} \rightarrow 2\,N_{ad}$$

(see section 3.C.3). If synthesis of ammonia would proceed through the molecularly adsorbed species, $N_{2\,ad}$, (as sometimes stated in earlier work) the surface would necessarily be saturated with N_{ad} under steady-state conditions. If, on the other hand, NH_3 is formed through $N_{ad} + 3\,H_{ad} \rightarrow$ $\rightarrow NH_3$, then it can easily be shown [163] that the steady-state concentration $[N_{ad}]_{st}$ will vary with partial pressures as

$$[N_{ad}]_{st} = k' p_{N_2} / p_{H_2}^x \tag{25}$$

as long as one works far away from equilibrium where the decomposition of NH_3 is negligible. The variable x was introduced in order to relate the surface concentration of H_{ad} with the H_2 partial pressure as determined by the adsorption isotherm. Figure 44 shows for a Fe(111) surface (which is the most active plane) the variation of $[N_{ad}]_{st}$ with p_{H_2} at constant p_{N_2} (20 kPa) and constant temperature (580 K). These data demonstrate that the second prediction is correct, that is, the reaction proceeds indeed through N_{ad}, and furthermore that in a stoichiometric $N_2 + H_2$ mixture the N_{ad} concentration is fairly small (as was also found with an industrial catalyst [158]). From the latter observation it has to be concluded that indeed dissociative nitrogen chemisorption is the rate-limiting step (as long as the temperature is not too low, say for $T \geqq 500$ K). This conclusion is in

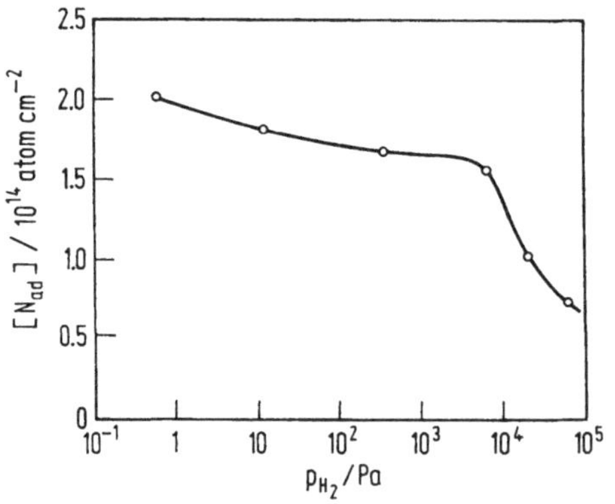

Figure 44. Variation of the steady-state surface concentration of atomic nitrogen, $[N_{ad}]$, on a Fe(111) surface with hydrogen pressure at constant $p_{N_2} = 20$ kPa, $T = 580$ K. (Reproduced with permission from ref. [158])

agreement with most of the suggestions in the earlier literature [147] and also explains why far from equilibrium the rate is just simply proportional to the N_2 partial pressure [164]. For example it was shown that hydrogenation proceeds much faster than nitrogen adsorption [165] and that on an industrial catalyst nitrogen chemisorption and ammonia synthesis proceed with about equal rate [166].

3. *Adsorption of Nitrogen*

Distinction between the two forms of surface nitrogen, either molecular ($N_{2\,ad}$) or atomic (N_{ad}) in nautre, can directly be made on the basis of UPS and XPS data [31, 167–170]. N_2 is isoelectronic with CO and therefore the bonding mechanism to the surface is qualitatively similar. It is bonded with its molecular axis perpendicular to the surface through coupling of its two highest σ-orbitals (5σ and 4σ) and by 'back-donation' of metallic *d*-electrons into the lowest empty ($2\pi^*$) level [171–174]. Experimental confirmation of this structure was obtained (for other surfaces) by means of angular resolved photoelectron spectroscopy [172] as well as from vibrational spectroscopy [175–179]. Figure 45 shows the changes of the orbital energies of N_2 in a Fe—N—N cluster as derived from *ab initio* calculations [174], demonstrating the stabilization of the 4σ and 5σ orbitals upon bond formation. Electron population analysis revealed a net electron transfer from

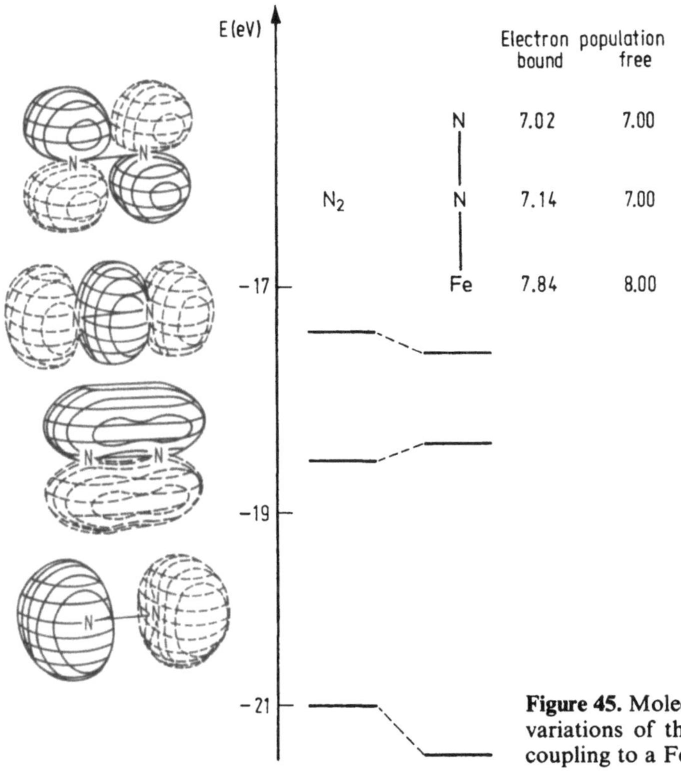

Figure 45. Molecular orbitals for N_2 and variations of the orbital energies upon coupling to a Fe atom [174]

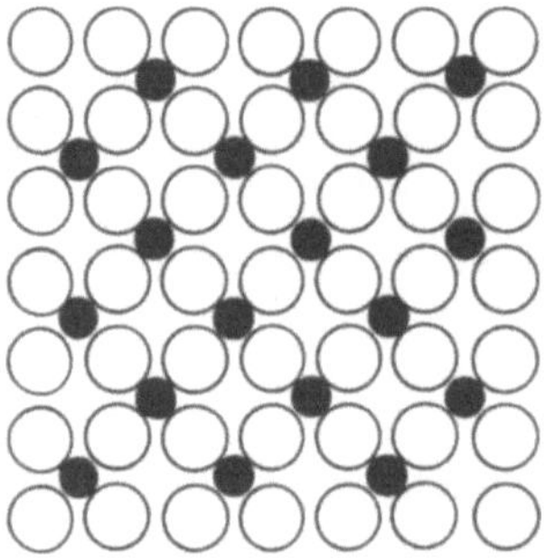

Figure 46. Structure model for the c2 × 2-structure formed by atomic nitrogen (dark circles) on a Fe(100) surface. (Reproduced with permission from ref. [31])

the metal to the ligand (due to the overcompensating $d \rightarrow \pi^*$ 'backdonation' effect) which is consistent with the observed *increase* of the work function [22]. These results show that adsorbed N_2 is an overall electron acceptor — a conclusion which will be of importance for the discussion of the promoter action of potassium. On Fe(111) the adsorption energy of $N_{2\,ad}$ is 32 kJ mol^{-1} giving rise to a maximum in the desorption spectra at about 160 K [22]. Similar adsorption energies were found for polycrystalline Fe surfaces [180, 181].

Adsorbed atomic nitrogen, N_{ad}, is formed through dissociation of $N_{2\,ad}$ as for example followed by XPS [170]. Bond formation with the surface occurs through the N 2*p* levels which show up in UPS as a pronounced maximum at about 5 eV below E_F. The activation energy for desorption ranges between 210 and 240 kJ mol^{-1} (depending on surface orientation) from which the strength of the M—N bond results to be about 580 kJ mol^{-1} [31, 167] which has to be compared with the 950 kJ mol^{-1} dissociation energy of the N_2 molecule. A model for the c2 × 2-structure formed on Fe(100) is reproduced in Figure 46 [31]. The atomic arrangement is quite similar to that in the (002) plane of (bulk) Fe_4N. By contrast, the Fe(111) and (110) planes reconstruct under the influence of the strong bond formation with the N atoms. A series of complicated LEED patterns was observed in these cases which could be reconciled with slightly distorted (111) planes of Fe_4N on top of the iron substrate. That is why the notation 'surface nitride' is preferred over that of a chemisorbed overlayer [31, 167]. It has to be pointed out, however, that bulk iron nitrides can never be formed by interaction of N_2 with Fe for thermodynamic reasons. This is only possible in an NH_3 atmosphere which offers a very high 'virtual' nitrogen pressure [182].

The initial sticking coefficient for dissociative nitrogen chemisorption was found to be very low, *viz.* in the range of 10^{-7} to 10^{-6} with clean surfaces, whereby the activity varies in the sequence (111) > (100) > (110) by about two orders of magnitude [31, 167]. This has to be compared with the kinetic data on the rate of NH_3 synthesis [160] mentioned in the preceding section and is strong further support for the rate-limiting nature of dissociative nitrogen adsorption. The effective activation energy is fairly low at zero coverage, ranging between 0 and 30 kJ mol^{-1}, but increases continuously with increasing coverage. Such a linear increase (from 20

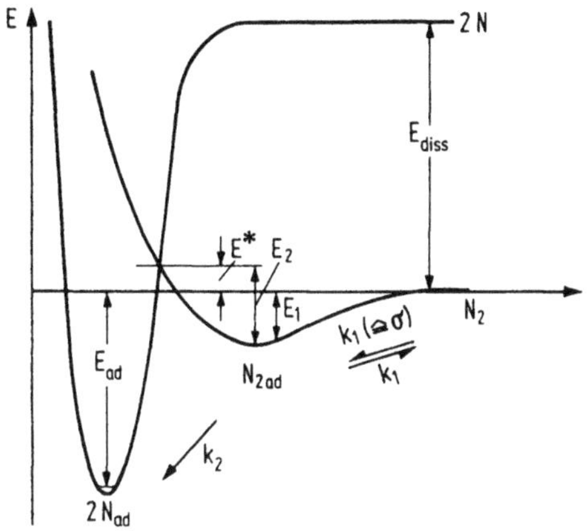

Figure 47. Lennard-Jones diagram illustrating the dissociative chemisorption of nitrogen on iron

to 100 kJ mol^{-1}) was also found in adsorption experiments with 'real' (singly promoted) catalysts [183]. Parallel to this increase of the activation energy also the preexponential increases markedly (compensation effect). As a consequence the sticking coefficient is still appreciable at higher coverages where the activation energy reaches values in the range 40 to 80 kJ mol^{-1}.

The detailed steps of dissociative nitrogen adsorption were recently studied with a Fe(111) surface and the results are best illustrated by the Lennard-Jones type potential diagram reproduced in Figure 47 [189]. The overall rate of dissociative chemisorption (at zero coverage) via

$$N_2 \underset{k_{-1}}{\overset{k_1}{\rightleftharpoons}} N_{2\,ad} \xrightarrow{k_2} 2N_{ad} \tag{26}$$

is given by

$$\frac{d[N_{ad}]}{dt} = k_2[N_{2\,ad}] = 2s_0 p_{N_2}(2\pi MRT)^{-\frac{1}{2}} \tag{27}$$

where s_0 is the sticking coefficient which can be written as

$$s_0 = \nu_s \exp(-E^*/RT)\,. \tag{28}$$

Here, ν_s is the effective preexponential and E^* the effective activation energy. It turned out that $k_{-1} \gg k_2$ so that the equilibrium $N_2 \rightleftharpoons N_{2\,ad}$ will be established. This leads to

$$s_0 = \frac{\sigma_0}{2}\frac{k_2}{k_{-1}} = \frac{\sigma_0}{2}\frac{\nu_2}{\nu_{-1}}\exp[(E_1 - E_2)/RT] \tag{29}$$

so that

$$\nu_s = \frac{\sigma_0}{2}\frac{\nu_2}{\nu_{-1}} \tag{30}$$

and $E^* = E_2 - E_1$ (see Figure 47), whereby $\sigma_0(\hat{=} k_1)$ is the probability that an impinging N_2 molecule is trapped in the molecular state. The resulting data were $\sigma_0 = 10^{-2}$, $\nu_{-1} \approx 10^{11}\ s^{-1}$, $\nu_2 \approx 10^7\ s^{-1}$, yielding $\nu_s \approx 10^{-6}$ and $E_2 \approx 28\ kJ\ mol^{-1}$, $E_1 = 31\ kJ\ mol^{-1}$ and thereby $E^* = 3\ kJ\ mol^{-1}$. The unusually small preexponential factor for dissociative chemisorption at zero coverage ($\nu_s \approx 10^{-6}$) is thus caused by two effects: i) The sticking coefficient into the molecular state is small ($\sigma_0 \approx 10^{-2}$) which is unusual but is presumably related with the shallow potential minimum for $N_{2\,ad}$, ii) ν_{-1} is about four orders of magnitude larger than ν_2, that means an adsorbed N_2 molecule has a much higher chance for desorption (k_{-1}) than for dissociation (k_2). While the transition state for the former process is the free N_2 molecule with all its degrees of freedom, the process $N_{2\,ad} \rightarrow 2\,N_{ad}$ obviously requires a rather restricted transition state (where the molecular axis has to be parallel to the surface while $N_{2\,ad}$ is standing upright). In view of transition state theory, this makes the large difference in partition functions and therefore preexponentials at least qualitatively plausible. The observed differences in the initial sticking coefficients on the (111), (100), and (110) single crystal planes can be traced back to different activation energies, *i.e.* ν_s (at $\theta = 0$) values appear to be essentially identical ($\approx 10^{-6}$). Even the kinetic data by Scholten *et al.* [183] obtained with a well reduced, singly promoted Fe catalyst turn out to be characterised by a quite similar value, $\nu_s = 3 \times 10^{-6}$ at $\theta = 0$, while also at $\theta = 0$ $E^* = 22\ kJ\ mol^{-1}$ being between the values found for the (100) and (110) single crystal planes. No conclusive explanation can, however, be offered for the observed compensation effect, *i.e.* the parallel increase of ν_s with E^* by increasing the coverage, although this is a quite generally observed phenomenon.

4. *Adsorption of Hydrogen*

Adsorption of hydrogen on Fe(110), (100) and (111) single crystal planes revealed to be rather 'normal' [36]. Dissociative adsorption is non-activated and proceeds with initial sticking coefficients in the range between 0.1 and 0.2. The initial adsorption energy varies between 88 and 110 $kJ\ mol^{-1}$ on the various crystal planes, that is, the strength of the M—H bond is of the order of about 250 $kJ\ mol^{-1}$. The adsorption energy decreases with increasing coverage due to the operation of lateral interactions and/or occupation of various adsorption sites, giving rise to multiple desorption states in the thermal desorption spectra (cf. Figure 18). As another example Figure 48 shows a series of TDS data for H_2/Fe(110) where the growth of a low-temperature state above $\theta = 0.5$ can be recognized. This plane is interesting in so far as below room temperature a series of ordered overlayer structures can be observed by LEED. Order-disorder transitions take place by raising the temperature, and these observations have led to the establishment of a two-dimensional phase diagram [184]. Under the conditions of ammonia synthesis the adsorbed hydrogen atoms are highly mobile and their concentration will be determined by the adsorption/desorption equilibrium (which is probably influenced by the presence of other surface species). Due to the relatively low adsorption energy the hydrogen coverage

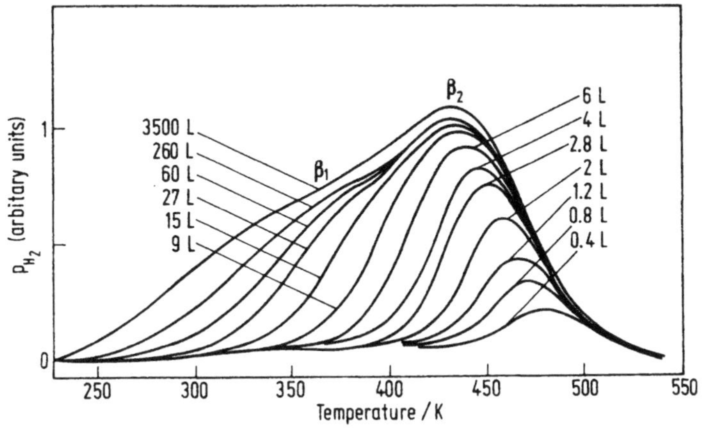

Figure 48. Series of thermal desorption spectra of hydrogen from Fe(110), with exposures in the range 0.4–3500 L. (Reproduced with permission from ref. [36])

will be fairly small at $T \geqq 670$ K even at high H_2 pressures, so that inhibition effects (see 3.C.5) will have no marked influence.

5. Coadsorption

Earlier reports in the literature indicated that the presence of hydrogen might accelerate the rate of dissociative nitrogen chemisorption [185]. Some observations with a Fe(111) surface tended to support this suggestion [186]. Recent more extended studies demonstrated, however, that in fact just the opposite is the case [163]. The earlier observation of a more rapid increase of the N_{ad} concentration in the presence of a hydrogen atmosphere was obviously caused by the fact that hydrogen enhances surface segregation of nitrogen atoms which were initially dissolved in the bulk. Tamaru's [185] measurements were not performed in UHV and it appears therefore highly probable that H_2 produces a higher degree of reduction of the iron surface. (Since oxygen strongly poisons nitrogen chemisorption, this would explain the apparent increase of the adsorption rate).

It was found [163] that hydrogen may be dissociatively adsorbed on a Fe surface precovered with *molecular* nitrogen whereby, however, $N_{2\,ad}$ is displaced from the surface. This becomes plausible in view of the much higher strength of the M—H bond (250 kJ mol^{-1}) if compared with that of the M—N_2 bond (35 kJ mol^{-1}). Preadsorbed H_{ad}, on the other hand, inhibits the adsorption of molecular nitrogen. Since the N-atoms needed for the reaction are formed through $N_2 \rightleftharpoons N_{2\,ad} \rightarrow 2\,N_{ad}$, these results indicate clearly that hydrogen inhibits dissociative nitrogen adsorption and therefore also ammonia synthesis. This effect is illustrated by Figure 49 which shows the variation of the N_{ad} concentration with N_2 exposure on Fe(111) with and without hydrogen in the gas phase. The temperature was rather low (433 K) in this case so that even with a small H_2 pressure (7×10^{-3} Pa) an appreciable fraction of the surface was covered with H_{ad}. As mentioned

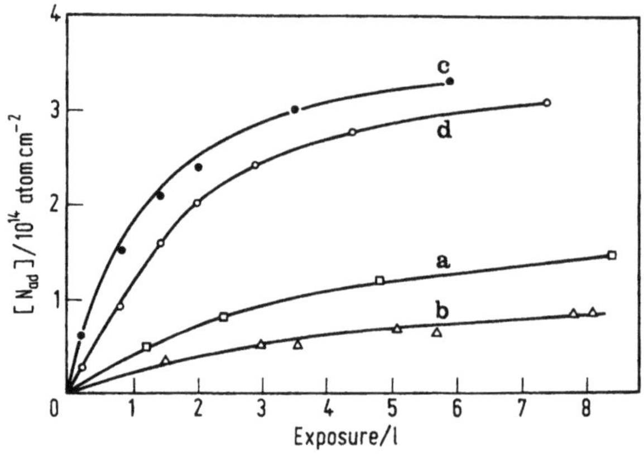

Figure 49. Variation of the surface concentration of atomic nitrogen, $[N_{ad}]$, on Fe(111) with N_2 exposure at 430 K. (a) Clean surface, pure N_2 ($p_{N_2} = 6.7 \times 10^{-3}$ Pa). (b) Clean surface, $p_{N_2} = 6.7 \times 10^{-3}$ Pa + $p_{H_2} = 6.7 \times 10^{-3}$ Pa. (c) Surface precovered with 1.8×10^{14} K atom cm^{-2}, $p_{N_2} = 6.7 \times 10^{-3}$ Pa. (d) Surface as c), but $p_{N_2} = 6.7 \times 10^{-3}$ + $p_{H_2} = 6.7 \times 10^{-3}$ Pa. (Reproduced with permission from ref. [163])

in the preceding section, under actual synthesis conditions ($T \gtrsim 670$ K) this effect will, however, play no major role. Experiments with Ru catalysts at somewhat lower temperatures demonstrated, however, the inhibiting effect of hydrogen on the rate of ammonia synthesis [187].

The presence of adsorbed *atomic* nitrogen, on the other hand, inhibits the adsorption of hydrogen [163]. (Now the M—N bond is much stronger, *viz.* 580 kJ mol^{-1}, than the M—H bond).

The overall conclusion from these observations is that ammonia synthesis is most favorably performed under conditions where both the H_{ad} and N_{ad} concentrations are not too high which will usually be fulfilled at actual catalyst operation. The temperature is high enough to shift the $H_2 \rightleftharpoons 2\,H_{ad}$ equilibrium to the left-hand side, and $[N_{ad}]$ will be small since its formation is rate-limiting.

6. *Adsorption and Decomposition of Ammonia*

At low temperatures NH_3 adsorbs non-dissociatively on the Fe(100), (111) [186] as well as (110) planes [188]. This becomes most directly evident from inspection of the UPS data as shown in Figure 50, curve *b*. The two emission maxima are derived from the $3a_1$ (= N lone pair) and $1e$ (= N—H bond) orbitals of NH_3. Comparison with gas-phase data shows that the energetic separation between these two levels has decreased by about 1 eV which has to be ascribed to the stabilization of the $3a_1$-level by donor-bond formation with the metal surface. The configuration of an adsorbed NH_3 molecule as well as the lateral order on Fe(110) as derived from LEED is reproduced in Figure 51. This electron donor mechanism is also responsible for the fact that the dipole moment of the NH_3 surface complex is even larger (~2.2 Debye) than that of free NH_3. *Ab initio* calculations for a Fe—NH_3 complex confirmed this picture of electronic interaction [189, 190]. The heat of adsorption depends again on surface orientation and coverage and attains a maximum value of 70 kJ mol^{-1}. As a consequence thermal desorption *in vacuo* is completed at 370 K and NH_3 formed under reaction conditions will almost instantaneously leave the surface.

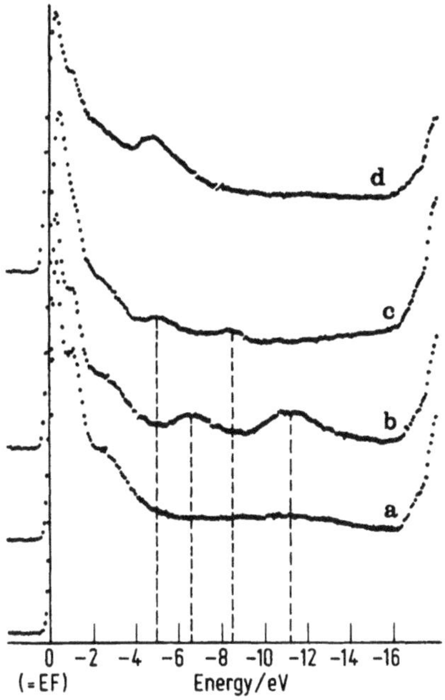

Figure 50. UPS data for NH_3/Fe(110). (a) Clean surface. (b) After NH_3 adsorption at 175 K (= $NH_{3\,ad}$). (c) After exposure to ammonia at 350 K (= NH_{ad}). (d) After complete dissociation (= N_{ad}). (Reproduced with permission from ref. [188])

Parallel to desorption, however, also dissociation may take place which, of course, is a necessary prerequisite for establishing the equilibrium $N_2 + 3\,H_2 \rightleftharpoons 2\,NH_3$. Decomposition occurs stepwise, *viz.* $NH_{3\,ad} \rightarrow NH_{2\,ad} + H_{ad} \rightarrow NH_{ad} + 2\,H_{ad} \rightarrow N_{ad} + 3\,H_{ad}$. The intermediate formation of $NH_{2\,ad}$ was concluded from the observed production of NH_2D between 300 and 400 K when a Fe(111) surface had been exposed to $NH_3 + D_2$ [186]. Similarities in the transient UPS data from NH_3 and N_2H_4 interacting with Fe(111) [191] also indicated the formation of $NH_{2\,ad}$. On the other hand it was possible to completely isolate and identify NH_{ad} on a Fe(110) surface. Curve *c* in Figure 50 shows how the photoelectron spectrum changed if NH_3 was interacting with this surface not at 175 K but at 350 K. Two maxima are now visible at energies which are quite different from those

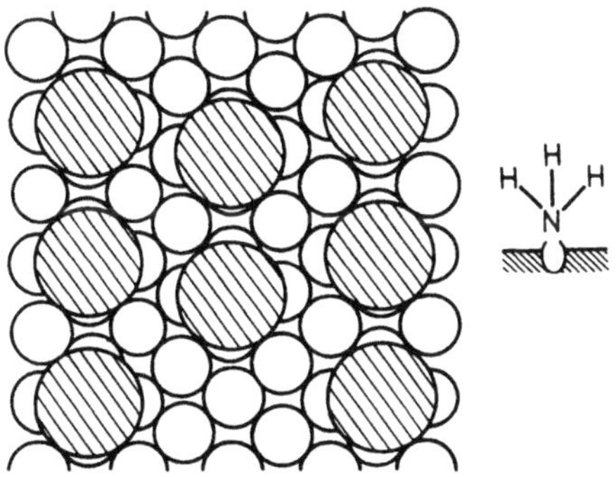

Figure 51. Structure model for NH_3 adsorbed on Fe(110). (Reproduced with permission from ref. [188])

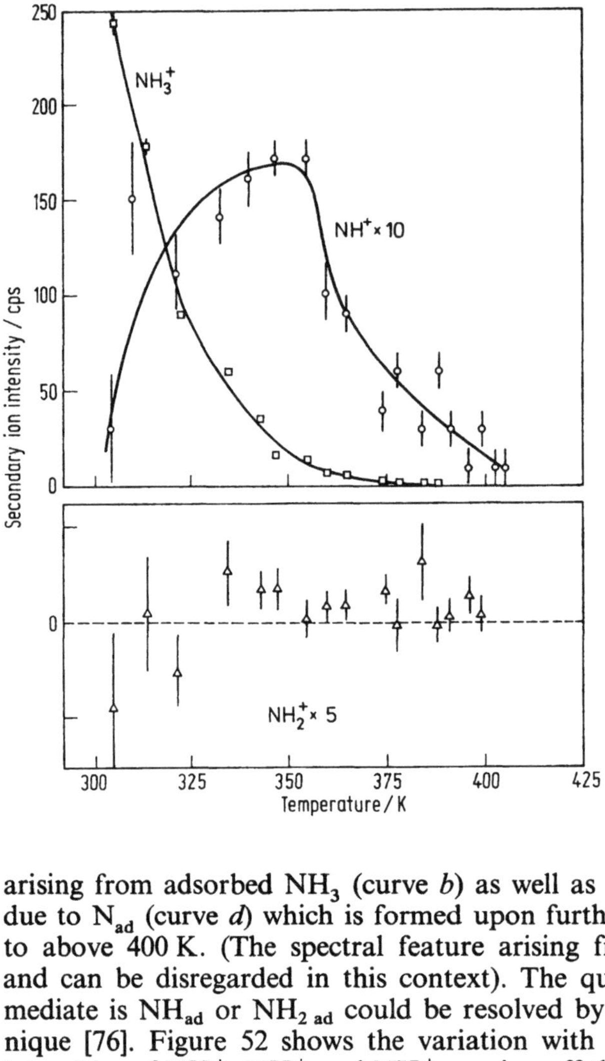

Figure 52. NH_3/Fe(110). Variation of the SIMS signals with surface temperature, demonstrating the intermediate formation of NH_{ad}. (Reproduced with permission from ref. [16])

arising from adsorbed NH_3 (curve *b*) as well as from the single maximum due to N_{ad} (curve *d*) which is formed upon further raising the temperature to above 400 K. (The spectral feature arising from H_{ad} is hardly visible and can be disregarded in this context). The question whether this intermediate is NH_{ad} or $NH_{2\,ad}$ could be resolved by means of the SIMS technique [76]. Figure 52 shows the variation with temperature of the signal intensities of NH_3^+, NH_2^+ and NH^+ coming off the surface under argon ion bombardment. The NH_3 concentration is continuously decreasing while that of NH_2 remains negligible. The NH concentration, on the other hand, increases and reaches a maximum at 350 K. At even higher temperature this species dissociates further. So both the UPS and the SIMS data indicate that at 350 K *in vacuo* (apart from H_{ad}) NH_{ad} is the only stable surface species.

The steady-state rate of the ammonia decomposition, $NH_3 \rightarrow N_2 + \frac{3}{2} H_2$, is limited by recombination and desorption of nitrogen at lower temperatures and becomes determined by the adsorption of NH_3 if the temperature is high enough ($\gtrsim$800 K). As a consequence the reaction order with respect to p_{NH_3} changes from zero to one with increasing temperature and the activation energy decreases from ~200 kJ mol^{-1} (desorption of nitrogen) to almost zero (sticking coefficient for NH_3) [192]. Since the rate of nitrogen desorption reaches appreciable values only above ~700 K this will be the rate-limiting step for ammonia decomposition under the ordinary conditions of synthesis.

7. Reaction Mechanism and Kinetic Considerations

Based on the information outlined so far the individual reaction steps can be formulated as follows:

$$\left.\begin{array}{lll} & H_2 & \rightleftharpoons 2\,H_{ad} \\ N_2 & \rightleftharpoons N_{2\,ad} & \rightleftharpoons 2\,N_{ad} \\ N_{ad} & + H_{ad} & \rightleftharpoons NH_{ad} \\ NH_{ad} & + H_{ad} & \rightleftharpoons NH_{2\,ad} \\ NH_{2\,ad} & + H_{ad} & \rightleftharpoons NH_{3\,ad} \\ & NH_{3\,ad} & \rightleftharpoons NH_3 \end{array}\right\} \qquad (31)$$

Indeed this mechanism had been suggested already in the earlier literature [147] whereby, however, no direct spectroscopic evidence for the various surface species was available. As has been shown, dissociative nitrogen chemisorption is rate-limiting at not too low temperatures. Since $N_{2\,ad}$ is only very weakly held, its surface concentration will always be very small and therefore proportional to p_{N_2} thus explaining that the reaction rate is given by $r = k'p_{N_2}$ under conditions far from equilibrium. If NH_3 decomposition is no longer negligible the steady-state N_{ad} concentration increases which parallels a continuous increase of the activation energy. This is also one of the basic assumptions underlying the derivation of the original Temkin rate law [81] as well as its various extensions [151, 193–195]. In this theory, the rate of ammonia synthesis is set equal to the rate of (dissociative) nitrogen adsorption, viz.

$$r = \frac{d[NH_3]}{dt} = \frac{d[N_{ad}]}{dt} = k_1 \cdot p_{N_2} = k_0 \exp\left(-g[N_{ad}]\right) p_{N_2} \qquad (32)$$

where k_0 corresponds to the sticking coefficient at zero coverage. The exponential decrease of r with $[N_{ad}]$ is mainly due to the linear increase of the activation energy with coverage which is partly compensated by the parallel increase of v_s (see section 3.C.3). The Langmuir term $(1 - \theta)$ or $(1 - \theta)^2$ is only of minor importance at lower converages (in view of the exponential variation of r) and is usually neglected in the discussion. An exponential decrease of the rate of nitrogen chemisorption over two orders of magnitude has for example been reported by Scholten et al. [197] for a singly-promoted Fe catalyst. Curve *a* in Figure 53 shows the variation of the relative sticking coefficient with $[N_{ad}]$ on a Fe(111) surface at $T = 508$ K which shows the same trend [198]. In the Temkin theory the steady-state concentration

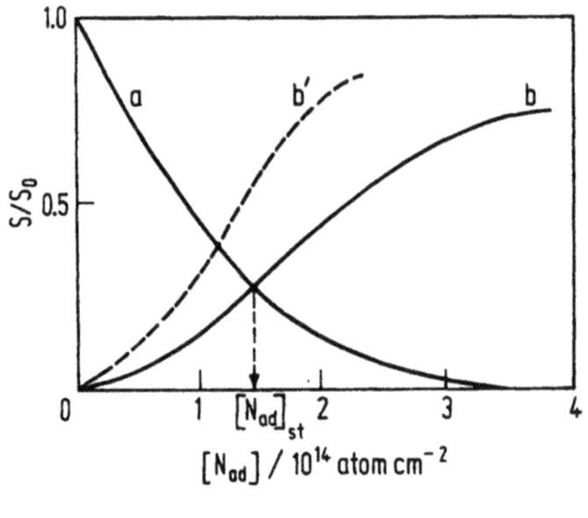

Figure 53. Curve a: Variation of the relative sticking coefficient, s/s_0, for dissociative nitrogen adsorption on Fe(111) at 508 K with surface concentration $[N_{ad}]$. Curve b: Schematic variation of the rate of N_{ad} hydrogenation with nitrogen coverage. The crossing point of curves a and b defines the steady-state N_{ad} concentration, $[N_{ad}]_{st}$, under the chosen conditions of synthesis. Curve b': Increasing p_{H_2} will change curve b into b' and thereby lower $[N_{ad}]_{st}$

$[N_{ad}]_r$ under reaction conditions is determined by the fugacity of N_2, p_{N_2}, *i.e.* the N_2 pressure which would be in equilibrium with ambient H_2 and NH_3, $(= p^2_{NH_3}/Kp^3_{H_2})$.

This concept of course breaks down under conditions far from equilibrium, *i.e.* if $p_{NH_3} \to 0$. In this case the simplest approximation consists in neglecting all backward reactions in the above reaction scheme yielding

$$[N_{ad}]_r \approx k_1 p_{N_2}/k_2[H_{ad}] \tag{33}$$

where k_2 is the rate constant for the subsequent step $N_{ad} + H_{ad} \to NH_{ad}$. If all reverse reactions except $2\,N_{ad} \to N_2$ and $NH_3 \to NH_{3\,ad}$ are admitted (*i.e.* still for conditions far from equilibrium), $[N_{ad}]_r$ will be determined by a much more complicated equation

$$[N_{ad}]_r = \frac{k_1 p_{N_2}}{k_2[H_{ad}]\left\{1 - k_{-2}\left[k_{-2} + k_3[H_{ad}] - \dfrac{k_3 k_{-3}[H_{ad}]}{k_{-3} + \left(k_4 - \dfrac{k_4 k_{-4}}{k_{-4} + k_5}\right)[H_{ad}]}\right]^{-1}\right\}} \tag{34}$$

This equation transforms into the simple form (33) as soon as $k_3[H_{ad}] \gg k_{-2}$, *i.e.* if hydrogenation of NH_{ad} proceeds faster than its decomposition.

$[H_{ad}]$ in equation (33) is determined by the adsorption equilibrium $H_2 \rightleftharpoons 2\,H_{ad}$ (which is also affected by N_{ad}) and increases with increasing p_{H_2} (in accordance with the results of Figure 44). Near saturation of the H_{ad}-layer additional complications due to inhibition of nitrogen adsorption will come into play.

The rate of hydrogenation of N_{ad} at a fixed temperature and p_{H_2} will vary schematically with $[N_{ad}]$ as shown by curve b in Figure 53, and the crossing point of curves a and b determines the concentration $[N_{ad}]_r$ under stationary reaction conditions (since there rates a and b have to be equal). Increasing p_{H_2} will cause an increase of the slope of curve b ($\to b'$) and thereby lower $[N_{ad}]_r$.

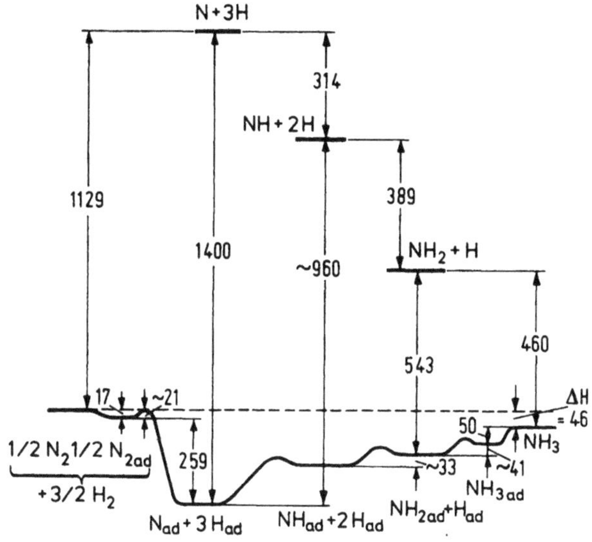

Figure 54. Schematic potential energy diagram for ammonia synthesis on iron at low coverages. Energies in kJ mol^{-1}. (Reproduced with permission from ref. [196])

These few remarks should illustrate how qualitative insights into the kinetics of this reaction can be obtained fairly easily, but how complicated, on the other hand, a complete quantitative kinetic analysis would be. It is quite remarkable that the rather simple rate laws following from the Temkin theory are indeed able to describe the actual reaction kinetics fairly well.

The energy values for the different steps either derived experimentally or based on plausible estimates [186] can now be used to construct a potential diagram which schematically illustrates the progress of the reaction (Figure 54) [196]. The energies vary of course with coverage and surface orientation and should therefore only be considered as representing the qualitative features far from equilibrium. This diagram demonstrates how dissociative chemisorption enables to overcome the high bond strengths of the reactants and that the surface reaction itself is even endothermic. Since dissociative nitrogen adsorption has a lower activation energy than the subsequent step (N_{ad} + H_{ad} → NH_{ad}) it becomes plausible that at low enough temperatures the latter reaction should become rate-limiting.

8. *Promoters and Poisons*

The surface of an actual practical catalyst is covered with a potassium/oxygen layer which has obviously to influence the rate of the slowest step, *viz.* $N_2 \rightarrow 2\,N_{ad}$. Model studies with adsorbed potassium (without oxygen) showed indeed very pronounced effects [62, 73]. With Fe(100) already small K concentrations increase the sticking coefficient for dissociative nitrogen adsorption, s_0, by more than two orders of magnitude up to a maximum value of 4×10^{-5} (at 430 K) which is reached with about $n_K = 1.5 \times 10^{14}$ K atom cm^{-2}. Figure 55 shows the variation of s_0 with n_K for a Fe(111) surface [62]: While the potassium-free Fe (100) and (111) surfaces differ

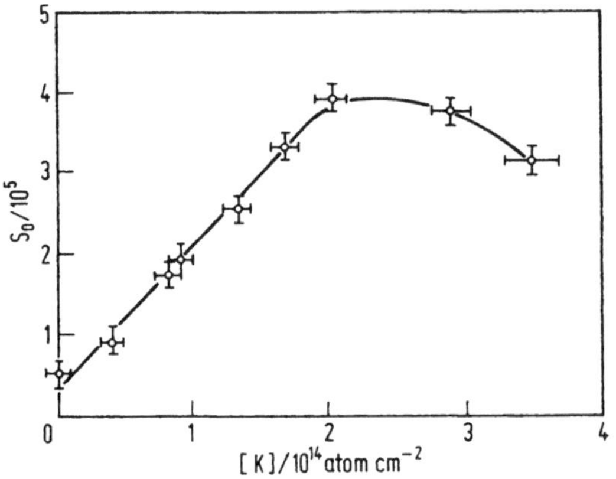

Figure 55. Variation of the initial sticking coefficient, s_0, for dissociative nitrogen adsorption on Fe(111) at 430 K with K concentration, [K], on the surface. (Reproduced with permission from ref. [62])

by about a factor 30 in activity this difference is removed by the addition of potassium. With Fe(111) the same maximum value for s_0 is reached. The decrease of s_0 with further increasing n_K parallels observations made with 'real' ammonia synthesis catalysts for which the catalytic activity as a function of K_2O content passes through a maximum [199].

It is quite evident that the nitrogen atoms once formed on the surface are not attached to the adsorbed K atoms but are distributed over the whole surface. This becomes obvious from LEED observations and follows also from the fact that the saturation concentration of N_s is independent of n_K (as long as the latter is not too high to cause considerable site blocking). On the other hand the effective activation energy for dissociative adsorption is markedly reduced. This effect in fact accounts for the promoter action [62].

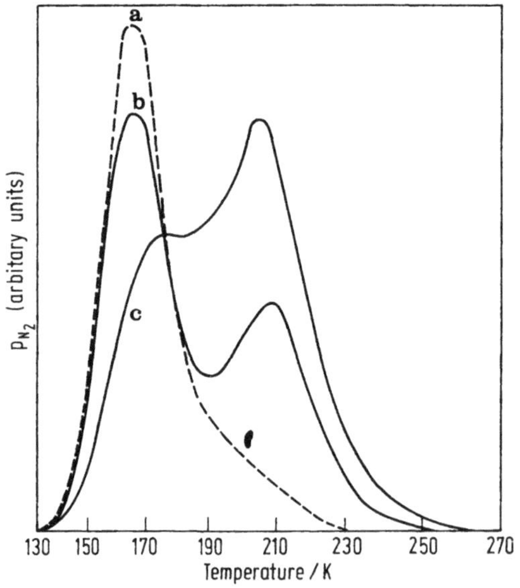

Figure 56. Thermal desorption spectra for molecularly adsorbed nitrogen from Fe(111) with various K concentrations. Curve *a*, K-free surface; curve *b*, 7×10^{13} K atom cm^{-2}; curve *c*, 1.5×10^{14} K atom cm^{-2}. (Reproduced with permission from ref. [63])

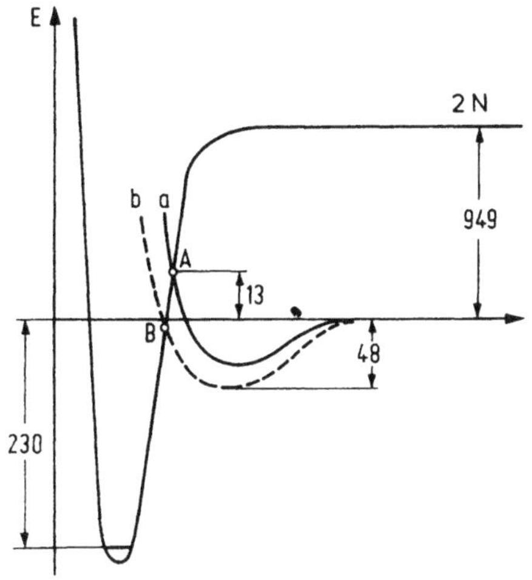

Figure 57. Potential energy diagram illustrating the effect of adsorbed K on nitrogen adsorption. Curve *a*, N_2 + Fe(100); curve *b*, N_2 + K/Fe(100). Energies in kJ mol^{-1}

In view of the discussion of the kinetics of dissociative nitrogen adsorption and the schematic energy diagram (Figure 47) illustrating this process, further explanation has to be sought in possible modifications of the adsorption properties of molecular nitrogen, $N_{2\,ad}$.

Figure 56 shows thermal desorption spectra for this species from Fe(111) with different K concentrations [62]. While with a potassium-free surface a single desorption maximum at 170 K appears, a second state at 210 K is growing up with increasing n_K. This indicates that N_2 is more tightly held in the vicinity of an adsorbed K atom. The adsorption energy is in fact increased from 30 kJ mol^{-1} to 45 kJ mol^{-1} while simultaneously the activation energy for dissociation is lowered by about 10 kJ mol^{-1}. The potential diagram of Figure 57 illustrates how these two effects are correlated with each other, so that the increased adsorption energy for $N_{2\,ad}$ is indeed the key for the promoter effect.

Adsorption of K on Fe surfaces causes a very pronounced lowering of the work function [200], which effect in turn is expected to enhance the 'back-donation' of metallic electrons to the antibonding N_2 2π level thus strengthening the M—N_2 bond. The MO calculations (Figure 45) discussed in section 3.C.3 indicate indeed a net electron transfer from the metal to N_2, that is, this ligand acts as an electron acceptor. Since the work function minimum caused by K adsorption reaches quite similar values for Fe(111), (100) and (110) it now also becomes plausible why the structural effects in dissociative nitrogen chemisorption are levelled off by the presence of potassium.

Since N_2 is isoelectronic with CO quite similar effects are to be expected with the adsorption of the latter molecule which indeed has been found to be the case [69, 201]. With a Pt(111) surface the adsorption energy of CO was observed to increase by as much as 50 kJ mol^{-1} while simultaneously the frequency of the C—O vibration (reflecting the degree of 'back-donation') is considerably shifted towards lower values [201].

If the surface of an actual ammonia synthesis catalyst would be covered by potassium *alone* only small surface concentrations could be maintained under reaction conditions, however. The heat of K adsorption on Fe decreases rapidly with increasing coverage so that a large fraction of the overlayer would not be thermally stable [200]. Addition of oxygen, however, increases the desorption temperature by more than 200 K to above 700 K [202]. The surface of the industrial catalysts is covered with such a composite K + O adlayer (which is not idential with one of the known bulk compounds between K and O) which is stable up to about 800 K [203]. The strong interaction between K and O also prevents complete reduction of a surface under reaction conditions while Fe, on the other hand, is in its metallic state [203]. Quite similar observations on such strong K + O interactions were found with Pt surfaces [204].

Adsorbed oxygen is also an electron acceptor and it is therefore not surprising that it acts as a poison in ammonia synthesis. It was found that the uptake of nitrogen decreases linearly with the concentration of O_{ad} [205]. As a consequence also a continuous decrease of the nitrogen sticking coefficient with increasing O-coverage of a Fe + K surface was observed [202] and therefore the promoter effect of an actual catalyst is much less pronounced than if only potassium alone would be present on the surface. This is consistent with observations made by Ozaki *et al.* [206] with Ru catalysts covered by either K or K + O.

Since excess oxygen is reduced under reaction conditions, poisoning caused by the presence of small O_2 contents in the reaction gas mixture can be removed. This is not the case with sulfur which has a similar poisoning effect. It appears as if the K + O adlayer also slows down the deposition of S (at least at room temperature) [203] which agrees with earlier observations [154]. A further effect of the promoter could come into play at reaction conditions closer to the equilibrium where the presence of NH_3 is no longer negligible, which, however, has still to be explored. The adsorption energy and kinetics of hydrogen on iron is influenced to some extent by the presence of K [207, 208] but will be without any major effect.

4. Conclusions

The three selected examples concerned reactions where only a single reaction product is formed and for which the individual steps appear to be relatively simple and well understood. Even there a series of questions still await their solution. The main goal was, however, to illustrate the basic principles which are underlying heterogeneously catalyzed reactions in general. Some effort was therefore also made to demonstrate how the gap between the 'surface science' approach and the 'real' world of catalysis can be bridged.

There exist of course numerous studies on other reactions which have been performed along similar lines [209]. These include, for example, studies on the interactions between CO and H_2 (methanation and Fisher-Tropsch) on Ni [210] or Fe [211], and in particular the wide field of hydrocarbon transformations [212], as well as model studies on the properties of bimetallic surfaces [213]. Although also in these cases considerable insight into the microscopic processes has been obtained, the situation is in general more complicated that with the examples presented here. Carbonaceous overlayers built up during the reaction may play a major role, direct identification of reaction intermediates has so far been scarce [214] and the picture becomes much more complex if more than one product is formed through parallel or consecutive routes. Very small changes in the energetics may then effect profoundly the selectivity, and much additional work will certainly be needed in order to clarify these processes in more detail.

Acknowledgements

This chapter was written while the author spent his sabbatical at the Department of Chemistry, University of California, Berkeley. The generous hospitality of Professor G. A. Somorjai as well as the stimulating interaction with him and members of his research group are most gratefully acknowledged. The author is also indebted to G. Fisher, J. Gland, R. J. Madix, E. L. Muetterties, W. H. Weinberg and J. T. Yates for fruitful discussions and for information about their work prior to publication.

Financial support was obtained through the Lawrence Radiation Laboratory as well as from the Stiftung Volkswagenwerk.

References

1. a) Somorjai, G. A.: Chemistry in two dimensions: Surfaces. Ithaca: Cornell Univ. Press 1981
   b) Methods of surface analysis (A. W. Czanderna, ed.). Amsterdam: Elsevier 1975
   c) Ertl, G., Küppers, J.: Low energy electrons and surface chemistry. Weinheim: Verlag Chemie 1974
   d) Characterisation of catalysts (J. M. Thomas and R. M. Lambert, eds.). New York: Wiley 1980
   e) Experimental methods of surface physics (R. L. Park, ed.). To be published
2. Somorjai, G. A.: Adv. Catalysis **26**, 2 (1977)
3. Boudart, M.: ibid. **20**, 153 (1969) and references therein
4. High-pressure cells combined with UHV techniques have been developed in various laboratories
5. a) Wedler, G.: Chemisorption: An experimental approach, 2nd ed. London: Butterworths 1976
   b) Roberts, M. W., McKee, C. S.: Chemistry of the metal-gas interface. Oxford: Clarendon Press 1978
   c) The nature of the surface chemical bond (T. N. Rhodin and G. Ertl, eds.). Amsterdam: North Holland 1979
   d) The chemical physics of solid surface and heterogeneous catalysis (D. A. King and D. P. Woodruff, eds.). In press
6. Muetterties, E. L., Rhodin, T. N., Band, E., Brucker, C. F., Pretzer, W. R.: Chem. Rev. **79**, 91 (1979)
7. Behm, R. J., Christmann, K., Ertl, G., Van Hove, M. A.: J. Chem. Phys. **73**, 2984 (1980)
8. Behm, R. J., Christmann, K., Ertl, G., Van Hove, M. A., Thiel, P. A., Weinberg, W. H.: Surface Sci. **88**, L 59 (1979)
9. See e.g.
   a) Boer, J. H. de: The dynamic character of adsorption. Oxford: University Press 1953
   b) Clark, A.: Theory of adsorption and catalysis. New York: Academic Press 1970

10. Ehrlich, G.: Adv. Catalysis **14**, 255 (1963)
11. Christmann, K., Behm, R. J., Ertl, G., Van Hove, M. A., Weinberg, W. H.: J. Chem. Phys. **70**, 4168 (1979)
12. See e.g.
    a) Bradshaw, A. M., Scheffler, M.: J. Vac. Sci. Techn. **16**, 447 (1979)
    b) Ertl, G.: J. Vac. Sci. Techn. **14**, 435 (1977)
13. Hoffmann, R.: Acc. Chem. Res. **4**, 1 (1977)
14. Einstein, T. L.: Surface Sci. **84**, L 497 (1979)
15. Conrad, H., Ertl, G., Küppers, J., Latta, E. E.: Surface Sci. **57**, 475 (1976)
16. See e.g. Ponec, V., Knor, Z., Cerny, S.: Adsorption on solids, p. 326. London: Butterworths 1974
17. See e.g. Ertl, G.: Surface Sci. **89**, 361 (1979)
18. Campbell, C. T., Ertl, G., Segner, J.: Surface Sci. **115**, 309 (1982)
19. a) Gadzuk, J. W., Metiu, H.: J. Chem. Phys. **74**, 2641 (1980)
    b) Schönhammer, K., Gunnarsson, O.: Phys. Rev. B **22**, 1629 (1980)
20. Hoinkes, H.: Rev. Mod. Phys. **52**, 933 (1980)
21. Kleyn, A. W., Luntz, A. C., Auerbach, D. J.: Phys. Rev. Lett. **47**, 1169 (1981)
22. Ertl, G., Lee, S. B., Weiss, M.: Surface Sci. **114**, 515 (1982)
23. Ehrlich, G.: J. Phys. Chem. **59**, 473 (1955)
24. Kisliuk, P. J.: J. Phys. Chem. Solids **3**, 95 (1957); **5**, 78 (1958)
25. Frenkel, F., Häger, J., Krieger, W., Walther, H., Campbell, C. T., Ertl, G., Kuipers, H., Segner, J.: Phys. Rev. Lett. **46**, 152 (1981)
26. King, D. A., Wells, M. G.: Proc. Roy. Soc. (London) **A 339**, 245 (1974)
27. Lennard-Jones, J. E.: Trans Faraday Soc. **28**, 28 (1932)
28. Polanyi, J. C., Schreiber, J. L.: Faraday Disc. **62**, 267 (1977)
29. Balooch, M., Cardillo, M. J., Miller, D. R., Stickney, R. E.: Surface Sci. **46**, 358 (1974)
30. Gland, J. L.: Surface Sci. **93**, 487 (1980)
31. Bozso, F., Ertl, G., Grunze, M., Weiss, M.: J. Catal. **49**, 18 (1977)
32. Campbell, C. T., Ertl, G., Kuipers, H., Segner, J.: Surface Sci. **107**, 220 (1981)
33. Elovich, Yu. S., Zhabrova, G. M.: Zh. Fiz. Khim. **13**, 1761, 1775 (1939)
34. Redhead, P. A.: Trans. Faraday Soc. **57**, 641 (1961)
35. Redhead, P. A.: Vacuum **12**, 203 (1962)
36. Bozso, F., Ertl, G., Grunze, M., Weiss, M.: Appl. Surf. Sci. **1**, 103 (1977)
37. King, D. A.: Surface Sci. **47**, 384 (1975)
38. Yates, J. T.: In: The chemical physics of solid surfaces and heterogeneous catalysis (D. A. King and D. P. Woodruff, eds.). Amsterdam: Elsevier (in press)
39. Menzel, D.: In: Interactions on metal surfaces (R. Gomer, ed.), p. 102. Berlin, Heidelberg, New York: Springer 1975
40. Taylor, J. L., Weinberg, W. H.: Surface Sci. **78**, 259 (1978)
41. Chan, C. M., Aris, R., Weinberg, W. H.: Appl. Surface Sci. **1**, 360 (1978)
42. Pfnür, H., Feulner, P., Engelhardt, H. A., Menzel, D.: Chem. Phys. Lett. **59**, 481 (1978)
43. Barthes, M. G., Rhead, G. E.: Surf. Sci. **80**, 421 (1979)
44. See e.g.
    a) Kohrt, C., Gomer, R.: Surface Sci. **24**, 77 (1971); **40**, 71 (1973)
    b) Bienfait, M., Venables, J. A.: Surface Sci. **64**, 425 (1977)
45. a) Jones, R. H., Olander, D. R., Siekhaus, W. J., Schwarz, J. A.: J. Vac. Sci. Techn. **9**, 1429 (1972)
    b) Schwarz, J. A., Madix, R. J.: Surface Sci. **46**, 317 (1974)
    c) Chang, H. C., Weinberg, W. H.: J. Chem. Phys. **66**, 4176 (1977); Surface Sci. **65**, 153 (1977); **72**, 617 (1978)
    d) Olander, D. R., Ullmann, A.: Int. J. Chem. Kin. **8**, 625 (1976)
46. Tracy, J. C., Palmberg, P. W.: Surface Sci. **14**, 274 (1969)
47. Frenkel, I.: Z. Phys. **26**, 117 (1924)
48. Glasstone, S., Laidler, K. J., Eyring, H.: The theory of rate processes. New York: MacGraw Hill 1941
49. Ibach, H., Erley, W., Wagner, H.: Surface Sci. **92**, 29 (1980)

50. Campbell, C. T., Ertl, G., Kuipers, H., Segner, J.: Surface Sci. **107**, 207 (1981)
51. Dresser, M. J., Madey, T. E., Yates, J. T.: Surface Sci. **42**, 533 (1974)
52. a) Falconer, J., Madix, R. J.: Surface Sci. **48**, 393 (1975)
    b) Helms, C., Madix, R. J.: Surface Sci. **52**, 677 (1975)
53. Comrie, C., Weinberg, W. H.: J. Chem. Phys. **64**, 250 (1976)
54. Brenig, W., Schönhammer, K.: Z. Phys. **B 36**, 81 (1979)
55. Tully, J. C.: Personal communication
56. Doyen, G.: Vacuum **32**, 91 (1982)
57. Janda, K. C., Hurst, J. E., Becker, C. A., Cowin, J. P., Auerbach, D. J., Wharton, L.: J. Chem. Phys. **72**, 2403 (1980)
58. Leuthäusser, U., Brenig, W.: To be published
59. King, D. A.: Surface Sci. **64**, 43 (1977)
60. Gorte, R., Schmidt, L. D.: Surface Sci. **76**, 559 (1978)
61. Cassuto, A., King, D. A.: Surface Sci. **102**, 388 (1981)
62. Comsa, G.: Proc. 7th Int. Vacuum Congr., Vienna 1977, p. 1317
63. Comsa, G., David, R., Schumacher, B. J.: Proc. 8th Int. Vacuum Congr., Cannes 1980, p. 252
64. Christmann, K., Schober, O., Ertl, G., Neumann, M.: J. Chem. Phys. **60**, 4528 (1974)
65. Ertl, G.: In: Molecular processes at solid surfaces (Eds. Drauglis, Gretz, Jaffee), p. 147. New York: McGraw Hill 1969
66. Conrad, H., Ertl, G., Küppers, J.: Surface Sci. **76**, 323 (1978)
67. Thomas, G. E., Weinberg, W. H.: J. Chem. Phys. **70**, 954 (1979)
68. Ertl, G., Lee, S. B., Weiss, M.: Surface Sci. **114**, 527 (1982)
69. Broden, G., Gafner, G., Bonzel, H. P.: Surface Sci. **84**, 295 (1979)
70. Thiel, P. A., Weinberg, W. H., Yates, J. T.: J. Chem. Phys. **71**, 1643 (1979)
71. Crowell, J., Garfunkel, E., Somorjai, G. A.: Personal communication
72. Engel, T., Ertl, G.: Adv. Catalysis **28**, 1 (1979)
(73) Ertl, G., Weiss, M., Lee, S. B.: Chem. Phys. Lett. **60**, 391 (1979)
74. See e.g. Streitwieser, A.: Chem. Rev. **56**, 571 (1956)
75. Conrad, H., Ertl, G., Küppers, J., Latta, E. E.: Proc. VIth Int. Congr. on Catalysis, London 1976, p. 427
76. Drechsler, M., Hoinkes, H., Kaarmann, H., Wilsch, H., Ertl, G., Weiss, M.: Appl. Surf. Sci. **3**, 217 (1979)
77. See e.g. Bond, G. C., Catalysis by metals. New York: Academic Press 1962
78. Harris, J., Kasemo, B.: Surface Sci. **105**, L281 (1981)
79. Harris, J., Kasemo, B., Törnquist, E.: Surface Sci. **105**, L288 (1981)
80. See e.g. Löffler, D. G., Schmidt, L. D.: J. Catalysis **41**, 440 (1976)
81. Temkin, M. I., Pyzhev, V. M.: Acta Physicochim. USSR **12**, 327 (1940)
82. Sheintuch, M., Schmitz, R. A.: Catalysis Rev. **15**, 107 (1977)
83. Lagos, R. E., Sales, B. C., Suhl, H.: Surface Sci. **82**, 525 (1979)
84. Dagonnier, R., Nuyts, J.: J. Chem. Phys. **65**, 2061 (1976)
85. Turner, J. E., Sales, B. C., Maple, M. B.: Surface Sci. **103**, 54 (1981)
86. Engel, T., Ertl, G.: In: The chemical physics of surfaces and heterogeneous catalysis (P. Woodruff and D. A. King, eds.), Vol. 4, 73 (1982)
87. Madey, T. E., Yates, J. T., Bradshaw, A. M., Hoffmann, F. M.: Surface Sci. **89**, 370 (1979)
88. Sayers, M. J., McClellan, M. R., Shinn, N. D., Trenary, M., McFeely, F. R.: Chem. Phys. Lett. **80**, 521 (1981)
89. Baro, A. M., Ibach, H.: J. Chem. Phys. **71**, 4812 (1979)
90. Campbell, C. T., Doyen, G., Ertl, G., Segner, J.: To be published
91. Gland, J. L., Sexton, B. A., Fischer, G. B.: Surface Sci. **95**, 587 (1980)
92. Hopster, H., Ibach, H., Comsa, G.: J. Catalysis **46**, 37 (1977)
93. Taylor, J. L., Ibbotson, D. E., Weinberg, W. H.: Surface Sci. **79**, 349 (1979)
94. Engel, T.: J. Chem. Phys. **69**, 373 (1978)
95. Thomas, G. E., Weinberg, W. H.: J. Chem. Phys. **61**, 3611 (1978)
96. Chan, C. M., Weinberg, W. H.: J. Chem. Phys. **71**, 2788 (1979)

97. Segner, J., Vielhaber, W., Ertl, G.: Isr. J. Chem. (in press)
98. Campbell, C. T., Ertl, G., Kuipers, H., Segner, J.: J. Chem. Phys. **73**, 5862 (1980)
99. Norton, P. R., Creber, D. K., Goodale, J. W.: Surf. Sci. **119**, 411 (1982)
100. Ertl, G., Neumann, M., Streit, K.: Surface Sci. **64**, 393 (1977)
101. Engel, T., Ertl, G.: J. Chem. Phys. **69**, 1267 (1978)
102. Matsushima, T.: J. Catalysis **55**, 337 (1978); Surf. Sci. **79**, 63 (1979)
103. For references see ref. [98]
104. Palmer, R. L., Smith, J. N.: J. Chem. Phys. **60**, 1453 (1974)
105. Becker, C. A., Cowin, J. P., Wharton, L., Auerbach, D. J.: J. Chem. Phys. **67**, 3394 (1977)
106. Mantell, D. A., Ryali, S. B., Halpern, B. L., Haller, G. L., Fenn, J. B.: Chem. Phys. Lett. **81**, 185 (1981)
107. Bernasek, S. L., Leone, S. R.: Chem. Phys. Lett. (in press)
108. Cardillo, M. J., Ching, C. S. Y., Greene, E. F., Becker, G. E.: J. Vac. Sci. Techn. **15**, 423 (1978)
109. Taylor, J. L., Ibbotson, D. E., Weinberg, W. H.: J. Catalysis **62**, 1 (1980)
110. Zhdan, P. A., Boreskov, G. K., Egelhoff, W. F., Weinberg, W. H.: Surface Sci. **61**, 377 (1976)
111. For references see ref. [72]
112. Takoudis, C. G., Schmidt, L. D., Aris, R.: Surface Sci. **105**, 325 (1981); Chem. Eng. Sci. **36**, 377 (1981); **37**, 69 (1982)
113. Eigenberger, G.: Chem. Eng. Sci. **33**, 1263 (1978)
114. White, J. M., Golchet, A.: J. Catalysis **53**, 266 (1978)
115. Creighton, J. R., Tseng, F. H., White, J. M., Turner, J. S.: J. Phys. Chem. **85**, 703 (1981)
116. Adlhoch, W., Lintz, H. G., Weisker, T.: Surf. Sci. **103**, 576 (1981)
117. Ertl, G., Norton, P. R., Rüstig, J.: Phys. Rev. Lett. **49**, 177 (1982)
118. Turner, J. E., Sales, B. C., Maple, M. B.: Surface Sci. **109**, 591 (1981)
119. Ertl, G., Koch, J.: Proc. Vth Int. Congr. on Catalysis, Palm Beach 1972, p. 969
120. Ladas, S., Poppa, H., Boudart, M.: Surface Sci. **102**, 151 (1981)
121. Cant, N. W., Hicks, P. C., Lennon, B. S.: J. Catalysis **54**, 372 (1978)
122. For a review of recent work see Norton, P. R.: In: The chemical physics of surfaces and heterogeneus catalysis (P. Woodruff and D. A. King, eds.), Vol. 4, 27 (1982)
123. Behm, R. J., Christmann, K., Ertl, G.: Surface Sci. **99**, 320 (1980)
124. Christmann, K., Ertl, G., Pignet, T.: Surface Sci. **54**, 365 (1976)
125. Poelsema, B., Mechtersheimer, G., Comsa, G.: Surface Sci. **111**, 519 (1981)
126. Conrad, H., Ertl, G., Latta, E. E.: Surface Sci. **41**, 435 (1974)
127. Salmeron, M., Gale, R. J., Somorjai, G. A.: J. Chem. Phys. **70**, 2807 (1979)
128. Baro, A. M., Ibach, H., Bruchmann, H. D.: Surface Sci. **88**, 384 (1979)
130. Lewis, R., Gomer, R.: Surface Sci. **17**, 333 (1969)
131. Engel, T., Kuipers, H.: Surface Sci. **90**, 181 (1979)
132. Thiel, P. A., Hoffmann, F. M., Weinberg, W. H.: J. Chem. Phys. **75**, 5556 (1981)
133. Fisher, G. B., Gland, J. L.: Surface Sci. **94**, 446 (1980)
134. Sexton, B. A.: Surface Sci. **94** (1980)
135. Firment, L. E., Somorjai, G. A.: Surface Sci. **55**, 413 (1976)
136. Fisher, G. B., Sexton, B. A.: Phys. Rev. Lett. **44**, 683 (1980)
137. Tevault, D. E., Talley, L. D., Lin, M. C.: J. Chem. Phys. **72**, 3314 (1980)
138. Talley, L. D., Sanders, W. A., Bogan, D. J., Lin, M. C.: J. Chem. Phys. Lett. **66**, 500 (1981)
139. CRC Handbook of Chemistry and Physics
140. Fisher, G. B., Gland, J. L., Schmieg, S. J.: to be published
141. Gland, J. L., Fisher, G. B., Kolling, E. B.: J. Catalysis **77**, 263 (1982)
142. Collins, D. M., Lee, J. B., Spicer, W. E.: Phys. Rev. Lett. **35**, 592 (1975)
143. Monroe, D. R., Merrill, R. P.: J. Catalysis **65**, 461 (1980)
144. Ceyer, S. T., Siekhaus, W. J., Somorjai, G. A.: J. Vac. Sci. Techn. (in press)
145. Gland, J. L., Fisher, G. B.: To be published
146. Hanson, F. V., Boudart, M.: J. Catalysis **53**, 56 (1978)
147. Emmett, P. H.: In: The physical basis for heterogeneous catalysis (E. Drauglis and R. I. Jaffee, eds.), p. 3. New York: Plenum 1975

148. Ozaki, A., Aika, K.: In: Catalysis — Science and Technology, Vol. 1 (J. A. Anderson and M. Boudart, eds.), p. 87. Berlin Heidelberg, New York: Springer 1981
149. Tamaru, K.: In: New Trends in the chemistry of nitrogen fixation (J. Chatt et al., eds.), p. 13. New York: Academic Press 1980
150. Grunze, M.: In: The chemical physics of surfaces and heterogeneous catalysis (P. W. Woodruff and D. A. King, eds.), Vol. 4, 143 (1982)
151. Nielsen, A.: An investigation of promoted iron catalysts for the synthesis of ammonia. Copenhagen: J. Gjellerups Forlag 1968
152. Ludwiczek, H., Preisinger, A., Fischer, A., Hosenmann, R., Schönfeld, A., Vogel, W.: J. Catal. **51**, 326 (1978)
153. Nielsen, A.: Catal. Rev. **4**, 1 (1970)
154. Chen, H. C., Anderson, R. B.: J. Catal. **28**, 161 (1973)
155. Emmett, P. H., Brunauer, S.: J. Am. Chem. Soc. **59**, 1553 (1937)
156. Brunauer, S., Emmett, P. H.: J. Am. Chem. Soc. **62**, 1732 (1940)
157. Solbakken, V., Solbakken, A., Emmett, P. H.: J. Catal. **15**, 90 (1969)
158. Ertl, G., Thiele, N.: Appl. Surface Sci. **3**, 99 (1970)
159. Ertl, G., Prigge, D., Schloegl, R., Weiss, M.: J. Catal. (in press)
160. Spencer, N. D., Schoonmaker, R. C., Somorjai, G. A.: J. Cat. **74**, 129 (1982)
161. Brill, R., Richter, E. L., Ruch, E.: Angew. Chem. **79**, 905 (1967)
162. a) Dumesic, J. A., Topsoe, H., Khammouma, S., Boudart, M.: J. Catal. **37**, 503 (1975)
b) Dumesic, J. A., Topsoe, H., Boudart, M.: J. Catal. **37**, 513 (1975)
163. Ertl, G., Huber, M., Lee, S. B., Paal, Z., Weiss, M.: Appl. Surf. Sci. **8**, 373 (1981)
164. Kiperman, S., Granovskaya, V. Sh.: J. Phys. Chem. USSR **26**, 1615 (1952)
165. Kozhenova, K. T., Kagan, M. Ya.: J. Phys. Chem. USSR **14**, 1250 (1940)
166. Emmett, P. H., Brunauer, S.: J. Am. Chem. Soc. **56**, 35 (1934)
167. Bozso, F., Ertl, G., Weiss, M.: J. Catal **50**, 519 (1977)
168. Kishi, K., Roberts, M. W.: Surface Sci. **62**, 252 (1977)
169. Gay, I. D., Textor, M., Mason, R., Iwasawa, Y.: Proc. R. Soc. London **A 356**, 25 (1977)
170. Johnson, D. W., Roberts, R. W.: Surf. Sci. **87**, 1255 (1979)
171. Doyen, G., Ertl, G.: Surface Sci. **69**, 157 (1977)
172. Umbach, E., Schichl, A., Menzel, D.: Solid State Comm. **36**, 93 (1980)
173. Bagus, P. A., Brundle, C. R., Hermann, K., Menzel, D.: J. Electr. Spectr. **20**, 253 (1980)
174. Itoh, H., Ertl, G., Kunz, A. B.: Chem. Phys. **59**, 149 (1981)
175. Grunze, M., Driscoll, R. K., Burland, G. N., Cornish, J. C. L., Pritchard, J.: Surf. Sci. **89**, 381 (1979)
176. Ho, W., Willis, R. F., Plummer, E. W.: Surf. Sc. **95**, 171 (1980)
177. Eischens, R. P., Jacknow, J.: Proc. 3rd Int. Congr. on Catalysis, p. 627. Amsterdam: North Holland 1965
178. Van Hardeveld, R. A., Van Montfoort, A.: Surf. Sci. **4**, 396 (1966)
179. Shigeishi, R. A., King, D. A.: Surf. Sci. **62**, 379 (1977)
180. Beeck, O., Cole, W. A., Wheeler, A.: Disc. Faraday Soc. **52**, 96 (1956)
181. Wedler, G., Borgmann, D., Geuss, K. P.: Surf. Sci. **47**, 592 (1975)
182. Ertl, G., Huber, M., Thiele, N.: Z. Naturforsch. **34a**, 30 (1979)
183. Scholten, J. J. F., Zwietering, P., Konvalinka, J. A., de Boer, J. H.: Trans. Faraday Soc. **55**, 2166 (1959)
184. Imbihl, R., Behm, R. J., Christmann, K., Ertl, G., Matsushima, T.: Surf. Sci. **117**, 257 (1982)
185. Tamaru, K.: Trans. Faraday Soc. **59**, 979 (1963)
186. Grunze, M., Bozso, F., Ertl, G., Weiss, M.: Appl. Surf. Sci. **1**, 241 (1978)
187. Amariglio, H.: Symposium on Nitrogen Fixation, Tokyo, July 1980
188. Weiss, M., Ertl, G., Nitschke, F.: Appl. Surf. Sci. **2**, 614 (1979)
189. Hermann, K.: Proc. 4th Int. Conf. on Solid Surfaces, Cannes 1980, p. 196
190. Itoh, H., Ertl, G., Kunz, A. B.: Z. Naturforsch. **36a**, 347 (1981)
191. Grunze, M.: Surf. Sci. **81**, 217 (1979)
192. Ertl, G., Huber, M.: J. Catalysis **61**, 537 (1980)
193. Temkin, M.: Zhur. Fiz. Khim. **24**, 1312 (1950)

194. Brunauer, S., Love, K. S., Keenan, R. G.: J. Am. Chem. Soc. **64**, 751 (1942)
195. Ozaki, A., Taylor, H. A., Boudart, M.: Proc. Roy. Soc. **A258**, 47 (1960)
196. Ertl, G.: Cat. Rev. **21**, 201 (1980)
197. Scholten, J. J. F., Konvalinka, J. A., Zwietering, P.: Trans. Faraday Soc. **56**, 262 (1960)
198. Lee, S. B., Ertl, G.: Unpublished results
199. Krabetz, R., Peters, C.: Angew. Chem. **77**, 333 (1965)
200. Lee, S. B., Weiss, M., Ertl, G.: Surf. Sci. **108**, 357 (1981)
201. Garfunkel, E. L., Crowell, J. E., Somorjai, G. A.: J. Phys. Chem. **86**, 310 (1982)
202. Paál, Z., Ertl, G., Lee, S. B.: Appl. Surf. Sci. **8**, 231 (1981)
203. Prigge, D., Ertl, G.: In preparation
204. Garfunkel, E. L., Somorjai, G. A.: Surf. Sci. **115**, 441 (1982)
205. Ertl, G., Huber, M.: Z. Phys. Chem. N.F. **119**, 97 (1980)
206. Ozaki, A., Aika, K., Morikawa, Y.: Proc. 5th Int. Congr. on Catalysis, p. 1251. Amsterdam: North Holland 1973
207. Benziger, J., Madix, R. J.: Surf. Sci. **94**, 119 (1980)
208. Ertl, G., Lee, S. B., Weiss, M.: Surf. Sci. **111**, 711 (1981)
209. Somorjai, G. A.: Science **201**, 489 (1978)
210. Kelley, R. D., Goodman, D. W.: In: The chemical physics of surfaces and heterogeneous catalysis (P. Woodruff and D. A. King, eds.), Vol. 4, 427 (1982)
211. a) Bonzel, H. P., Krebs, H. J.: Surf. Sci. **91**, 499 (1980)
b) Dubois, L., Somorjai, G. A.: J. Chem. Phys. **72**, 5234 (1980)
c) Krebs, H. J., Bonzel, H. P., Gafner, G.: Surf. Sci. **88**, 269 (1979)
d) Dwyer, D. J., Somorjai, G. A.: J. Catal. **56**, 249 (1979)
212. Somorjai, G. A.: Chemistry in the two dimensions: Surfaces, ch. 9. Ithaca: Cornell University Press 1981
213. a) Christmann, K., Ertl, G., Shimizu, H.: J. Catal. **61**, 397, 412 (1980)
b) Vickermann, J. C., Christmann, K., Ertl, G.: J. Cat. **71**, 175 (1981)
c) Shi, S. K., Lee, H. I., White, J. M.: Surface Sci. **102**, 56 (1981)
d) Sachtler, J. W., A., Biberian, J. P., Somorjai, G. A.: Surf. Sci. **110**, 43 (1981)
214. a) Ibach, H., Lehwald, S.: J. Vac. Sci. Techn. **15**, 407 (1978)
b) Demuth, J. E.: Surf. Sci. **80**, 367 (1979)
c) Kesmodel, L. L., Dubois, L. H., Somorjai, G. A.: J. Chem. Phys. **70**, 2180 (1979)

Subject Index

Author Index Volume 1–4

Catalysis
Science and Technology

Volume 1

Editors: **J.R. Anderson, M. Boudart**

1981. 107 figures. X, 309 pages. ISBN 3-540-10353-8

Contents:

Volume 2

Editors: **J.R. Anderson, M. Boudart**

1981. 145 figures. X, 282 pages. ISBN 3-540-10593-X

Contents/Information:

G.-M. Schwab: **History of Concepts in Catalysis.** The concept of catalysis can be attributed to J. Berzelius (1838), whose formulation was based on the manifold observations made in the 17th and 18th centuries. This article traces the development of this and related theories along with the scientific research and empirical material from which they are drawn.

J. Haber: **Crystallography of Catalyst Types.** Structural properties of metals and their substitutional and interstitial alloys, transition metal oxides as well as alumina, silica, aluminosilicates and phosphates are discussed. Implications of point and extended defects for catalysis are emphasized and the problem of the structure and composition of the surface as compared to the bulk is considered.

G. Froment, L. Hosten: **Catalytic Kinetics: Modelling.** The text reviews the methodology of kinetic analysis for simple as well as complex reactions. Attention is focused on the differential and integral methods of kinetic modelling. The statistical testing of the model and the parameter estimates required by the stochastic character of experimental data is described in detail and illustrated by several practical examples. Sequential experimental design procedures for discrimination between rival models and for obtaining parameter estimates with the greatest attainable precision are developed and applied to real cases.

A.J. Lecloux: **Texture of Catalysts.** Useful guidelines and methods for a systematic investigation and a coherent description of catalyst texture are proposed in this contribution. Such a description requires the specification of a very large number of parameters and implies the use of "models" involving assumptions and simplifications. The general approach for determining the porous texture of solids is based on techniques, whose results are cross analyzed in such a way that a self-consistent picture of the porous texture of solids is obtained.

K. Tanabe: **Solid Acid and Base Catalysts.** This chapter deals with the types of solid acids and bases, the acidic and basic properties, and the structure of acidic and basic sites. The chemical principles of the determination of acid-base properties and the mechanism for the generation of acidity and basicity are also described. How acidic and basic properties are controlled chemically is discussed in connection with the preparation method of solid acids and bases.

Catalysis
Science and Technology

Volume 3

Editors: **J.R. Anderson, M. Boudart**

1982. 91 figures. Approx. X, 290 pages. ISBN 3-540-11634-6

Contents/Information:

E.E. Donath: **History of Catalysis in Coal Liquefaction.** Coal liquefaction is a process of immense future significance for the production of alternative liquid fuels and chemical feedstocks. It is a process with a long technical history, and it is important for current research to consider at least the catalytic component from the historical viewpoint. This chapter summarizes the industrial experience in coal hydrogenation up to the end of the second World War (20 references).

G.K. Boreskov: **Catalytic Activation of Dioxygen.** Dioxygen is the most common oxidative agent, and many oxidation reactions with dioxygen proceed via heterogeneous catalysis. This chapter reviews these catalytic reactions which form the basis for important industrial processes such as production of sulphuric or nitric acids. They are further used for detoxication of organic substances and carbon monoxide from industrial and motor-transport exhaust gases. Attempts have recently been made to utilize the catalytic oxidation of fuels for energy production (212 references).

M.A. Vannice: **Catalytic Activation of Carbon Monoxide on Metal Surfaces.** Renewed interest in the production of fuels and chemical from synthetic gas has resulted in much research directed towards chemistry of CO/H_2 reactions. This chapter concentrates on studies of CO adsorption and the reactions of CO with H_2 and H_2O on metal surfaces. Recent work is reviewed which describes the adsorbed states of CO, its interaction with hydrogen on surfaces, and its subsequent reaction in the methanation and Fischer-Tropsch reactions. The water gas shift reaction is also discussed (230 references).

S.R. Morrison: **Chemisorption on Nonmetallic Surfaces.** This chapter describes ways in which atoms can be bonded to the surface of a nonmetallic solid. It shows both adsorption with local bonding only, where the band model is not necessary to describe the process, and the opposite extreme, where tha band model dominates the adsorption and provides the only reasonable description of the adsorbed species. Finally, the author discusses the effect of adsorption on the properties of the solid, emphasizing effects that lead to techniques for studying and understanding adsorption on nonmetals (37 references).

Z. Knor: **Chemisorption of Dihydrogen.** Chemisorption is the process of central importance in the catalytic activation of dihydrogen. This chapter emphasized the nature of the chemisorption bond, and the relation between theory and experiment, as theory seems at present best suited to providing an essentially qualitative framework within which the experimentalist may be able to consider the significance of his results (309 references).

P.N. Rylander: **Catalytic Processes in Organic Conversions.** Catalytic conversions of organic compounds are the heart of the modern chemical industry. This chapter reviews heterogeneous catalysts and chemicals of industrial importance having functionality beyond those of simple olefins and aromatics. It illustrates the diversity of chemical transformations that can be achieved and discusses the interplay of catalyst and chemical properties of the organic reactants, to show how various intrinsic problems can be minimized, and to suggest the type of catalyst suitable for various reactions (182 references).

Bei Fragen zur Produktsicherheit wenden Sie sich bitte an:
If you have any questions regarding product safety,
please contact:

Walter de Gruyter GmbH
Genthiner Straße 13
10785 Berlin
productsafety@degruyterbrill.com